Bioactive Gel Films and Coatings Applied in Active Food Packaging

Bioactive Gel Films and Coatings Applied in Active Food Packaging

Editors

Aris Giannakas
Constantinos Salmas
Charalampos Proestos

Basel • Beijing • Wuhan • Barcelona • Belgrade • Novi Sad • Cluj • Manchester

Editors
Aris Giannakas
Department of Food Science & Technology
University of Patras
Agrinio
Greece

Constantinos Salmas
Department of Materials Science Engineering
University of Ioannina
Ioannina
Greece

Charalampos Proestos
Department of Chemistry
National and Kapodistrian University of Athens
Athens
Greece

Editorial Office
MDPI
St. Alban-Anlage 66
4052 Basel, Switzerland

This is a reprint of articles from the Special Issue published online in the open access journal *Gels* (ISSN 2310-2861) (available at: www.mdpi.com/journal/gels/special_issues/food_films_coatings).

For citation purposes, cite each article independently as indicated on the article page online and as indicated below:

Lastname, A.A.; Lastname, B.B. Article Title. *Journal Name* **Year**, *Volume Number*, Page Range.

ISBN 978-3-0365-9719-5 (Hbk)
ISBN 978-3-0365-9718-8 (PDF)
doi.org/10.3390/books978-3-0365-9718-8

Contents

Editorial

Editorial for Special Issue: Gel Films and Coatings Applied in Active Food Packaging

Aris E. Giannakas

Department of Food Science and Technology, University of Patras, 30100 Agrinio, Greece; agiannakas@upatras.gr

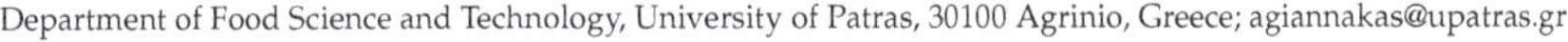

1. Introduction

Nowadays, the global trends of bioeconomy and sustainability require the use of biobased raw materials in all scientific and application fields to reduce the global carbon dioxide fingerprint. In this direction, natural biopolymers such as chitosan, sodium alginate, gelatin, pectin, and xanthan are potential candidates for replacing polymers derived from mineral oils in new food packaging applications. At the same time, nanotechnology provides innovative applications in the field of food packaging, similar to those provided in the biomedical and pharmaceutical fields. By incorporating biobased bioactive compounds such as essential oils, derivatives of essential oils, propolis, anthocyanin, aloe vera, acacia gum, and collagen in biopolymers, novel gels and hydrogels can be developed with the control release properties of enriched bioactive compounds, which can be applied as smart films and coatings in the food industry. Moreover, nanomaterials such as nanoclays and natural zeolites can be used as new, cheap, and natural abundant nanocarriers for such bioactive compounds and enhance their release control properties from the biopolymer matrix. These bioactive films and coating materials enhance solubility, improve bioavailability, facilitate controlled release, and protect bioactive ingredients during manufacture and storage. The current Special Issue provides a fresh bouquet of articles on the bioactive gel films and coatings applied to active food packaging.

2. Contributions

In the article of Bhatia et al. [1], hydrogel-based films loaded with varying concentrations of Melissa officinalis (MOEO) (0.1%, 0.15%, and 0.2%) were prepared using the solvent-casting method, and their physicochemical and antioxidant properties were examined. The incorporation of MOEO into the films gave films with a higher elongation at break (EAB) (30.24–36.29%), lower tensile strength (TS) (3.48–1.25 MPa), lower transparency (87.30–82.80%), higher water barrier properties, and higher antioxidant properties, indicating that MOEO has the potential to be used in active food packaging material for potential applications.

The same research group developed sodium alginate (SA) and acacia gum (AG) hydrogel-based films loaded with cinnamon essential oil (CEO) and revealed that, as the oil concentration in the films increased, the thickness and elongation at break (EAB) increased, while the transparency, tensile strength (TS), water vapor permeability (WVP), and moisture content (MC) decreased. As the concentration of CEO increased, the hydrogel-based films demonstrated a significant improvement in their antioxidant properties [2]. Overall, it was concluded that the incorporation of CEO into edible SA–AG composite films presented a promising strategy for the production of hydrogel-based films with the potential to serve as food-packaging materials.

Elsebaie et al. [3] studied the preparation, characterization (physical, mechanical, optical, and morphological properties, as well as antioxidant and antimicrobial activities), and packaging application of chitosan (CH)-based gel films containing varying empty green pea pod extract (EPPE) concentrations (0, 1, 3, and 5% *w*/*w*). Their studies revealed that

Citation: Giannakas, A.E. Editorial for Special Issue: Gel Films and Coatings Applied in Active Food Packaging. *Gels* **2023**, *9*, 743. https://doi.org/10.3390/gels9090743

Received: 7 September 2023
Accepted: 12 September 2023
Published: 13 September 2023

adding EPPE to CH increased the thickness (from 0.132 ± 0.08 to 0.216 ± 0.08 mm), density (from 1.13 ± 0.02 to 1.94 ± 0.02 g/cm^3), and opacity (from 0.71 ± 0.02 to 1.23 ± 0.04), while decreasing the water vapor permeability, water solubility, oil absorption ratio, and whiteness index from 2.34 to 1.08 × 10^{-10} g^{-1} s^{-1} pa^{-1}, from 29.40 ± 1.23 to 18.75 ± 1.94%, from 0.31 ± 0.006 to 0.08 ± 0.001%, and from 88.10 ± 0.43 to 77.53 ± 0.48, respectively. The EPPE films had a better tensile strength (maximum of 26.87 ± 1.38 MPa), elongation percentage (maximum of 58.64 ± 3.00%), biodegradability (maximum of 48.61% after 3 weeks), and migration percentages than the pure CH gel film. By adding EPPE, the antioxidant and antibacterial activities of the film were improved. Furthermore, compared to control samples, corn oil packaging in CH-based EPPE gel films slowed increases in thiobarbituric acid and superoxide. As an industrial application, CH-based EPPE films have the potential to be beneficial in food packaging.

Salmas et al. [4] presented a novel adsorption process for producing a thymol-halloysite nanohybrid and dispersed it in a chitosan/poly-vinyl-alcohol matrix. The results indicated that the increased fraction of thymol from thyme oil significantly enhanced the antimicrobial and antioxidant activities of the prepared chitosan-poly-vinyl-alcohol gel. The use of halloysite increased the mechanical and water–oxygen barrier properties, leading to a controlled release process of thymol, which extended the preservation and shelf-life of kiwi fruits. Finally, the results indicated that the halloysite improved the properties of the chitosan/poly-vinyl-alcohol films, and the thymol made them further advantageous.

The same research group [5] developed a green method for producing natural zeolite (TO@NZ) nanostructure rich in thymol. This material was used to prepare sodium-alginate/glycerol/xTO@NZ (ALG/G/TO@NZ) nanocomposite active films for the packaging of soft cheese to extend its shelf-life. The water vapor transmission rate, oxygen permeation analyzer, tensile measurements, antioxidant measurements, and antimicrobial measurements were used to estimate the film's water and oxygen barrier and mechanical properties and the nanostructure's nanoreinforcement activity, antioxidant activity, and antimicrobial activity. The findings from this study revealed that the ALG/G/TO@NZ nanocomposite film could be used as an active packaging film for foods with enhanced mechanical properties, oxygen and water barrier, and antioxidant and antimicrobial activities, and it is capable of extending food shelf life.

Puscaselu et al. [6] developed biopolymer-based films (agar and sodium alginate) and tested their properties following additions of 7.5% and 15% (w/v) of various essential oils (lemon, orange, grapefruit, cinnamon, clove, chamomile, ginger, eucalyptus, or mint). The samples were tested immediately after their development and after one year of storage to examine the possible long-term property changes. All the films showed reductions in mass, thickness, and microstructure, as well as their mechanical properties. The most considerable variations in their physical properties were observed in the 7.5% lemon oil sample and 15% grapefruit oil sample, with the largest reductions in mass (23.13%), thickness (from 109.67 μm to 81.67 μm), and density (from 0.75 g/cm^3 to 0.43 g/cm^3). However, the microstructures of the samples were considerably improved. Although the addition of lemon essential oil prevented a reduction in mass during the storage period, it favored the degradation of the microstructure and a loss of elasticity (from 16.7% to 1.51% for the sample with 7.5% lemon EO and from 18.28% to 1.91% for the sample with 15% lemon EO). Although the addition of the essential oils of mint and ginger resulted in films with a more homogeneous microstructure, the increase in concentration favored the appearance of pores and modifications in the color parameters. Apart from the films with added orange, cinnamon, and clove EOs, the antioxidant capacity of the films decreased during storage. The most obvious variations were identified in the samples with lemon, mint, and clove EOs. The most unstable samples were those with added ginger (95.01%), lemon (92%), and mint (90.22%). The same research group [7] furthered their research on the development of such films and tested the modification of the properties of the previously developed biopolymeric materials, by adding 10 and 20% w/v essential oils, respectively, of lemon, grapefruit, orange, cinnamon, clove, mint, ginger, eucalypt, and chamomile. Films with a thickness between 53 and 102 μm were obtained, with a roughness ranging between 147 and 366 nm. Most films had a water activity index significantly below what is required for microorganism growth, as low as 0.27, while all the essential oils induced a microbial growth reduction or 100% inhibition. Tested for an evaluation of their physical, optical, microbiological, or solubility properties, all the films with an addition of essential oil to their composition showed improved properties compared to the control sample.

Chit et al. [8] developed a single-layer coating using chitosan (Ch) and sodium alginate (SA) solutions, and their gel coating (ChCSA) was formed via layer-by-layer (LbL) electrostatic deposition using calcium chloride (C) as a cross-linking agent to improve the storage qualities and shelf-life of fresh-cut purple-flesh sweet potatoes (PFSP). The preservative effects of the single-layer coating in comparison to LbL on the quality parameters of the fresh-cut PFSPs, including color change, weight loss, firmness, microbial analysis, CO_2 production, pH, solid content, total anthocyanin content (TAC), and total phenolic content (TPC), were evaluated during 16 days of storage at 5 °C. Uncoated samples were applicable as a control. The result established the effectiveness of coating in reducing microbial proliferation (~2 times), color changes (~3 times), and weight loss (~4 times), with negligible firmness losses after the storage period. In addition, the TAC and TPC were better retained in the coated samples than in the uncoated samples. In contrast, quality deterioration was observed in the uncoated fresh cuts, which progressed with storage time. Relatively, gel coating ChCSA showed superior effects on preserving the quality of the fresh-cut PFSPs and could be suggested as a commercial method for preserving fresh-cut purple-flesh sweet potato and other similar roots.

Al Hilifi et al. [9] developed aloe vera gel (AVG)-based edible coatings enriched with anthocyanin and investigated the effect of different formulations of these aloe-vera-based edible coatings, such as neat AVG (T1), AVG with glycerol (T2), aloe vera with 0.2% anthocyanin + glycerol (T3), and AVG with 0.5% anthocyanin + glycerol (T4), on the postharvest quality of fig (Ficus carica L.) fruits under refrigerated conditions (4 °C) for up to 12 days of storage with 2-day examination intervals. The results of the present study revealed that the T4 treatment was the most effective for reducing the weight loss in the fig fruits throughout the storage period (~4%), followed by T3, T2, and T1. The minimum weight loss after 12 days of storage (3.76%) was recorded for the T4 treatment, followed by T3 (4.34%), which was significantly higher than that of uncoated fruit (~11%). The best quality attributes, such as the total soluble solids (TSS), titratable acidity (TA), and pH, were also demonstrated in the T3 and T4 treatments. The T4 coating caused a marginal change of 0.16 in the fruit titratable acidity, compared to the change of 0.33 in the untreated fruit control after 12 days of storage at 4 °C. Similarly, the total soluble solids in the T4-coated fruits increased marginally (0.43 °Brix) compared to that in the uncoated control fruits (>2 °Brix) after 12 days of storage at 4 °C. The results revealed that the incorporation of anthocyanin content into AVG is a promising technology for the development of active edible coatings to extend the shelf life of fig fruits.

Sheikha et al. [10] studied the effect of a xanthan coating containing various concentrations (0, 1, and 2%; w/v) of ethanolic extract of propolis (EEP) on the physicochemical, microbial, and sensory quality indices in mackerel fillets stored at 2 °C for 20 days. The pH, peroxide value, K-value, TVB-N, TBARS, microbiological, and sensory characteristics were determined every 5 days over the storage period (20 days). The samples treated with xanthan (XAN) coatings containing 1 and 2% of EEP were shown to have the highest level of physicochemical protection and maximum level of microbial inhibition ($p < 0.05$) compared to the uncoated samples (control) over the storage period. Furthermore, the addition of EEP to XAN was more effective in notably preserving ($p < 0.05$) the taste and odor of the coated samples compared to the controls.

3. Conclusions

In conclusion, we are thankful to have edited this Special Issue, as it collects relevant contributions that reflect the increasingly widespread interest in gel films and coatings applied to food packaging. Overall, it was revealed that such films and gel coatings can be used to extend the shelf life of various food products such as fruits, dairy products, and fish products. We look forward to this Special Issue reaching the highest scientific audience and promoting the second edition.

Conflicts of Interest: The authors declare no conflict of interest.

References

1. Bhatia, S.; Al-Harrasi, A.; Alhadhrami, A.S.; Shah, Y.A.; Kotta, S.; Iqbal, J.; Anwer, M.K.; Nair, A.K.; Koca, E.; Aydemir, L.Y. Physical, Chemical, Barrier, and Antioxidant Properties of Pectin/Collagen Hydrogel-Based Films Enriched with Melissa Officinalis. *Gels* **2023**, *9*, 511. [CrossRef]
2. Bhatia, S.; Al-Harrasi, A.; Shah, Y.A.; Altoubi, H.W.K.; Kotta, S.; Sharma, P.; Anwer, M.K.; Kaithavalappil, D.S.; Koca, E.; Aydemir, L.Y. Fabrication, Characterization, and Antioxidant Potential of Sodium Alginate/Acacia Gum Hydrogel-Based Films Loaded with Cinnamon Essential Oil. *Gels* **2023**, *9*, 337. [CrossRef]
3. Elsebaie, E.M.; Mousa, M.M.; Abulmeaty, S.A.; Shaat, H.A.Y.; Elmeslamy, S.A.-E.; Asker, G.A.; Faramawy, A.A.; Shaat, H.A.Y.; Abd Elrahman, W.M.; Eldamaty, H.S.E.; et al. Chitosan-Based Green Pea (*Pisum sativum* L.) Pod Extract Gel Film: Characterization and Application in Food Packaging. *Gels* **2023**, *9*, 77. [CrossRef]
4. Salmas, C.E.; Giannakas, A.E.; Moschovas, D.; Kollia, E.; Georgopoulos, S.; Gioti, C.; Leontiou, A.; Avgeropoulos, A.; Kopsacheili, A.; Avdylaj, L.; et al. Kiwi Fruits Preservation Using Novel Edible Active Coatings Based on Rich in Thymol Halloysite Nanostructures and Chitosan/Polyvinyl Alcohol Gels. *Gels* **2022**, *8*, 823. [CrossRef]
5. Giannakas, A.E.; Salmas, C.E.; Moschovas, D.; Zaharioudakis, K.; Georgopoulos, S.; Asimakopoulos, G.; Aktypis, A.; Proestos, C.; Karakassides, A.; Avgeropoulos, A.; et al. The Increase of Soft Cheese Shelf-Life Packaged with Edible Films Based on Novel Hybrid Nanostructures. *Gels* **2022**, *8*, 539. [CrossRef] [PubMed]
6. Gheorghita Puscaselu, R.; Lobiuc, A.; Sirbu, I.O.; Covasa, M. The Use of Biopolymers as a Natural Matrix for Incorporation of Essential Oils of Medicinal Plants. *Gels* **2022**, *8*, 756. [CrossRef] [PubMed]
7. Puscaselu, R.G.; Lobiuc, A.; Gutt, G. The Future Packaging of the Food Industry: The Development and Characterization of Innovative Biobased Materials with Essential Oils Added. *Gels* **2022**, *8*, 505. [CrossRef] [PubMed]
8. Chit, C.-S.; Olawuyi, I.F.; Park, J.J.; Lee, W.Y. Effect of Composite Chitosan/Sodium Alginate Gel Coatings on the Quality of Fresh-Cut Purple-Flesh Sweet Potato. *Gels* **2022**, *8*, 747. [CrossRef]
9. Al-Hilifi, S.A.; Al-Ali, R.M.; Al-Ibresam, O.T.; Kumar, N.; Paidari, S.; Trajkovska Petkoska, A.; Agarwal, V. Physicochemical, Morphological, and Functional Characterization of Edible Anthocyanin-Enriched Aloevera Coatings on Fresh Figs (*Ficus carica* L.). *Gels* **2022**, *8*, 645. [CrossRef] [PubMed]
10. Sheikha, A.F.E.; Allam, A.Y.; Oz, E.; Khan, M.R.; Proestos, C.; Oz, F. Edible Xanthan/Propolis Coating and Its Effect on Physicochemical, Microbial, and Sensory Quality Indices in Mackerel Tuna (*Euthynnus affinis*) Fillets during Chilled Storage. *Gels* **2022**, *8*, 405. [CrossRef] [PubMed]

Article

Physical, Chemical, Barrier, and Antioxidant Properties of Pectin/Collagen Hydrogel-Based Films Enriched with *Melissa officinalis*

Saurabh Bhatia [1,2,3,*,†], Ahmed Al-Harrasi [1,*,†], Aysha Salim Alhadhrami [1], Yasir Abbas Shah [1], Sabna Kotta [4,5], Javed Iqbal [6], Md Khalid Anwer [7], Anjana Karunakaran Nair [8], Esra Koca [9] and Levent Yurdaer Aydemir [9]

1 Natural and Medical Sciences Research Center, University of Nizwa, Birkat Al Mauz, P.O. Box 33, Nizwa 616, Oman; ayshalhadhrami@gmail.com (A.S.A.); yasir.shah@unizwa.edu.om (Y.A.S.)
2 School of Health Science, University of Petroleum and Energy Studies, Dehradun 248007, India
3 Saveetha Institute of Medical and Technical Sciences, Saveetha University, Chennai 600077, India
4 Department of Pharmaceutics, Faculty of Pharmacy, King Abdulaziz University, Jeddah 21589, Saudi Arabia; skotta@kau.edu.sa
5 Center of Excellence for Drug Research and Pharmaceutical Industries, King Abdulaziz University, Jeddah 21589, Saudi Arabia
6 Center of Nanotechnology, King Abdulaziz University, Jeddah 21589, Saudi Arabia; iqbaljavedch@gmail.com
7 Department of Pharmaceutics, College of Pharmacy, Prince Sattam Bin Abdulaziz University, Al-Kharj 11942, Saudi Arabia; m.anwer@psau.edu.sa
8 Department of Biomedical Sciences, College of Pharmacy, Shaqra University, Shaqra 11961, Saudi Arabia; anju@su.edu.sa
9 Department of Food Engineering, Faculty of Engineering, Adana Alparslan Turkes Science and Technology University, Adana 01250, Turkey; esrakoca.tr@outlook.com (E.K.); lyaydemir@atu.edu.tr (L.Y.A.)
* Correspondence: sbsaurabhbhatia@gmail.com (S.B.); aharrasi@unizwa.edu.om (A.A.-H.)
† These authors contributed equally to this work.

Citation: Bhatia, S.; Al-Harrasi, A.; Alhadhrami, A.S.; Shah, Y.A.; Kotta, S.; Iqbal, J.; Anwer, M.K.; Nair, A.K.; Koca, E.; Aydemir, L.Y. Physical, Chemical, Barrier, and Antioxidant Properties of Pectin/Collagen Hydrogel-Based Films Enriched with *Melissa officinalis*. *Gels* **2023**, 9, 511. https://doi.org/10.3390/gels9070511

Academic Editors: Aris Giannakas, Constantinos Salmas and Charalampos Proestos

Received: 16 May 2023
Revised: 14 June 2023
Accepted: 15 June 2023
Published: 25 June 2023

Abstract: The essential oil extracted from *Melissa officinalis* (MOEO) exhibits a wide range of therapeutic properties, including antioxidant, antibacterial, and antifungal activities. The current research aimed to analyze the mechanical, barrier, chemical, and antioxidant properties of pectin and collagen-based films. Hydrogel-based films loaded with varying concentrations of MOEO (0.1%, 0.15%, and 0.2%) were prepared by solvent-casting method, and their physicochemical as well as antioxidant properties were examined. GC-MS analysis revealed the presence of major components in MOEO such as 2,6-octadienal, 3,7-dimethyl, citral, caryophyllene, geranyl acetate, caryophyllene oxide, citronellal, and linalool. Fourier transform infrared (FTIR) results revealed the interaction between components of the essential oil and polymer matrix. Scanning electron microscopy (SEM) revealed that films loaded with the highest concentration (0.2%) of MOEO showed more homogeneous structure with fewer particles, cracks, and pores as compared to control film sample. MOEO-incorporated films exhibited higher elongation at break (EAB) (30.24–36.29%) and thickness (0.068–0.073 mm); however, they displayed lower tensile strength (TS) (3.48–1.25 MPa) and transparency (87.30–82.80%). MOEO-loaded films demonstrated superior barrier properties against water vapors. According to the results, the incorporation of MOEO into pectin–collagen composite hydrogel-based films resulted in higher antioxidant properties, indicating that MOEO has the potential to be used in active food packaging material for potential applications.

Keywords: biopolymers; edible films; composite material; food packaging; essential oil

1. Introduction

The food packaging material in the form of polymeric edible films is gaining more attention these days due to its numerous benefits, such as being edible, biodegradable, nontoxic, biocompatible, etc. These films are made at a laboratory scale to improve the safety, quality, and shelf life of food products [1]. There are several manufacturers (Innoteq, NewGem Foods, Watson, Biofilm Limited, ODF Pharma, Proinec, MonoSol, Umang Pharmatech, etc.) that make edible films at large scale. Pectin, a major constituent of plant cells, is primarily used in the manufacturing of fruit juices, jams, and bread fillings [2,3]. The utilization of pectin as a polymer to produce hydrogel-based films is a viable option owing to its gelation characteristics, lack of toxicity, biodegradability, and accessibility. Collagen is one of the natural polymers that hold significant potential as a vital industrial raw material [4]. Its physicochemical characteristics, biodegradability, and non-toxicity have made it increasingly important and useful in the food sector [5].

Several research studies have shown that incorporating various plant-based bioactive components into pectin-based films can result in to effective food packaging material [6,7]. Blending two polymers (pectin/collagen) is an effective way to synthesize the edible film with favorable barrier and mechanical attributes. Hydrogel-based films have been formulated with essential oils due to their possible antibacterial, antioxidant, and antifungal characteristics [8]. Essential oils are widely used in the food industries due to their composition of volatile compounds, including a combination of terpenes, terpenoids, alpha-aromatic and aromatic compounds, as well as non-volatile compounds [9].

Melissa officinalis is a member of the Lamiaceae family and possesses diverse therapeutic properties, and Melissa oil exhibits antibacterial, antifungal, and antioxidant properties [10,11]. Melissa essential oil (MOEO) primarily comprises citronella, but it also contains other components such as citral, geraniol, linalool, and beta-caryophyllene [12]. In earlier research, researchers developed active films using a combination of carboxymethyl chitosan and locust bean gum. These films contained nanoemulsions of essential oil extracted from *Melissa officinalis* L. Following the analysis of the results, it was noted that incorporating MOEO nanoemulsion into the active films resulted in an elevation of their elasticity and resistance to water [13]. Another research revealed the development of active edible films using sodium caseinate, which contained a combination of zinc oxide nanoparticles (ZnONPs) and microcapsules of MOEO. The properties of the films were studied, and it was observed that they exhibited strong antioxidant activity [14].

To date, there has been no investigation into the impact of MOEO on the physiochemical characteristics of composite hydrogel-based films composed of pectin and collagen. Present study aims to investigate the effect of MOEO on the structural, mechanical, thermal, chemical, optical and barrier properties of hydrogel films based on pectin/collagen. In addition to investigating the physicochemical changes to determine the compatibility of the polymeric material with MOEO, the antioxidant activity was also evaluated as a potential food packaging material.

2. Results and Discussion

2.1. Visual Appearance

The prepared edible film samples including B (Control), PC20, PC30, and PC40 were visually examined, and it was observed that control (B) and PC20 film samples had superior visual characteristics and was easy to peel from the Petri plate. Moreover, the PC20 film exhibited relatively improved characteristics regarding brittleness, flexibility, and stiffness. Film samples PC30 and PC40 containing 0.15% and 0.2% MOEO, respectively, were less transparent while exhibiting uniformity and flexibility. The better visual characteristics of the PC30 and PC40 film samples could be due to the addition of higher concentrations of MOEO and Tween 80, resulted in the uniform dispersion of the oil in the film matrix (Figure 1).

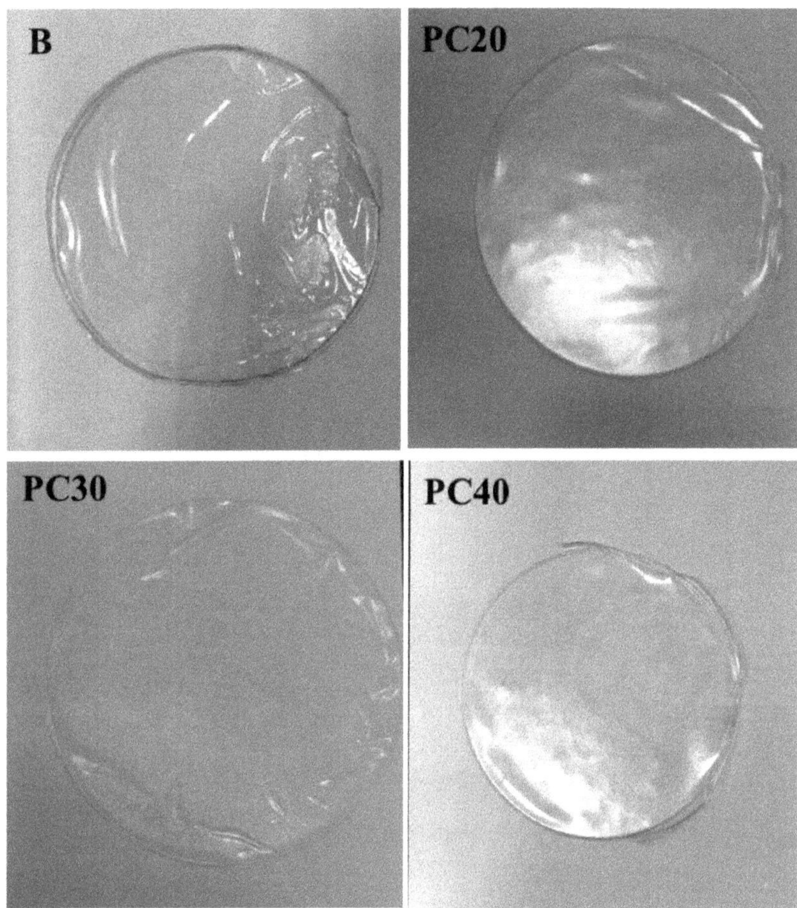

Figure 1. Visual appearance of B (control), PC20 (containing 0.1% MOEO), PC30 (containing 0.15% of MOEO), and PC40 (containing 0.2% MOEO).

2.2. GC-MS Analysis of Melissa Essential Oil

The chromatogram of Melissa essential oil is shown in Figure 2. The main compounds detected in MOEO were 2,6-octadienal, 3,7-dimethyl (20.40%), citral (19.64%), caryophyllene (9.28%), geranyl acetate (7.33%), caryophyllene oxide (6.85%), citronellal (6.58%), and linalool (6.00%) (Table 1). As per Sorensen's research [15], the primary constituents of *M. officinalis* obtained through laboratory distillation of verified plant material are the isomers of citral, geranial, and neral. Previous studies have reported that the essential oil of lemon balm contains citronellol and linalool as their primary chemical constituents [15,16]. Park and Lee [17] reported that the major compound of essential oil in lemon was 2,6-octadienal and 3,7-dimethyl-. The varied chemical components in the literature could be attributed to several factors influencing the chemical diversity of essential oils, such as light, precipitation, growth location, and soil [18].

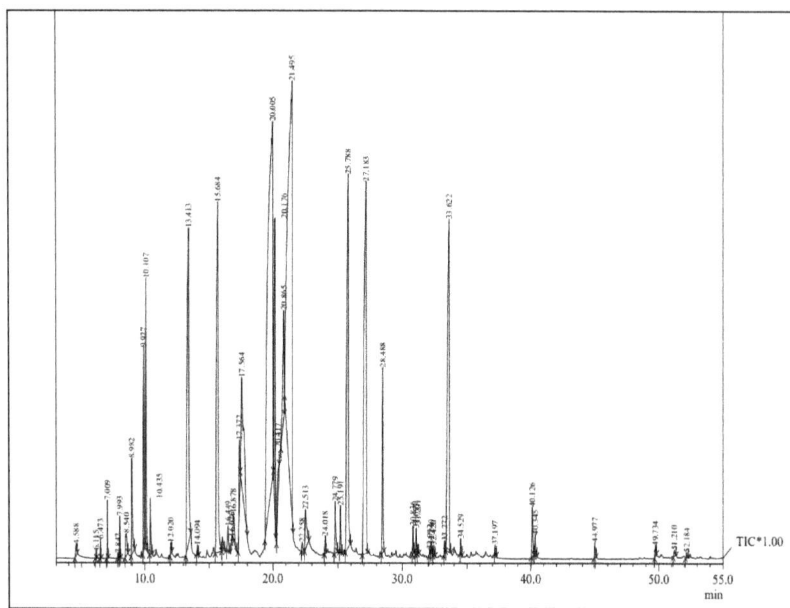

Figure 2. Chromatogram of Melissa essential oil.

Table 1. Primary components detected in Melissa essential oil.

No.	NAME	R. Time	Composition %
1	2,6-OCTADIENAL,3,7-DIMETHYL	20.005	20.40
2	Citral	21.495	19.64
3	Caryophyllene	27.183	9.28
4	Geranyl acetate	25.788	7.33
5	Caryophyllene oxide	33.622	6.85
6	Citronellal	15.68	6.58
7	Linalool	13.413	6.00

2.3. *Thickness*

Table 2 presents the results that were obtained for the average thickness of the edible film samples. The thickness values varied between 0.061 and 0.073 mm by the incorporation of MOEO. Control film (B) had the lowest value of thickness (0.061 mm); however, the highest value (0.073 mm) was observed in PC40 with a maximum concentration (0.2%) of MOEO. The results demonstrate that incorporating MOEO in pectin–collagen-based hydrogel films had a significant ($p > 0.05$) effect on the resultant film thickness. The observed increase in thickness may be attributed to the incorporation of oil into the film matrix, which consequently led to the generation of microdroplets owing to the oil's hydrophobic properties [19]. In our prior research, the incorporation of ginger essential oil had an impact on the thickness of hydrogel films composed of chitosan and porphyran [20].

Table 2. Thickness, mechanical, and barrier properties of edible film samples loaded with MOEO.

Film Samples	Thickness (mm)	Elongation at Break (%)	Tensile Strength (MPa)	Water Vapor Permeability (g mm/m^2 h kPa)	Moisture Content (%)
B	0.061 ± 0.005 ^{a}	27.36 ± 1.94 ^{a}	8.12 ± 0.68 ^{a}	0.676 ± 0.02 ^{a}	29.92 ± 0.53 ^{a}
PC20	0.068 ± 0.005 ^{ab}	30.24 ± 2.23 ^{bc}	3.48 ± 0.11 ^{b}	0.569 ± 0.04 ^{b}	28.30 ± 0.57 ^{b}
PC30	0.072 ± 0.008 ^{b}	34.97 ± 3.76 ^{cd}	2.37 ± 0.21 ^{c}	0.536± 0.05 ^{b}	26.55 ± 0.23 ^{c}
PC40	0.073 ± 0.006 ^{b}	36.29 ± 2.80 ^{d}	1.25 ± 0.16 ^{d}	0.327 ± 0.01 ^{c}	25.16 ± 0.83 ^{c}

The values with different letters (a, b, c, and d) inside a column indicate significant differences ($p < 0.05$).

2.4. Mechanical Attributes

The mechanical properties of biopolymer films are commonly evaluated through the measurement of TS and EAB. The mechanical resistance of a film is attributed to cohesive forces between chains, which is represented by its tensile strength. On the other hand, the flexibility of a film, or its capacity to elongate before breaking, is evaluated through its elongation at break. The concurrent study of tensile strength and elongation at break is necessary due to the structural characteristics of films, whereby those with high tensile strength typically exhibit low elongation at break [21].

Table 2 represents the values for the mechanical properties, including TS and EAB of pectin–collagen-based composite films loaded with MOEO. The study found that the tensile strength varied between 8.12 and 1.25MPa, with the control film exhibiting the highest value and PC40 exhibiting the lowest value. The values of elongation at the break exhibited significant variation ($p < 0.05$) ranging from 27.36 to 36.29%. The control film (B) demonstrated the lowest value, while the PC40 film showed the highest value. The incorporation of MOEO into the biopolymer network decreased TS caused mostly by intra- and inter-molecular interactions [13]. The observed enhancement in the EAB can be attributed to the incorporation of MOEO, which acted as a plasticizer and facilitated the mobility of polymer chains, thereby imparting flexibility to the films [22]. The results of the present study are consistent with those reported by Al-Harrasi et al. [20], wherein the incorporation of ginger essential oil into chitosan and porphyran-based composite film resulted in a reduction in tensile strength.

2.5. Barrier Properties

The water vapor permeability (WVP) and moisture content of hydrogel-based films made of pectin and collagen and loaded with different concentrations of MOEO were assessed. The results indicated that the control film sample (B) without the addition of MOEO showed higher values for WVP as compared to the film samples loaded with oil (Table 2). The WVP of the film samples decreased from 0.676 to 0.327 (g mm)/(m^2 h kPa) with increasing concentrations of MOEO. This behavior of the films could be due to the hydrophobicity of the MOEO causing the inhibition of water permeability, since the Tween 80 has a hydrophilic character. The water WVP of the film could be affected by the hydrophilic–lipophilic ratio of the film matrix, as water diffuses through the hydrophilic section of the matrix [23]. In addition, the reduced WVP of the films could be ascribed to the crosslinking between the polymer chains and reduced chain mobility with the addition of MOEO and Tween 80, resulting from the filling of empty spaces in the film matrix [24]. Similar findings were reported by Almasi et al. [23], wherein the pectin-based hydrogel films resulted in to decrease in the permeability of water vapors when loaded with marjoram essential oil nanoemulsions.

Assessment of the moisture content is a crucial parameter because it affects the mechanical, barrier, and sensory characteristics of the films. The film samples that were loaded with MOEO exhibited a lower level of moisture content in comparison to the control film, as indicated in Table 2. The moisture content decreased from 29.92 to 25.16% with increasing concentrations of the oil in the film samples. The findings of the present investigation are consistent with those obtained by Nisar et al. [25], who found that the incorporation of clove bud essential oil into edible films based on citrus pectin resulted in a decrease in the amount of water present in the films.

2.6. Transparency and Color Analysis

The transparency of the edible films is an important parameter as it affects the visual appearance of the food products. The percentage transparency values varied significantly ($p < 0.05$) from 90.37 to 82.80% with the addition of oil in the films. The transparency of the control film (B) was highest, while the film sample, PC40 with the highest concentration of oil exhibited the lowest transparency. The reduction in transparency after the inclusion of MOEO may be attributed to the presence of colorful constituents within the oil [20]. The results of the current investigation are consistent with those reported by Scartazzini et al. [26], wherein a decrease in transparency was noted upon the incorporation of mint essential oil into films.

The color parameters of different film samples loaded with varying concentrations of MOEO are presented in Table 3. The lightness (L^*) of the film samples slightly increased with higher MOEO concentration, but there were no significant differences observed in the values of the lightness of different samples. The inclusion of MOEO resulted in notable changes in the yellowness (b^*) (0.95–1.57) of the film samples. The increase in the yellow color may be associated with the existence of various pigments within MOEO. Furthermore, a significant difference was detected in the a^* redness values of the film samples. In addition, a higher ΔE^* value was associated with a higher amount of MOEO; for instance, PC40 had the highest value of ΔE (1.90), and B film (control film) had the lowest value (0.85). These results demonstrated that the addition of MOEO changed the color characteristics of pectin-collagen-based composite films. Similar findings were reported by Al-Harrasi [20] et al., indicating that the incorporation of essential oil increased the yellowness of the edible films.

Table 3. Color parameters of different edible film samples loaded with MOEO.

Film Samples	Transparency (%)	L	a^*	b^*	ΔE^*
B	90.37 ± 0.55 a	95.82 ± 0.05 a	-0.10 ± 0.01 a	0.95 ± 0.07 a	0.85 ± 0.07 a
PC20	87.30 ± 0.28 b	96.07 ± 0.02 a	-0.15 ± 0.02 b	1.32 ± 0.13 b	1.16 ± 0.08 b
PC30	84.33 ± 0.55 c	96.52 ± 0.34 a	-0.17 ± 0.01 b	1.57 ± 0.27 bc	1.63 ± 0.01 c
PC40	82.80 ± 0.75 c	96.54 ± 0.46 a	-0.23 ± 0.02 c	1.57 ± 0.06 c	1.90 ± 0.01 d

The values with different letters (a, b, c, and d) inside a column indicate significant differences ($p < 0.05$). L: lightness, a^*: green-red color, b^*: blue-yellow color, and ΔE^*: overall color variation.

2.7. Scanning Electron Microscopy (SEM)

SEM analysis can offer a comprehensive depiction of the surface texture of the film, which can help to determine its roughness, porosity, and other surface properties. SEM images of the prepared film samples with and without the addition of MOEO are presented in Figure 3. The surface morphology of the control film sample was observed as homogeneous, with cracks and microscopic particles all over the surface. PC20 film sample with 0.1% concentration of MOEO showed a rough surface with fewer cracks and particles across the surface. However, PC30 and PC40 film samples with the higher MOEO concentrations of 0.15 and 0.2%, relatively had homogeneous structures, and the oil was uniformly dispersed in the film matrix. Tween 80 stabilizes the oil-in-water emulsion, and the uniform dispersion of the oil in the film matrix could be due to the addition of the highest concentrations of Tween 80 in the PC 40 film sample. Overall, the SEM analysis revealed that films loaded with different concentrations of MOEO showed a more homogeneous structure with less numbers of particles, cracks, and pores as compared to the control films sample. The observed surface properties of the films may be attributed to the intermolecular interaction among the constituents that contribute to film formation, such as polymers, oil, and plasticizer.

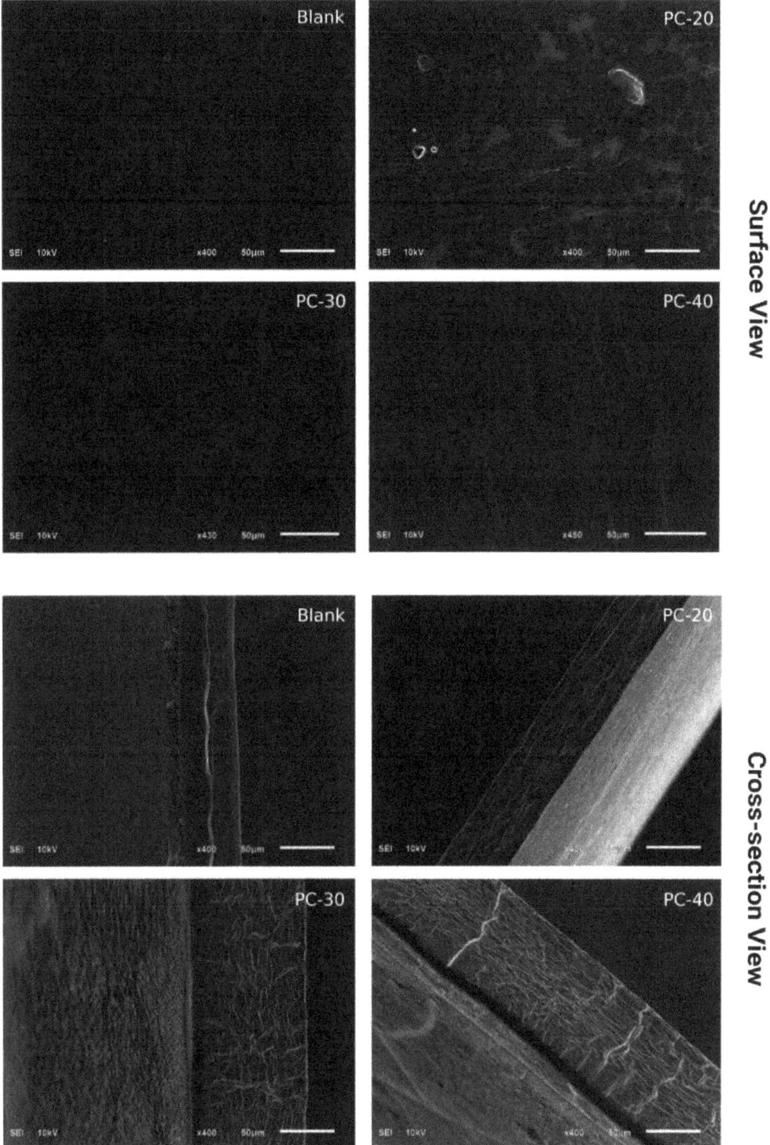

Figure 3. Morphological characteristics of B (control), PC20 (containing 0.1% MOEO), PC30 (containing 0.15% of MOEO), and PC40 (containing 0.2% MOEO).

2.8. X-ray Diffraction (XRD) Analysis

XRD analysis is an important tool for the characterization of edible films, as it can provide valuable information about their crystalline structure and properties, which is essential for their development and optimization. Figure 4 depicts an overlay of XRD diffraction peaks and intensities for different edible film samples loaded with MOEO. The XRD analysis indicated that the samples exhibit peaks at similar positions, although with differing intensities. The distinctive peaks of film samples loaded with MOEO and control were observed at position 21 of 2θ. It was observed that each of the film samples displayed a significant amorphous phase. The present research indicates that there is a reduction in the crystalline nature of the films as the concentration of MOEO increased. The tensile strength of edible film samples can be significantly influenced by their crystallinity. In general, a higher degree of crystallinity in the film leads to a higher tensile strength, while a more amorphous structure leads to a lower tensile strength.

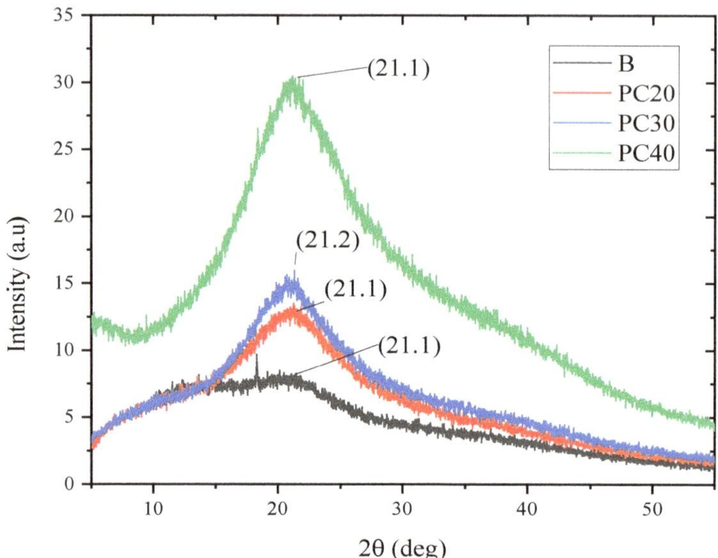

Figure 4. X-ray diffraction analysis of B (control), PC20 (containing 0.1% MOEO), PC30 (containing 0.15% of MOEO), and PC40 (containing 0.2% MOEO).

2.9. Fourier Transform Infrared Spectroscopy (FTIR) Analysis

FTIR analysis provides information about the interactions among functional groups of film constituents. The FTIR pattern of the film samples based on pectin and collagen containing different concentrations of MOEO is shown in Figure 5. The 3300 cm^{-1} region band peaks reflected OH stretching bonds. Inter-molecular or intramolecular hydrogen bonds express the stretching vibration of OH [13]. The O=C=O stretching vibrations were responsible for the peak at the 2389 cm^{-1} region. The active films containing MOEO exhibited bands in those areas, suggesting that MOEO was linked to the biopolymers comprising film matrix, as depicted in Figure 5 [27]. The band at 2274 cm^{-1} was linked to the N=C=O isocyanate group, whereas the band at 1170–1220 cm^{-1} was linked to the C–O stretching vibrations. The stretching vibrations of –OH and hydrogen bonding among the hydroxyl groups of the film-forming components can be indicated by a broad peak observed in the 3600–3200 cm^{-1} range [28]. Films loaded with MOEO showed an increase in the peak intensities of (–OH) stretching as compared to the control film sample, which indicates the formation of new hydrogen bonds with the addition of oil in the film matrix. Furthermore, alterations in the FTIR spectrum were noted in oil-loaded samples, resulting in the appearance of new characteristic peaks (Figure 5). In general, the FTIR examination demonstrated the interaction among the functional groups of the film-forming constituents.

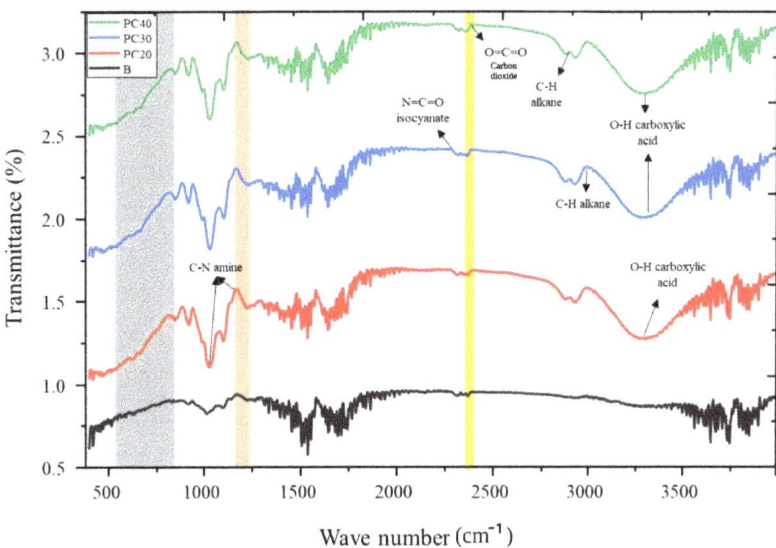

Figure 5. Fourier transform infrared patterns of B (control), PC20 (containing 0.1% MOEO), PC30 (containing 0.15% of MOEO), and PC40 (containing 0.2% MOEO).

2.10. Antioxidant Activities of the Hydrogel-Based Films

Edible films that contain essential oils or plant extracts exhibit promising antioxidant characteristics and offer a natural substitute for synthetic antioxidants. Additionally, these films enhance the sensory attributes, life expectancy, and overall quality of food products. The prepared hydrogel-based films containing MOEO were assessed for their antioxidant properties by using DPPH and ABTS assays. The results of the present investigation are shown in Figure 6, which indicate a rise in the DPPH and ABTS radical scavenging activity of the hydrogel-based films when loaded with MOEO. The highest DPPH and ABTS radical scavenging activity was shown by the PC40 film sample loaded with the highest concentration of MOEO; however, the lowest antioxidant activity was observed in the control film sample. The DPPH and ABTS radical scavenging activity in the film samples was enhanced with the addition of MOEO, increasing from 18.56 to 59.03% and from 64.31 to 73.75%, correspondingly. The improvement observed in the antioxidant activity of the films may be attributed to the existence of phenolic compounds in the MOEO.

Sani et al. [14] also reported similar results in which the sodium caseinate-based edible films possessed higher antioxidant activity when incorporated with a combination of zinc oxide nanoparticles (ZnONPs) and microcapsules of MOEO. Furthermore, many researchers have shown an increase in the antioxidant potency of edible films upon the incorporation of essential oils [28–30].

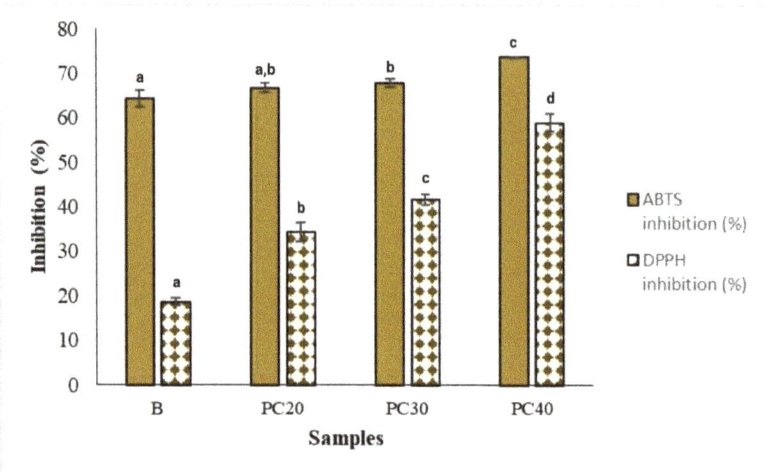

Figure 6. ABTS and DPPH inhibition percentage of B (control), PC20 (containing 0.1% MOEO), PC30 (containing 0.15% of MOEO), and PC40 (containing 0.2% MOEO). Different letters (a, b, c, and d) indicate significant differences ($p < 0.05$).

3. Conclusions

The *Melissa officinalis* essential oil (MOEO) appears as an interesting ingredient for the formulation of pectin and collagen composite-based hydrogel films. MOEO-loaded films showed more elasticity and thickness; however, the tensile strength and transparency were lower. The addition of the highest concentration (0.2%) of MOEO improved the barrier and antioxidant properties of the hydrogel-based films. The findings of the present investigation indicate that composite hydrogel films composed of pectin and collagen loaded with MOEO exhibit promising prospects as a viable food packaging material. However, more research is needed to enhance the mechanical and optical properties of the prepared films.

4. Materials and Methods

4.1. Material Procurement

Biopolymers including pectin and collagen with a purity of 95% were purchased from SRL Pvt. Ltd., Mumbai, India. The plasticizer (glycerol) was obtained from BDH Laboratory located in London, England. The Melissa oil used in this study was sourced from Nature Natural, a company based in Ghaziabad, India, and was identified by the batch number NNIMEEO/154/0821.

4.2. Film Preparation

The film-forming solutions of pectin (1.5% *w*/*v*) and collagen (1% *w*/*v*) were prepared separately by using distilled water. After the complete dissolution of polymers in distilled water, both solutions were blended. The solution was distributed in four labeled beakers (B, PC-20, PC-30, and PC-40), and concentrations of Tween 80, MOEO, and glycerol were added in the film-forming solutions, as shown in Table 4. The resultant solutions were eventually transferred into the respective Petri dishes and then subjected to drying at suitable room temperature conditions. Finally, the films were peeled and stored for further analysis.

Table 4. The composition of the film-forming solution for different samples.

Film Samples	Film Forming Components
B/Control	(1.5%) Pectin (*w*/*v*) + 1% Collagen (*w*/*v*) + (1.25%) Glycerol (*v*/*v*)
PC20	(1.5%) Pectin (*w*/*v*) + 1% Collagen (*w*/*v*) + (1.25%) Glycerol (*v*/*v*) + (0.2%) Tween 80 (*v*/*v*) + (0.1%) MOEO (*v*/*v*)
PC30	(1.5%) Pectin (*w*/*v*) + 1% Collagen (*w*/*v*) + (1.25%) Glycerol (*v*/*v*) + (0.3%) Tween 80 (*v*/*v*) + (0.15%) MOEO (*v*/*v*)
PC40	(1.5%) Pectin (*w*/*v*) + 1% Collagen (*w*/*v*) + (1.25%) Glycerol (*v*/*v*) + (0.4%) Tween 80 (*v*/*v*) + (0.2%) MOEO (*v*/*v*)

4.3. Chromatographic (GC/MS) Analysis

Gas Chromatography–Mass Spectrometry (GCMS) analysis was performed to determine the composition of MOEO using GCMS-QP-2010 Plus Gas chromatograph Mass Spectrometer, Shimadzu, Japan. Sample size and GCMS conditions such as oven temperature, ramp rate, split ratio, the flow rate of Helium, and temperature of the ion source and injector were used as reported in our previous study [31]. The analysis was performed by similarity searches in the National Institute of Standards and Technology (NIST) mass spectra database with the obtained pure mass spectrum of each component.

4.4. The Thickness of the Films

The sample thickness (in mm) was determined by digital micrometer (Yu-Su 150, Yu-Su Tools, Shanghai, China). Five different measurements from different positions were obtained from each film sample, and the mean value was calculated.

4.5. Mechanical Testing

A texture analyzer (XT plus, Stable Micro Systems, Godalming, UK) was used to assess the mechanical strength of the films using the ASTM D882 standard method [32]. Tensile strength (MPa) and elongation at break (%) of the prepared samples were evaluated. The equations utilized in the determination of TS and EAB are as follows.

$$TensileStrength(TS) = (\frac{F}{A}) \tag{1}$$

The variable F denotes the maximum force, while A stands for the cross-sectional area of the film.

$$Elongationatbreak(EAB)(\%) = \frac{Lf - Li}{Li} \times 100 \tag{2}$$

The symbol Li represents the initial length of the film, while Lf denotes the length of the film in millimeters at the point of break.

4.6. Water Vapor Permeability (WVP)

The WVP of the film samples was determined by Erdem et al. [33]. The measured WVP of the films was calculated using Equation (3):

$$WaterVaporPermeability = \frac{\Delta m}{\Delta t \times \Delta P \times A} \times d \tag{3}$$

The above equation involves the representation of $\Delta m/\Delta t$, which denotes the amount of moisture gained per unit of time. The variable A represents the surface area of the film in square meters. The symbol ΔP denotes the disparity in water vapor pressure across the film, measured in kPa. Lastly, the letter d signifies the thickness of the film in millimeters.

4.7. Moisture Content (MC)

The gravimetric method was used to determine the MC of the film samples by subjecting them to drying at 105 °C. The weight variation among films was determined using the following equation:

$$MoistureContent = \frac{W1 - W2}{W1} \times 100 \quad (4)$$

where $W1$ and $W2$ represent the weight before and after drying of the films.

4.8. Transparency and Color Analysis

The light transmission of the prepared films was quantified at 550 nm wavelength using an ONDA-Vis spectrophotometer, V-10 Plus, ONDA, Padova, Italy. Film strips were placed in a spectrophotometer test cell, and the transparency of the films was determined as per the methodology described by Zhao et al. [34].

A Konica Minolta colorimeter (Konica Minolta, Tokyo, Japan) was utilized to perform color analysis of the film samples. The L^*, a^*, and b^* parameters were measured on a reference plate with a lightness value of 100. The analysis was conducted at different points on the surface of the films. The equation utilized for the computation of the overall color variation is as follows:

$$\Delta E^* = \left[(\Delta L^*)^2 + (\Delta a^*)^2 + (\Delta b^*)^2\right]^{1/2} \quad (5)$$

L: lightness, a^*: green-red color, b^*: blue-yellow color, and ΔE^*: overall color variation.

4.9. Scanning Electron Microscopy (SEM)

The surface, as well as cross-sectional microstructural features of the films, was investigated using scanning electron microscopy (SEM) (JSM6510LA, Analytical SEM, Jeol, Tokyo, Japan) following the procedure as per our previous study [31].

4.10. X-ray Diffraction (XRD) Analysis

The percentage of crystallinity of the samples was recorded by an X-ray diffractometer (Bruker D8 Discover, Billerica, MA, USA) running with copper (Kα) radiation (1.5418 Å), 40 kV. Samples were analyzed between $2\theta = 5°$ and $2\theta = 50°$ at the rate of 0.500 s/point.

4.11. FTIR Spectra Analysis

The study employed an Infrared Tensor 37 instrument (InfraRed Bruker, Ettlingen, Germany) to evaluate the chemical interaction among the constituents of the film matrix. The Fourier Transform Infrared Spectroscopy (FTIR) instrument was configured to conduct 32 scans across a wide spectral region spanning from 400 to 4000 cm^{-1}.

4.12. Antioxidant Activity

The free radical scavenging capability of the film samples was determined by using the DPPH and ABTS radical scavenging activity. The antioxidant activity of DPPH was studied as per the procedure described earlier by Brand-Williams et al. [35], utilizing 50 mg of film samples. The measurement of absorbance was conducted at a wavelength of 517 nm utilizing an ONDA Vis spectrophotometer. The outcomes were expressed as a percentage of inhibition. The antioxidant potential was assessed by using the chromogenic compound ABTS as per the procedure followed by Re et al. [36], with some minor modifications. The sample amount of 25 mg was used in the ABTS assay. The measurement of absorbance was conducted at a wavelength of 734 nm. The outcomes were expressed as a percentage of inhibition and were derived from an average of three measurements.

4.13. Statistical Analysis

The results were presented as mean ± SD of the triplicated determinations and were analyzed through a one-way analysis of variance using a statistical software package (SPSS ver. 17.0, SPSS Inc., Chicago, IL, USA). The current study presents results as the average and standard deviation (SD) of three separate measurements. The means were compared using Duncan's test at a significant level of 5%.

Author Contributions: Conceptualization, S.B.; formal analysis, S.K., J.I., M.K.A., A.K.N., E.K. and L.Y.A.; methodology, A.S.A., E.K. and L.Y.A.; software, A.S.A., Y.A.S. and E.K.; supervision, S.B. and A.A.-H.; validation, S.K., J.I., M.K.A. and A.K.N.; writing—original draft, S.B., A.S.A. and Y.A.S.; writing—review and editing, Y.A.S. and L.Y.A. All authors have read and agreed to the published version of the manuscript.

Funding: This research work was funded by Institutional Fund Projects under grant number IFPIP: 1777-249-1443.

Informed Consent Statement: Not applicable.

Data Availability Statement: Not applicable.

Acknowledgments: The authors gratefully acknowledge the technical and financial support provided by the Ministry of Education and Abdulaziz University, DSR, Jeddah, Saudi Arabia. The authors are also thankful to the Natural and Medical Sciences Research Center, University of Nizwa, Oman, for providing the research facilities used to conduct the current study.

Conflicts of Interest: The authors declare no conflict of interest.

References

1. de Morais Lima, M.; Carneiro, L.C.; Bianchini, D.; Dias, A.R.G.; Zavareze, E.d.R.; Prentice, C.; Moreira, A.d.S. Structural, thermal, physical, mechanical, and barrier properties of chitosan films with the addition of xanthan gum. *J. Food Sci.* **2017**, *82*, 698–705. [CrossRef]
2. Casas-Orozco, D.; Villa, A.L.; Bustamante, F.; González, L.-M. Process development and simulation of pectin extraction from orange peels. *Food Bioprod. Process.* **2015**, *96*, 86–98. [CrossRef]
3. Gutierrez-Pacheco, M.; Ortega-Ramirez, L.; Cruz-Valenzuela, M.; Silva-Espinoza, B.; Gonzalez-Aguilar, G.; Ayala-Zavala, J. Combinational Approaches for Antimicrobial Packaging: Pectin and Cinnamon Leaf Oil. *Antimicrob. Food Packag.* **2016**, 609–617. [CrossRef]
4. Borges, J.G.; Silva, A.G.; Cervi-Bitencourt, C.; Vanin, F.M.; Carvalho, R.A.d. Lecithin, gelatin and hydrolyzed collagen orally disintegrating films: Functional properties. *Int. J. Biol. Macromol.* **2016**, *86*, 907–916. [CrossRef] [PubMed]
5. Wang, Z.; Hu, S.; Wang, H. Scale-up preparation and characterization of collagen/sodium alginate blend films. *J. Food Qual.* **2017**, *2017*, 4954259.
6. Jridi, M.; Abdelhedi, O.; Salem, A.; Kechaou, H.; Nasri, M.; Menchari, Y. Physicochemical, antioxidant and antibacterial properties of fish gelatin-based edible films enriched with orange peel pectin: Wrapping application. *Food Hydrocoll.* **2020**, *103*, 105688. [CrossRef]
7. Lei, Y.; Wu, H.; Jiao, C.; Jiang, Y.; Liu, R.; Xiao, D.; Lu, J.; Zhang, Z.; Shen, G.; Li, S. Investigation of the structural and physical properties, antioxidant and antimicrobial activity of pectin-konjac glucomannan composite edible films incorporated with tea polyphenol. *Food Hydrocoll.* **2019**, *94*, 128–135. [CrossRef]
8. Ali, A.; Noh, N.M.; Mustafa, M.A. Antimicrobial activity of chitosan enriched with lemongrass oil against anthracnose of bell pepper. *Food Packag. Shelf Life* **2015**, *3*, 56–61. [CrossRef]
9. Sánchez-González, L.; Vargas, M.; González-Martínez, C.; Chiralt, A.; Chafer, M. Use of essential oils in bioactive edible coatings: A review. *Food Eng. Rev.* **2011**, *3*, 1–16. [CrossRef]
10. Sani, I.K.; Pirsa, S.; Tağı, Ş. Preparation of chitosan/zinc oxide/Melissa officinalis essential oil nano-composite film and evaluation of physical, mechanical and antimicrobial properties by response surface method. *Polym. Test.* **2019**, *79*, 106004. [CrossRef]
11. Santos-Neto, L.L.d.; de Vilhena Toledo, M.A.; Medeiros-Souza, P.; de Souza, G.A. The use of herbal medicine in Alzheimer's disease—A systematic review. *Evid.-Based Complement. Altern. Med.* **2006**, *3*, 441–445. [CrossRef] [PubMed]
12. Karimi Sani, I.; Alizadeh, M.; Pirsa, S.; Moghaddas Kia, E. Impact of operating parameters and wall material components on the characteristics of microencapsulated *Melissa officinalis* essential oil. *Flavour Fragr. J.* **2019**, *34*, 104–112. [CrossRef]
13. Yu, H.; Zhang, C.; Xie, Y.; Mei, J.; Xie, J. Effect of Melissa officinalis L. essential oil nanoemulsions on structure and properties of carboxymethyl chitosan/locust bean gum composite films. *Membranes* **2022**, *12*, 568. [CrossRef] [PubMed]

14. Sani, I.K.; Marand, S.A.; Alizadeh, M.; Amiri, S.; Asdagh, A. Thermal, mechanical, microstructural and inhibitory characteristics of sodium caseinate based bioactive films reinforced by ZnONPs/encapsulated Melissa officinalis essential oil. *J. Inorg. Organomet. Polym. Mater.* **2021**, *31*, 261–271. [CrossRef]
15. Sorensen, J.M. Melissa officinalis. *Int. J. Aromather.* **2000**, *10*, 7–15. [CrossRef]
16. Mrlianova, M.; Tekel'ova, D.; Felklova, M.; Toth, J.; Musil, P.; Grancai, D. Comparison of the quality of *Melissa officinalis* L. cultivar Citra with Mellissas of European origin. *Ceska A Slov. Farm. Cas. Ceske Farm. Spol. Slov. Farm. Spol.* **2001**, *50*, 299–302.
17. Park, J.-H.; Lee, H.-S. Acaricidal target and mite indicator as color alteration using 3,7-dimethyl-2,6-octadienal and its derivatives derived from *Melissa officinalis* leaves. *Sci. Rep.* **2018**, *8*, 8129. [CrossRef]
18. Barra, A. Factors affecting chemical variability of essential oils: A review of recent developments. *Nat. Prod. Commun.* **2009**, *4*, 1934578X0900400827. [CrossRef]
19. Valizadeh, S.; Naseri, M.; Babaei, S.; Hosseini, S.M.H.; Imani, A. Development of bioactive composite films from chitosan and carboxymethyl cellulose using glutaraldehyde, cinnamon essential oil and oleic acid. *Int. J. Biol. Macromol.* **2019**, *134*, 604–612. [CrossRef]
20. Al-Harrasi, A.; Bhtaia, S.; Al-Azri, M.S.; Makeen, H.A.; Albratty, M.; Alhazmi, H.A.; Mohan, S.; Sharma, A.; Behl, T. Development and characterization of chitosan and porphyran based composite edible films containing ginger essential oil. *Polymers* **2022**, *14*, 1782. [CrossRef]
21. Galus, S.; Lenart, A. Development and characterization of composite edible films based on sodium alginate and pectin. *J. Food Eng.* **2013**, *115*, 459–465. [CrossRef]
22. Zhang, Y.; Zhou, L.; Zhang, C.; Show, P.L.; Du, A.; Fu, J.; Ashokkumar, V. Preparation and characterization of curdlan/polyvinyl alcohol/thyme essential oil blending film and its application to chilled meat preservation. *Carbohydr. Polym.* **2020**, *247*, 116670. [CrossRef] [PubMed]
23. Almasi, H.; Azizi, S.; Amjadi, S. Development and characterization of pectin films activated by nanoemulsion and Pickering emulsion stabilized marjoram (*Origanum majorana* L.) essential oil. *Food Hydrocoll.* **2020**, *99*, 105338. [CrossRef]
24. Zhu, J.-Y.; Tang, C.-H.; Yin, S.-W.; Yang, X.-Q. Development and characterization of novel antimicrobial bilayer films based on Polylactic acid (PLA)/Pickering emulsions. *Carbohydr. Polym.* **2018**, *181*, 727–735. [CrossRef] [PubMed]
25. Nisar, T.; Wang, Z.-C.; Yang, X.; Tian, Y.; Iqbal, M.; Guo, Y. Characterization of citrus pectin films integrated with clove bud essential oil: Physical, thermal, barrier, antioxidant and antibacterial properties. *Int. J. Biol. Macromol.* **2018**, *106*, 670–680. [CrossRef]
26. Scartazzini, L.; Tosati, J.; Cortez, D.; Rossi, M.; Flôres, S.; Hubinger, M.; Di Luccio, M.; Monteiro, A. Gelatin edible coatings with mint essential oil (*Mentha arvensis*): Film characterization and antifungal properties. *J. Food Sci. Technol.* **2019**, *56*, 4045–4056. [CrossRef] [PubMed]
27. Behjati, J.; Yazdanpanah, S. Nanoemulsion and emulsion vitamin D3 fortified edible film based on quince seed gum. *Carbohydr. Polym.* **2021**, *262*, 117948. [CrossRef]
28. Bhatia, S.; Al-Harrasi, A.; Shah, Y.A.; Jawad, M.; Al-Azri, M.S.; Ullah, S.; Anwer, M.K.; Aldawsari, M.F.; Koca, E.; Aydemir, L.Y. The Effect of Sage (*Salvia sclarea*) Essential Oil on the Physiochemical and Antioxidant Properties of Sodium Alginate and Casein-Based Composite Edible Films. *Gels* **2023**, *9*, 233. [CrossRef]
29. Xu, T.; Gao, C.; Feng, X.; Yang, Y.; Shen, X.; Tang, X. Structure, physical and antioxidant properties of chitosan-gum arabic edible films incorporated with cinnamon essential oil. *Int. J. Biol. Macromol.* **2019**, *134*, 230–236. [CrossRef]
30. Tongnuanchan, P.; Benjakul, S.; Prodpran, T. Properties and antioxidant activity of fish skin gelatin film incorporated with citrus essential oils. *Food Chem.* **2012**, *134*, 1571–1579. [CrossRef]
31. Bhatia, S.; Al-Harrasi, A.; Jawad, M.; Shah, Y.A.; Al-Azri, M.S.; Ullah, S.; Anwer, M.K.; Aldawsari, M.F.; Koca, E.; Aydemir, L.Y. A Comparative Study of the Properties of Gelatin (Porcine and Bovine)-Based Edible Films Loaded with Spearmint Essential Oil. *Biomimetics* **2023**, *8*, 172. [CrossRef] [PubMed]
32. *ASTM D882-02*; Standard Test Method for Tensile Properties of Thin Plastic Sheeting. ASTM International: West Conshohocken, PA, USA, 2002.
33. Erdem, B.G.; Dıblan, S.; Kaya, S. Development and structural assessment of whey protein isolate/sunflower seed oil biocomposite film. *Food Bioprod. Process.* **2019**, *118*, 270–280. [CrossRef]
34. Zhao, J.; Wang, Y.; Liu, C. Film Transparency and Opacity Measurements. *Food Anal. Methods* **2022**, *15*, 2840–2846. [CrossRef]
35. Brand-Williams, W.; Cuvelier, M.; Berset, C. Antioxidative activity of phenolic composition of commercial extracts of sage and rosemary. *LWT* **1995**, *28*, 25–30. [CrossRef]
36. Re, R.; Pellegrini, N.; Proteggente, A.; Pannala, A.; Yang, M.; Rice-Evans, C. Antioxidant activity applying an improved ABTS radical cation decolorization assay. *Free Radic. Biol. Med.* **1999**, *26*, 1231–1237. [CrossRef] [PubMed]

Article

Fabrication, Characterization, and Antioxidant Potential of Sodium Alginate/Acacia Gum Hydrogel-Based Films Loaded with Cinnamon Essential Oil

Saurabh Bhatia [1,2,3,*,†], Ahmed Al-Harrasi [1,*,†], Yasir Abbas Shah [1,†], Halima Waleed Khalifa Altoubi [1], Sabna Kotta [4,5], Priyanka Sharma [6], Md. Khalid Anwer [7], Deepa Sreekanth Kaithavalappil [8], Esra Koca [9] and Levent Yurdaer Aydemir [9]

1 Natural and Medical Sciences Research Center, University of Nizwa, Birkat Al Mauz, P.O. Box 33, Nizwa 616, Oman; yasir.shah@unizwa.edu.om (Y.A.S.); 7halima5757@gmail.com (H.W.K.A.)
2 School of Health Science, University of Petroleum and Energy Studies, Dehradun 248007, India
3 Center for Transdisciplinary Research, Department of Pharmacology, Saveetha Dental College and Hospital, Saveetha Institute of Medical and Technical Sciences, Saveetha University, Chennai 600077, India
4 Department of Pharmaceutics, Faculty of Pharmacy, King Abdulaziz University, Jeddah 21589, Saudi Arabia; skotta@kau.edu.sa
5 Center of Excellence for Drug Research and Pharmaceutical Industries, King Abdulaziz University, Jeddah 21589, Saudi Arabia
6 Center for Innovation in Personalized Medicine, King Fahad Medical Research Center, King Abdulaziz University, Jeddah 21589, Saudi Arabia; aamrahs@kau.edu.sa
7 Department of Pharmaceutics, College of Pharmacy, Prince Sattam Bin Abdulaziz University, Al-Kharj 11942, Saudi Arabia; m.anwer@psau.edu.sa
8 Department of Pharmaceutical Sciences, College of Pharmacy, Shaqra University, Shaqra 11961, Saudi Arabia; deepatv@su.edu.sa
9 Department of Food Engineering, Faculty of Engineering, Adana Alparslan Turkes Science and Technology University, Adana 01250, Turkey; esrakoca.tr@outlook.com (E.K.); lyaydemir@atu.edu.tr (L.Y.A.)
* Correspondence: sbsaurabhbhatia@gmail.com (S.B.); aharrasi@unizwa.edu.om (A.A.-H.)
† These authors contributed equally to this work.

Citation: Bhatia, S.; Al-Harrasi, A.; Shah, Y.A.; Altoubi, H.W.K.; Kotta, S.; Sharma, P.; Anwer, M.K.; Kaithavalappil, D.S.; Koca, E.; Aydemir, L.Y. Fabrication, Characterization, and Antioxidant Potential of Sodium Alginate/Acacia Gum Hydrogel-Based Films Loaded with Cinnamon Essential Oil. *Gels* **2023**, *9*, 337. https://doi.org/10.3390/gels9040337

Academic Editors: Aris Giannakas, Constantinos Salmas and Charalampos Proestos

Received: 12 March 2023
Revised: 4 April 2023
Accepted: 10 April 2023
Published: 15 April 2023

Abstract: Several studies have reported the advantages of incorporating essential oils in hydrogel-based films for improving their physiochemical and antioxidant attributes. Cinnamon essential oil (CEO) has great potential in industrial and medicinal applications as an antimicrobial and antioxidant agent. The present study aimed to develop sodium alginate (SA) and acacia gum (AG) hydrogel-based films loaded with CEO. Scanning Electron Microscopy (SEM), X-ray diffraction (XRD), Fourier-transform infrared spectroscopy (FTIR), Differential scanning calorimetry (DSC), and texture analysis (TA) were performed to analyze the structural, crystalline, chemical, thermal, and mechanical behaviour of the edible films that were loaded with CEO. Moreover, the transparency, thickness, barrier, thermal, and color parameters of the prepared hydrogel-based films loaded with CEO were also assessed. The study revealed that as the concentration of oil in the films was raised, the thickness and elongation at break (EAB) increased, while transparency, tensile strength (TS), water vapor permeability (WVP), and moisture content (MC) decreased. As the concentration of CEO increased, the hydrogel-based films demonstrated a significant improvement in their antioxidant properties. Incorporating CEO into the SA–AG composite edible films presents a promising strategy for producing hydrogel-based films with the potential to serve as food packaging materials.

Keywords: sodium alginate; hydrogel polymer; acacia gum; cinnamon essential oil; edible hydrogel-based films; food packaging

1. Introduction

Recent developments in the food packaging industry have revealed that biopolymer-based edible films have great potential to replace plastics. The different characteristics of edible films, such as their non-toxicity, biodegradability, and safety, make them a feasible option for food packaging material [1]. Hydrogels are used in a broad range of applications due to their characteristic properties such as hydrophilicity, non-toxicity, and biocompatibility [2–4]. Sodium alginate (SA) is a natural hydrophilic polysaccharide and has been used in the fabrication of biopolymer films due to its excellent film-forming properties [5–7]. Sodium alginate-based edible films have demonstrated superior mechanical properties along with good transparency [8]. Sodium alginate is a common hydrogel polymer that tends to form hydrogel via substituting sodium ions of the guluronic acid residues. Despite their good mechanical properties, edible films made from sodium alginate are limited by their high hydrophilicity and poor heat stability [9]. Combining SA with various other polymers is one approach that has been investigated extensively to overcome these challenges.

Acacia gum, also known as gum Arabic, is a gummy exudate obtained from the branches and trunk of the Acacia senegal and Acacia seyal plant species [10]. AG is a water-soluble, highly branched polysaccharide extensively used as a thickener, stabilizer, and emulsifier in the food industry [11]. It has been employed in the production of edible films due to its biocompatibility, renewability, non-toxicity, pH stability, low cost, high solubility, and gelling properties [12]. AG can be used to make edible films; however, it poses several challenges, such as its poor mechanical attributes, its high hydrophilicity, and its poor barrier features [13]. However, the preparation of composite films consisting of SA and AG could be an option to enhance the physiochemical properties of the resulting films. Over the past few years, numerous studies have highlighted the advantages of incorporating bioactive compounds such as essential oils in edible films for improved physiochemical and antioxidant attributes [14–16]. In the current study, cinnamon essential oil (CEO) was incorporated in the SA–AG composite hydrogel-based films due to its vast industrial and medicinal applications, such as antimicrobial and antioxidant agents [17]. The essential oil of cinnamon has a variety of significant chemical constituents, the most prominent of which are the compounds aldehyde and alcohol, along with trans-cinnamaldehyde and eugenol [17–19]. Zhou et al. [20] studied the physiochemical properties of cassava starch-based edible films that were loaded with cinnamon essential oil. After the assessment, it was found that the films' barrier properties, crystallinity, and elongation at break (EAB) were considerably enhanced and that there was also an improvement in their thermal stability [20].

Given its antimicrobial and antioxidant characteristics, cinnamon is widely acknowledged as a safe food preservative [21]. The objective of the current study is to assess the physiochemical and antioxidant properties of composite hydrogel-based films based on SA and AG, which have been loaded with varying concentrations of cinnamon essential oil.

2. Results and Discussion

2.1. Visual Assessment of the Films

The visual assessment of edible films is necessary to determine the overall quality, appearance, and uniformity of the film structure. This is a crucial aspect for visually assessing hydrogel-based films in terms of their functionality and acceptance by consumers. The visual appearance of the films can impact the consumer perception of the product's quality and visual defects such as tears, pores, poor mechanical strength, and inconsistencies in thickness. The visual appearance of the SA–AG composite edible films loaded with CEO is shown in Figure 1. In general, as the amount of CEO in the film increased, a less transparent appearance of the film was observed in comparison with the control film. The incorporation of oil resulted in a slightly yellow appearance of the films. The control film sample showed good uniformity and was more transparent compared with film samples that were loaded with oil. However, the CEO-loaded films had good flexibility and physical attributes compared with the control. Moreover, different factors can impact the visual characteristics of the films such as the film composition, the preparation technique, handling, and storage conditions.

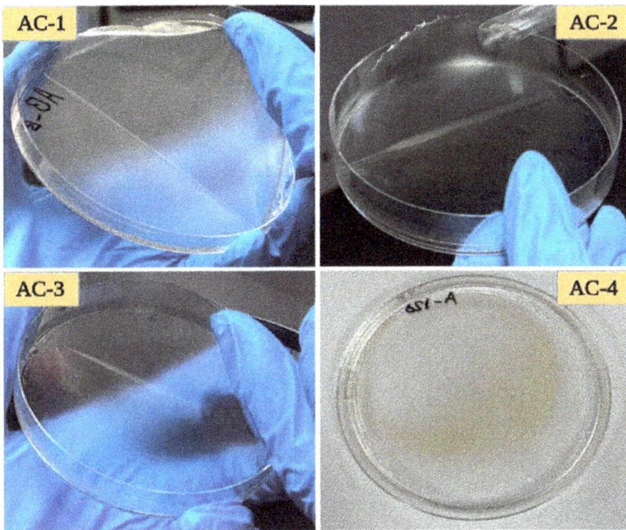

Figure 1. Visual assessment of SA–AG-based composite edible films; AC-1/Control, AC-2 contains 15 μL of CEO, AC-3 contains 20 μL of CEO, and AC-4 contains 30 μL of CEO.

2.2. Thickness of the Films

The thickness of edible films can vary depending on the type of material and the process used for the preparation. The composite hydrogel-based films based on SA and AG were examined for thickness measurement and the results are presented in Table 1. The control (AC-1) showed minimum thickness (0.065mm) compared with the samples loaded with CEO. The incorporation of the CEO in films significantly increased the thickness from 0.065 to 0.11 mm, with the maximum thickness in the AC-4 sample containing the maximum concentration (30 μL) of the CEO, followed by AC-3 and AC-2. The increase in the thickness could be due to the increase in the viscosity of the film-forming solution with the addition of CEO. Zhou et al. [20] also reported similar results, where the thickness of cassava starch-based films increased when the concentration of the CEO increased. Thickness also affects other parameters, such as mechanical and barrier properties, as well as the transparency and visual appearance of the films. Therefore, the thickness of biopolymer-based edible films is an important factor that needs to be carefully controlled during the preparation of the films to achieve optimal properties for their intended application.

Table 1. The thickness, tensile strength, elongation at break, moisture content and water vapor permeability of SA–AG hydrogel-based films.

Film Sample	Thickness (mm)	Tensile Strength (Mpa)	Elongation at Break (%)	Moisture Content (%)	Water Vapor Permeability ((g * mm)/(m^2 * h * kPa))
AC-1	0.065 ± 0.006 ^{a}	9.82 ± 0.41 ^{a}	7.57 ± 0.45 ^{a}	18.52 ± 0.41 ^{a}	0.424 ± 0.018 ^{a}
AC-2	0.075 ± 0.006 ^{a}	7.31 ± 0.65 ^{b}	12.24 ± 0.60 ^{b}	18.18 ± 0.39 ^{ab}	0.405 ± 0.007 ^{a}
AC-3	0.098 ± 0.005 ^{b}	5.59 ± 0.27 ^{c}	14.94 ± 0.79 ^{c}	17.89 ± 0.59 ^{bc}	0.390 ± 0.015 ^{b}
AC-4	0.110 ± 0.006 ^{c}	3.49 ± 0.02 ^{d}	18.41 ± 0.60 ^{d}	17.03 ± 0.08 ^{c}	0.353 ± 0.004 ^{c}

Means carrying the same letters are significantly identical ($p > 0.05$).

2.3. Mechanical Properties of the SA–AG-Based Films

Hydrogel-based films can be prepared from various materials such as proteins, carbohydrates, and lipids that have different mechanical properties. The mechanical properties of the edible film can be modified by changing the composition of the film or the processing conditions. In the current study, TS and EAB parameters were assessed to evaluate the mechanical properties of the SA–AG composite edible films. The TS of the edible films showed a significant decrease (9.82–3.49 Mpa) when increasing the concentration of the CEO. The minimum TS (3.49 Mpa) was observed in the AC-4 film sample with a maximum concentration (30 μL) of CEO, while the blank AC-1 had the maximum value (9.82 Mpa) (Table 1). Wu et al. [22] showed similar results, where the TS of the gelatin-based films decreased when increasing the concentration of the CEO. The tensile strength can be influenced by several factors such as polymer/plasticizer/other additive type and proportion, as well as their respective interaction with each other.

The results of the EAB of the analyzed SA–AG-based film samples are shown in Table 1. The EAB of the SA–AG-based films significantly increased from 7.57 to 18.41% with the addition of the CEO. The maximum and the minimum EAB values were found in AC-4 and AC-1, respectively. The increase in the EAB of the hydrogel-based films could be due to the addition of the CEO, as the oil acts as a plasticizer and makes films more flexible and less brittle. The results of the present study are in line with Wu et al. [23], who reported an increase in the EAB with the addition of CEO nanoliposomes in the gelatin films. Moreover, different factors affect the EAB of the edible films such as the concentration of the added bioactive compounds, the composition of the film-forming solution, and the conditions and method used for the preparations of films.

2.4. Moisture Content

The moisture content in edible films is an important factor to consider as it can affect the quality, safety, and shelf life of the food. Generally, the moisture content should be low to prevent microbial growth and maintain the structural integrity of the film. In the current study, the SA–AG-based film samples showed a slight decrease from 18.52% to 17.03% in moisture content with the addition of the CEO. The minimum MC (17.03%) was found in the AC-4 sample, followed by AC-3 (17.89%) and AC-2 (18.18%), while the maximum value for MC was observed in the control. The decrease in the moisture content of the films when increasing the concentration of the CEO could be due to the hydrophobic nature of the added oil. Jamróz et al. [24] reported a decrease in the moisture content of edible films, based on starch-turbellarian-gelatin, when incorporated with tea tree essential oil. Furthermore, sodium alginate-chitosan-based edible films showed similar behaviour when loaded with bitter orange oil [25].

2.5. Water Vapor Permeability

In edible films, water vapor permeability (WVP) significantly impacts the shelf life and the quality of the packed food product. The WVP of an edible film is dependent on the properties of the film-forming material, such as its composition and structure. The environmental conditions, such as the temperature and humidity of the storage environment, can also affect the WVP of the hydrogel-based films. A lower WVP is desirable for certain applications as it slows down moisture migration and increases the shelf life of the product. The water vapor permeability of the SA–AG composite films significantly decreased with the addition of the CEO (Table 1). The maximum WVP (0.424) was exhibited by the control compared with the film samples loaded with the CEO. The AC-4 sample showed the minimum WVP (0.353), followed by the AC-3 (0.390) and AC-2 (0.405) samples. The decrease in the WVP could be due to the hydrophobicity of the CEO, which resulted in better barrier properties for the hydrogel-based films regarding water vapor permeability. The results of the current study are in line with the findings of Suput et al. [26] and Sánchez-González et al. [27]. Furthermore, the WVP of the edible films also depends on the type of polymers used, the concentration of the oil, and the preparation technique and conditions.

2.6. Transparency and Color Parameters

Transparency and color parameters are important factors as they affect the visual appearance, freshness, and quality parameters of food products. In the current study, the SA–AG composite hydrogel films incorporated with the CEO were examined for transparency and color parameters, including Lightness (L), a*, b*, and delta E. The addition of oil had a significant impact on the transparency of the films (Table 2). The transparency decreased from 79 to 21% with the addition of oil; the maximum was observed in the control (AC-1) and the minimum was shown by the AC-4 sample, which contained the maximum concentration of the CEO. The results of the current study are in accord with the findings of a previous study, in which the transparency of starch-based films decreased with the addition of oregano and black cumin essential oil [26].

Table 2. Transparency and color parameters of SA–AG hydrogel-based films.

Film Sample	Transparency (%)	L	a*	b*	ΔE*
AC-1	79.99 ± 0.62 [a]	96.05 ± 0.12 [a]	−0.08 ± 0.02 [a]	0.94 ± 0.08 [a]	0.85 ± 0.09 [a]
AC-2	73.85 ± 2.04 [b]	94.40 ± 0.07 [b]	0.03 ± 0.03 [b]	2.68 ± 0.11 [b]	2.98 ± 0.10 [b]
AC-3	64.02 ± 1.05 [c]	92.39 ± 0.30 [c]	0.05 ± 0.02 [b]	3.42 ± 0.47 [c]	4.36 ± 0.43 [c]
AC-4	21.23 ± 2.27 [d]	91.64 ± 0.15 [d]	0.26 ± 0.05 [c]	4.97 ± 0.52 [d]	5.04 ± 0.54 [c]

Means carrying the same letters are significantly identical ($p > 0.05$). L: lightness, a*: green-red color, b*: blue-yellow color, ΔE*: overall color variation.

The SA–AG based films showed a slight decrease in Lightness (L), from 96.05 to 91.64%, with the addition of oil. The b* value of the films varied from 0.94 to 4.97, whereases the a* value ranged from 0.08 to 0.26. As mentioned in Table 2, the films showed (b*) yellowness as the concentration of the CEO increased, with a maximum b* value (4.97) in the AC-4 sample. The significant variation of ΔE values (0.85–5.04) also validates the overall color alterations in the films with the addition of oil. Essential oils contain a complex mixture of polyphenolic components that tends to absorb light, which can ultimately impact the color attributes of the film. Zhou et al. [20] reported a similar behavior in which the color of the cassava starch-based films changes to yellow with the addition of cinnamon essential oil. Furthermore, Tongnuanchan et al. [28] and Atarés et al. [29] reported similar results.

2.7. Scanning Electron Microscopy

Scanning Electron Microscopy (SEM) is a technique that is used to observe the surface structure of materials at high magnification. In the case of edible films, SEM is used to study their physical characteristics, such as their microstructure, uniformity, compactness and surface texture. SEM imaging can also help to identify any defects in the film, such as cracks, pores, or voids, that could impact its functionality and performance as food packaging. The SEM results of the SA–AG film samples with the added CEO are shown in Figure 2. The control film (AC-1) containing SA and AG without the CEO showed a structure with some pores on the film surface, compared with the films loaded with CEO. The protrusion can also be observed in the cross-section image of the AC-1 sample. The randomly distributed oil drops can be observed on the film surface of the AC-2 sample. The roughness of the AC-2 film samples could be linked to the agglomeration that resulted from the irregular distribution of hydrophobic components during the film-forming procedure. The most uniform dispersion was observed in the AC-4 sample, which contained the maximum concentration of the CEO. The film samples incorporated with the CEO showed smooth surfaces with a fewer number of particles, homogeneous structures, and no pores and cracks. Zhou et al. [20] also reported similar results, in which the CEO distributed homogeneously in the polymer matrix of cassava starch-based films. Moreover, the interaction between oil and the film-forming polymers can also influence the SEM appearance, as some polymers may be more compatible with oil than others.

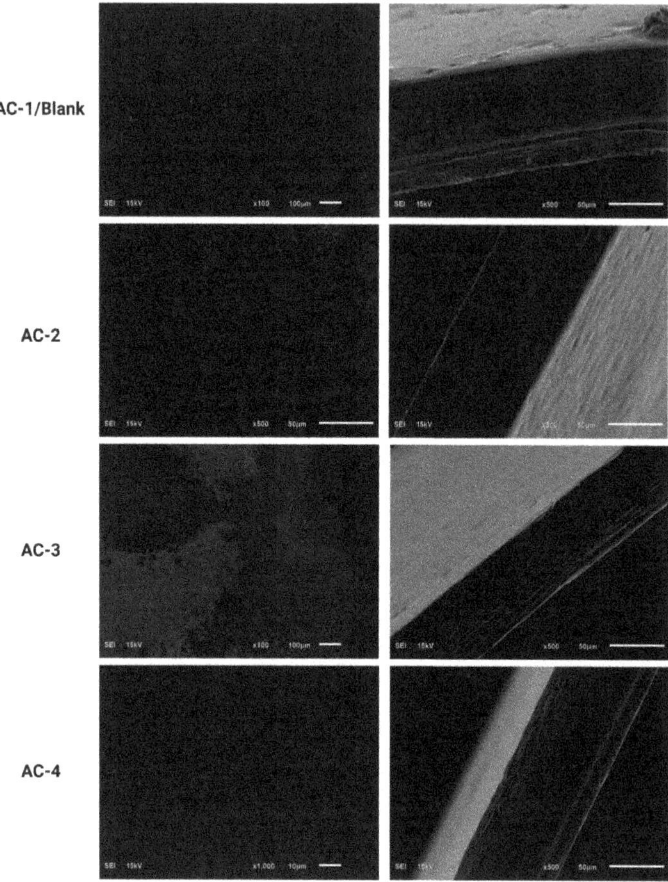

Figure 2. Scanning Electron Microscopy of SA–AG hydrogel-based edible films; AC-1/Control, AC-2 contains 15 µL of CEO, AC-3 contains 20 µL of CEO, and AC-4 contains 30 µL of CEO.

2.8. X-ray Diffraction Analysis

XRD analysis of edible films is important because it provides important information about the structure and composition of the material. XRD provides information about the crystalline structure of edible films, which provides insight into their mechanical properties, such as their tensile strength and flexibility. The SA–AG hydrogel-based composite films loaded with the CEO were examined for their structural characteristics and the resulting spectrum is shown in Figure 3. A Diffract Eva software package was used to calculate the crystallinity percentages of the film samples, and the crystallinity percentages of the AC-1, AC-2, AC-3, and AC-4 samples were found to be 38.1%, 38.7%, 35.4%, and 38.6%, respectively. All the samples showed characteristic peaks at similar positions with different intensities due to the variations in the concentration of the oil. However, there was no difference observed in the crystallinity of the SA–AG composite hydrogel films with the addition of the CEO. Overall, the XRD patterns showed good compatibility between the film-forming polymers including SA, AG and CEO.

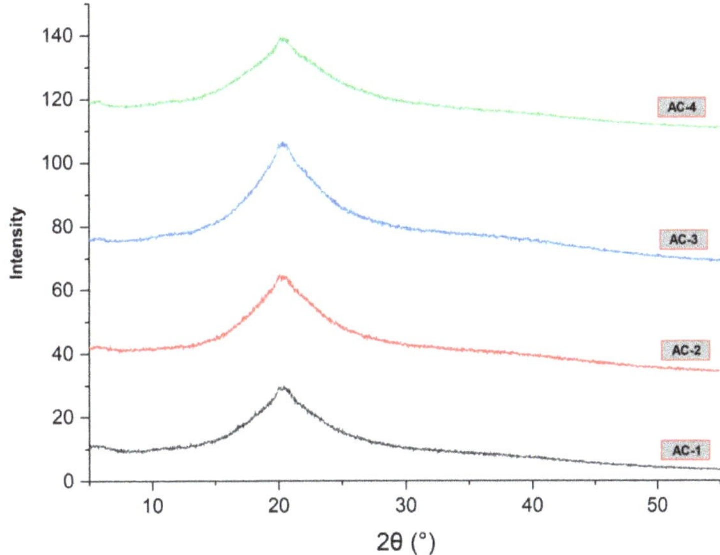

Figure 3. X-ray diffraction pattern of SA–AG hydrogel-based edible films; AC-1/blank, AC-2 contained 15 μL of CEO, AC-3 contained 20 μL of CEO, and AC-4 contained 30 μL of CEO.

2.9. Fourier Transform Infrared Spectroscopy

FTIR provides information about the molecular structure and composition of the film, it also provides valuable information about the degree of cross-linking in the film, which affects its mechanical properties and durability. SA and AG composite edible films loaded with the CEO were analyzed for FTIR and the absorption spectrum is shown in Figure 4. The broad spectrum at 3305 cm^{-1} indicates the stretching vibration of the secondary N-H amide bonds [30]. The characteristic peaks identified at 1410 cm^{-1} and 1600 cm^{-1} could have been due to the presence of sodium alginate in the film matrix, representing the symmetric and asymmetric stretching vibration of the COO-group respectively [31]. Additionally, studies have also reported that characteristic peaks identified at 1400 cm^{-1} and 2925 cm^{-1} positions could be due to the presence of sodium alginate [25]. In a previous study, it was reported that AG exhibited a characteristic peak at 2929 cm^{-1} in the FTIR analysis [32]. A study reported the -C=O stretching vibration of the tween 80 carbonyl group at 1735 cm^{-1} and 1736 cm^{-1} during the FTIR analysis [20]. In our current study, the film matrix exhibited a characteristic peak at 1733 cm^{-1} that could be ascribed to the presence of tween 80 in the film matrix; however, a little variation (from 1735 cm^{-1} to 1733 cm^{-1}) could have been due to the difference in the concentration. A distinctive peak at 1456 cm^{-1} corresponding to the phenolic group of cinnamaldehyde was observed in the FTIR spectrum that indicates the presence of CEO in the film matrix [20]. Overall, the FTIR analysis showed a good intermolecular interaction between the CEO, SA and AG.

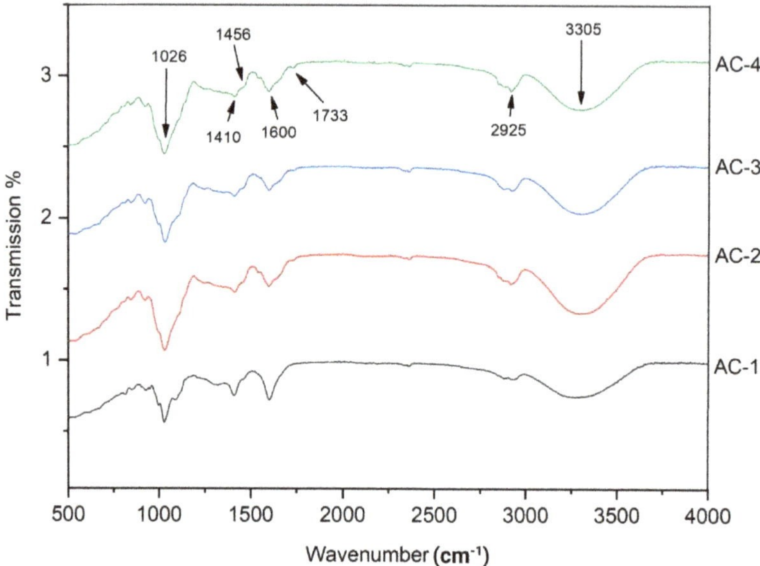

Figure 4. FTIR spectrum of SA–AG hydrogel-based edible films; AC-1/Control, AC-2 contains 15 µL of CEO, AC-3 contains 20 µL of CEO and AC-4 contains 30 µL of CEO.

2.10. DSC Analysis

The thermal stability of edible films refers to their ability to maintain their physical, chemical, and structural properties when exposed to elevated temperatures. Thermal stability is a critical factor for the performance and functionality of edible films, as it affects their shelf life and suitability for various food packaging applications. DSC analysis can be used to evaluate the thermal stability of edible films by measuring the temperature and heat flow during heating or cooling cycles. The DSC thermograms of the SA–AG-based composite films are shown in Figure 5. The AC-2, AC-3, and AC-4 films incorporated with CEO presented one broad endothermic peak at 70–128 °C, 72–126 °C, and 51–118 °C, respectively. This endothermic peak can be attributed to the evaporation of the residual solvent (water) that was used during the production of composite films [33,34]. After the incorporation of oil into the films, the temperature of the endothermic peak increased significantly in AC-2 and AC-3 samples, but the temperature slightly reduced in the AC-4 sample with a maximum (30 µL) of CEO. As the concentration of CEO was increased, the peak area of the AC-2, AC-3, and AC-4 samples also increased, indicating that the thermal stability of the composite films improved due to the addition of EOs. This shift could have been due to the plasticization effect of oil, which would have raised the free volume within the polymeric network and mobility of the polymeric chains as reported in the previous studies [35].

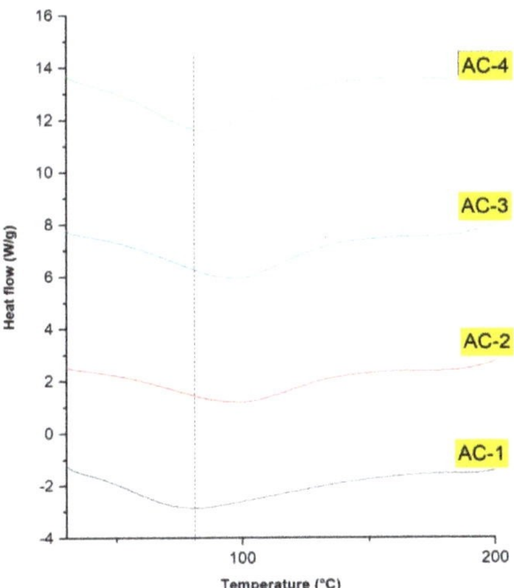

Figure 5. DSC analysis of SA–AG-based composite edible films; AC-1/Control, AC-2 contains 15 µL of CEO, AC-3 contains 20 µL of CEO, and AC-4 contains 30 µL of CEO.

The improved thermal stability of the composite hydrogel-based films indicates that there were strong intermolecular interactions between the CEO and composite material, which could potentially influence the mechanical properties of the films. The previous studies suggested that the incorporation of *Origanum vulgare* L. and *Matricaria recutita* essential oil caused a change in endothermic peaks, indicating the thermal stability of the polymeric films [36,37].

As per previous reports, the incorporation of *Origanum onites* L. essential oil reduced the heat transitions due to evaporation, as evidenced by changes in endothermic peaks. This could be due to the molecular structure of the essential oil that caused the changes in the overall chain mobility of the polymer matrix, as reported in previous studies [38,39].

2.11. Antioxidant Potential

The antioxidant activity of hydrogel-based films can help to protect the food packaged within the film from oxidation, which can cause the food to spoil or lose quality. The addition of essential oils to edible films can increase their antioxidant potential, and can help to protect the packaged food from oxidation and extend its shelf life. Essential oils are concentrated plant extracts that contain a variety of antioxidants, such as phenolic compounds, terpenes, and flavonoids [40]. In the current study, the antioxidant activity of the different samples of edible films based on SA and AG was assessed and the results are shown in Figure 6. The hydrogel-based films showed a significant increase in the DPPH radical scavenging activity with increasing the concentration of the CEO. Similar results were obtained for the ABTS radical scavenging activity of the SA–AG-based composite film samples. The films incorporated with CEO exhibited more antioxidant activity compared with the control. This could have been due to the presence of different bioactive compounds in the CEO, primarily cinnamaldehyde, contributing to the overall antioxidant potential of the edible films [41]. Xu et al. [42] found similar results in which the antioxidant activity of the chitosan-gum arabic films increased with the addition of CEO. Furthermore, many studies have reported an increase in the antioxidant activity of edible films when incorporated with essential oils [43–45].

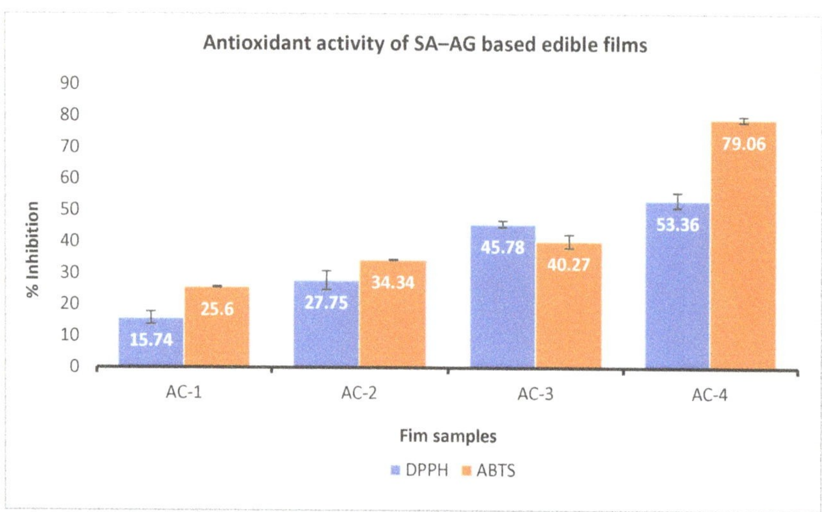

Figure 6. Antioxidant activity of SA–AG-based composite edible films; AC-1/Control, AC-2 contains 15 µL of CEO, AC-3 contains 20 µL of CEO, and AC-4 contains 30 µL of CEO.

3. Conclusions

The current study revealed that CEO has the potential to be used for the development of sodium alginate and acacia gum-based composite films with improved physicochemical properties at optimal concentrations. The findings of the study are likely to be a valuable resource for the development of future edible packaging formulations, as well as for identifying potential applications for different food products. However, further investigations can reveal the impact of adding the CEO on the antimicrobial properties of sodium alginate-acacia gum-based films.

4. Materials and Methods

4.1. Film-Formation

SA and AG were provided by Sisco Research Laboratories (SRL), Mumbai, India. 1.5% (*w*/*v*) solution of SA and a 0.5% (*w*/*v*) solution of AG were prepared separately by dissolving the polymers in distilled water and stirring them overnight for complete solubilization at the magnetic stirrer. After complete solubilization, both solutions were subjected to mixing with the gradual addition of 1% glycerol (BDH Laboratory, London, England) as a plasticizer. The resultant solution was subjected to stirring for 3 h at a magnetic stirrer. The obtained film-forming solution was divided equally into four beakers and labelled as AC-1, AC-2, AC-3, and AC-4. Subsequently, different concentrations (15 µL, 20 µL and 30 µL) of the CEO (Nature Natural, Ghaziabad, India) were added to AC-2, AC-3, and AC-4 samples, respectively, while the AC-1 sample without the addition of the CEO was used as a control sample. Additionally, different concentrations of 30 µL, 40 µL, and 80 µL of tween 80 (Sisco Research Laboratories, Mumbi, India) were also added as a surfactant to AC-2, AC-3, and AC-4 samples, respectively, for the uniform dispersion of oil in the film-forming solution. The obtained solutions were poured onto the labelled petri plates and left to dry for 48 h at room temperature. After drying, the films were evaluated visually and subjected to further examination.

4.2. Thickness

To determine the thickness of the films produced, a digital micrometre (Yu-Su 150, Yu-Su Tools, China) was utilized. The thickness measurements were taken at 5 distinct positions of the film and an average was calculated.

4.3. Mechanical Properties

The mechanical properties of the hydrogel films, specifically their tensile strength (TS) and percentage elongation at break (EAB), were determined using a texture analyzer (XT plus, Stable Micro Systems, Godalming, England) following the standard ASTM D882 method [46]. For the present test, film strips that were 60 mm long and 7 mm wide were used. The equations presented below were used to determine the values of the TS and EAB for the film samples.

$$\mathrm{TS} = \left(\frac{\mathrm{F}}{\mathrm{A}}\right) \quad (1)$$

F is the maximum force,
A is the cross-sectional area of the film.

$$\mathrm{EAB}\ (\%) = \frac{\mathrm{Lf} - \mathrm{Li}}{\mathrm{Li}} \times 100 \quad (2)$$

Lf is the final length at a break,
Li is the initial length of the film.

4.4. Moisture Content

The hydrogel film strips (3 cm × 4 cm) were examined for their moisture content (MC). The film strips were dried at 105 °C and the difference in weight was calculated before (W1) and after drying (W2), according to Equation (3).

$$\mathrm{MC} = \frac{\mathrm{W1} - \mathrm{W2}}{\mathrm{W1}} \times 100 \quad (3)$$

4.5. Water Vapor Permeability

The procedure followed by Erdem et al. [47] was used to measure the water vapor permeability (WVP) of the films. The relative humidity (RH) of the apparatus was adjusted to 100% and 0% by using water and silica gel, respectively. Hydrogel films were sealed firmly over the glass test cups (5cm of internal diameter and 3 cm depth) containing silica gel. The cups were weighed at specific intervals to calculate the weight-gain within a day. Equation (4) was used to calculate the WVP of the films and represented in g mm/(m^2)(d)(kPa).

$$\mathrm{WVP} = \frac{\Delta \mathrm{m}}{\Delta \mathrm{t} \times \Delta \mathrm{P} \times \mathrm{A}} \times \mathrm{d} \quad (4)$$

In Equation (4), the $\Delta m/\Delta t$ is the weight of moisture-gain per unit of time. A is the film area in m^2; ΔP is the water-vapor pressure difference between the two sides of the film in kPa; d is the film thickness in mm.

4.6. Transparency and Color Parameters

The transparency of the SA–AG hydrogel film samples was measured at a wavelength of 550 nm by using a spectrophotometer (ONDA-Vis spectrophotometer, V-10 Plus, ONDA, Padova, Italy) according to the method described by Zhao et al. [48].

The surface color analysis of the SA–AG-based films was carried out by using a colorimeter (Konica Minolta, Tokyo, Japan), and represented as L* (lightness), a* (red/green), and b* (yellow/blue). The film samples were placed on the surface of a standard plate (L* = 100) and the color parameters L*, a*, and b* were calculated. Equation (5) was used to calculate the ΔE (the overall color difference).

$$\Delta \mathrm{E}^* = \left[(\Delta \mathrm{L}^*)^2 + (\Delta \mathrm{a}^*)^2 + (\Delta \mathrm{b}^*)^2\right]^{1/2} \quad (5)$$

4.7. Microstructure of the Films

The microstructure analysis of the prepared hydrogel film samples based on SA–AG was carried out by using Scanning Electron Microscopy (SEM) (JSM6510LA from Analytical

SEM, Jeol, Tokyo, Japan). The films were initially mounted on double-sided tape on an aluminum stub that was coated with a thin layer of gold before the analysis.

4.8. X-ray Diffraction (XRD) Analysis

To determine the crystallinity of the samples, X-ray diffractometer (Bruker D8 Discover) was used by applying 40 kV voltage with 2 theta ranging from 5–50° at a rate of 0.500 s/point, employing copper (Kα) radiation (1.5418 Å).

4.9. Fourier Transforms Infrared Spectroscopy (FT-IR)

FTIR analysis of the SA–AG hydrogel-based films was performed by using an InfraRed Bruker Tensor 37, Ettlingen, Germany. The obtained spectrum was used to examine the functional groups and the interactions between film-forming components. The test was performed with an average of 32 scans, with a wavenumber range from 400 to 4000 cm^{-1}.

4.10. Differential Scanning Calorimetry

DSC measurements were performed using DSC-Q20 instrument (TA instruments, New Castle, DE, USA). 10 mg of film was hermetically encapsulated in aluminum capsules and placed in the sampling unit. The sample was heated from 25 °C to 200 °C at a heating rate of 10 °C/min in a nitrogen-rich atmosphere.

4.11. Antioxidant Analysis

The antioxidant potential of the SA–AG-based films was analyzed using DPPH and ABTS radical scavenging activities. For the DPPH assay, the method described by Brand-Williams et al. [49] was followed to carry out the analysis for 50 mg of film samples. The absorbance was recorded at 517 nm using an ONDA-Vis spectrophotometer. The obtained results were presented as % inhibition. For the ABTS assay, the methodology of Re et al. [50] was employed to evaluate the antioxidant activity of the 25 mg of film samples. The absorbance was recorded at 734 nm and the results that were obtained were presented as % inhibition as an average of three measurements. The films were directly treated with radical solutions, vortexed for 30 s and incubated for radical scavenging activity.

4.12. Statistical Analysis

All data in the current study are reported as the mean and standard deviation (SD) of three distinct assessments. A one-way analysis of variance was conducted using statistical analysis software, followed by Duncan's test with a 5% significant level. The purpose of these analyses was to examine whether the mean values differed significantly.

Author Contributions: Conceptualization, S.B. and Y.A.S.; Methodology, Y.A.S., H.W.K.A., E.K. and L.Y.A.; Software, H.W.K.A.; Validation, S.K., P.S. and D.S.K.; Formal analysis, S.K., P.S. and E.K.; Investigation, M.K.A.; Data curation, D.S.K. and L.Y.A.; Writing–original draft, S.B. and Y.A.S.; Writing–review & editing, E.K. and L.Y.A.; Visualization, M.K.A.; Supervision, S.B. and A.A.-H. All authors have read and agreed to the published version of the manuscript.

Funding: This research work was funded by Institutional Fund Projects under grant number IFPIP: 1782-249-1443. The authors gratefully acknowledge the technical and financial support provided by the Ministry of Education and Abdulaziz University, DSR, Jeddah, Saudi Arabia.

Institutional Review Board Statement: This study does not contain any studies with human or animal subjects performed by any of the authors.

Informed Consent Statement: Not applicable.

Data Availability Statement: Not applicable.

Acknowledgments: The Authors are thankful to the Natural and Medical Sciences Research Center, University of Nizwa, Oman, for providing research facilities to conduct the current study.

Conflicts of Interest: The authors declare no conflict of interest.

References

1. Dhumal, C.V.; Sarkar, P. Composite edible films and coatings from food-grade biopolymers. *J. Food Sci. Technol.* **2018**, *55*, 4369–4383. [CrossRef]
2. Ibrahim, A.G.; Abdel Hai, F.; Abd El-Wahab, H.; Aboelanin, H. Methylene blue removal using a novel hydrogel containing 3-Allyloxy-2-hydroxy-1-propanesulfonic acid sodium salt. *Adv. Polym. Technol.* **2018**, *37*, 3561–3573. [CrossRef]
3. Ibrahim, A.G.; Sayed, A.Z.; Abd El-Wahab, H.; Sayah, M.M. Synthesis of a hydrogel by grafting of acrylamide-co-sodium methacrylate onto chitosan for effective adsorption of Fuchsin basic dye. *Int. J. Biol. Macromol.* **2020**, *159*, 422–432. [CrossRef]
4. Elkony, A.M.; Ibrahim, A.G.; Abu El-Farh, M.H.; Abdelhai, F. Synthesis of Acrylamide-co-3-Allyloxy-2-hydroxy-1-propanesulfonic acid sodium salt Hydrogel for efficient Adsorption of Methylene blue dye. *Int. J. Environ. Anal. Chem.* **2021**, *103*, 1–20. [CrossRef]
5. Galus, S.; Lenart, A. Development and characterization of composite edible films based on sodium alginate and pectin. *J. Food Eng.* **2013**, *115*, 459–465. [CrossRef]
6. Gheorghita, R.; Gutt, G.; Amariei, S. The use of edible films based on sodium alginate in meat product packaging: An eco-friendly alternative to conventional plastic materials. *Coatings* **2020**, *10*, 166. [CrossRef]
7. Mahcene, Z.; Khelil, A.; Hasni, S.; Akman, P.K.; Bozkurt, F.; Birech, K.; Goudjil, M.B.; Tornuk, F. Development and characterization of sodium alginate based active edible films incorporated with essential oils of some medicinal plants. *Int. J. Biol. Macromol.* **2020**, *145*, 124–132. [CrossRef] [PubMed]
8. Dou, L.; Li, B.; Zhang, K.; Chu, X.; Hou, H. Physical properties and antioxidant activity of gelatin-sodium alginate edible films with tea polyphenols. *Int. J. Biol. Macromol.* **2018**, *118*, 1377–1383. [CrossRef] [PubMed]
9. Liu, S.; Li, Y.; Li, L. Enhanced stability and mechanical strength of sodium alginate composite films. *Carbohydr. Polym.* **2017**, *160*, 62–70. [CrossRef]
10. Sanchez, C.; Nigen, M.; Tamayo, V.M.; Doco, T.; Williams, P.; Amine, C.; Renard, D. Acacia gum: History of the future. *Food Hydrocoll.* **2018**, *78*, 140–160. [CrossRef]
11. Suresh, S.N.; Puspharaj, C.; Natarajan, A.; Subramani, R. Gum acacia/pectin/pullulan-based edible film for food packaging application to improve the shelf-life of ivy gourd. *Int. J. Food Sci. Technol.* **2022**, *57*, 5878–5886. [CrossRef]
12. Ibrahim, A.G.; Elkony, A.M.; El-Bahy, S.M. Methylene blue uptake by gum arabic/acrylic amide/3-allyloxy-2-hydroxy-1-propanesulfonic acid sodium salt semi-IPN hydrogel. *Int. J. Biol. Macromol.* **2021**, *186*, 268–277. [CrossRef]
13. Kang, S.; Xiao, Y.; Guo, X.; Huang, A.; Xu, H. Development of gum arabic-based nanocomposite films reinforced with cellulose nanocrystals for strawberry preservation. *Food Chem.* **2021**, *350*, 129199. [CrossRef]
14. Pelissari, F.M.; Grossmann, M.V.; Yamashita, F.; Pineda, E.A.G. Antimicrobial, mechanical, and barrier properties of cassava starch– chitosan films incorporated with oregano essential oil. *J. Agric. Food Chem.* **2009**, *57*, 7499–7504. [CrossRef] [PubMed]
15. Gómez-Estaca, J.; De Lacey, A.L.; López-Caballero, M.; Gómez-Guillén, M.; Montero, P. Biodegradable gelatin–chitosan films incorporated with essential oils as antimicrobial agents for fish preservation. *Food Microbiol.* **2010**, *27*, 889–896. [CrossRef]
16. Wu, Y.; Luo, Y.; Wang, Q. Antioxidant and antimicrobial properties of essential oils encapsulated in zein nanoparticles prepared by liquid–liquid dispersion method. *LWT-Food Sci. Technol.* **2012**, *48*, 283–290. [CrossRef]
17. Nwanade, C.F.; Wang, M.; Wang, T.; Zhang, X.; Zhai, Y.; Zhang, S.; Yu, Z.; Liu, J. The acaricidal activity of cinnamon essential oil: Current knowledge and future perspectives. *Int. J. Acarol.* **2021**, *47*, 446–450. [CrossRef]
18. Wang, R.; Wang, R.; Yang, B. Extraction of essential oils from five cinnamon leaves and identification of their volatile compound compositions. *Innov. Food Sci. Emerg. Technol.* **2009**, *10*, 289–292. [CrossRef]
19. Deng, X.; Liao, Q.; Xu, X.; Yao, M.; Zhou, Y.; Lin, M.; Zhang, P.; Xie, Z. Analysis of essential oils from cassia bark and cassia twig samples by GC-MS combined with multivariate data analysis. *Food Anal. Methods* **2014**, *7*, 1840–1847. [CrossRef]
20. Zhou, Y.; Wu, X.; Chen, J.; He, J. Effects of cinnamon essential oil on the physical, mechanical, structural and thermal properties of cassava starch-based edible films. *Int. J. Biol. Macromol.* **2021**, *184*, 574–583. [CrossRef]
21. Gooderham, N.J.; Cohen, S.M.; Eisenbrand, G.; Fukushima, S.; Guengerich, F.P.; Hecht, S.S.; Rietjens, I.M.; Rosol, T.J.; Davidsen, J.M.; Harman, C.L. FEMA GRAS assessment of natural flavor complexes: Clove, cinnamon leaf and West Indian bay leaf-derived flavoring ingredients. *Food Chem. Toxicol.* **2020**, *145*, 111585. [CrossRef]
22. Wu, J.; Sun, X.; Guo, X.; Ge, S.; Zhang, Q. Physicochemical properties, antimicrobial activity and oil release of fish gelatin films incorporated with cinnamon essential oil. *Aquac. Fish.* **2017**, *2*, 185–192. [CrossRef]
23. Wu, J.; Liu, H.; Ge, S.; Wang, S.; Qin, Z.; Chen, L.; Zheng, Q.; Liu, Q.; Zhang, Q. The preparation, characterization, antimicrobial stability and in vitro release evaluation of fish gelatin films incorporated with cinnamon essential oil nanoliposomes. *Food Hydrocoll.* **2015**, *43*, 427–435. [CrossRef]
24. Jamróz, E.; Juszczak, L.; Kucharek, M. Development of starch-furcellaran-gelatin films containing tea tree essential oil. *J. Appl. Polym. Sci.* **2018**, *135*, 46754. [CrossRef]
25. Bhatia, S.; Al-Harrasi, A.; Al-Azri, M.S.; Ullah, S.; Bekhit, A.E.-D.A.; Pratap-Singh, A.; Chatli, M.K.; Anwer, M.K.; Aldawsari, M.F. Preparation and physiochemical characterization of bitter Orange oil loaded sodium alginate and casein based edible films. *Polymers* **2022**, *14*, 3855. [CrossRef]
26. Suput, D.; Lazic, V.; Pezo, L.; Markov, S.; Vastag, Z.; Popovic, L.; Rudulovic, A.; Ostojic, S.; Zlatanovic, S.; Popovic, S. Characterization of starch edible films with different essential oils addition. *Pol. J. Food Nutr. Sci.* **2016**, *66*, 277–285. [CrossRef]
27. Sánchez-González, L.; Vargas, M.; González-Martínez, C.; Chiralt, A.; Cháfer, M. Characterization of edible films based on hydroxypropylmethylcellulose and tea tree essential oil. *Food Hydrocoll.* **2009**, *23*, 2102–2109. [CrossRef]

28. Tongnuanchan, P.; Benjakul, S.; Prodpran, T. Physico-chemical properties, morphology and antioxidant activity of film from fish skin gelatin incorporated with root essential oils. *J. Food Eng.* **2013**, *117*, 350–360. [CrossRef]
29. Atarés, L.; Bonilla, J.; Chiralt, A. Characterization of sodium caseinate-based edible films incorporated with cinnamon or ginger essential oils. *J. Food Eng.* **2010**, *100*, 678–687. [CrossRef]
30. Marzbani, P.; Resalati, H.; Ghasemian, A.; Shakeri, A. Surface modification of talc particles with phthalimide: Study of composite structure and consequences on physical, mechanical, and optical properties of deinked pulp. *BioResources* **2016**, *11*, 8720–8738. [CrossRef]
31. Chan, H.; Nyam, K.; Yusof, Y.; Pui, L. Investigation of properties of polysaccharide-based edible film incorporated with functional melastoma malabathricum extract. *Carpathian J. Food Sci. Technol.* **2020**, *12*, 120–134.
32. Venkatesham, M.; Ayodhya, D.; Madhusudhan, A.; Veerabhadram, G. Synthesis of stable silver nanoparticles using gum acacia as reducing and stabilizing agent and study of its microbial properties: A novel green approach. *Int. J. Green Nanotechnol.* **2012**, *4*, 199–206. [CrossRef]
33. Altiok, D.; Altiok, E.; Tihminlioglu, F. Physical, antibacterial and antioxidant properties of chitosan films incorporated with thyme oil for potential wound healing applications. *J. Mater. Sci. Mater. Med.* **2010**, *21*, 2227–2236. [CrossRef] [PubMed]
34. Kaya, M.; Khadem, S.; Cakmak, Y.S.; Mujtaba, M.; Ilk, S.; Akyuz, L.; Salaberria, A.M.; Labidi, J.; Abdulqadir, A.H.; Deligöz, E. Antioxidative and antimicrobial edible chitosan films blended with stem, leaf and seed extracts of Pistacia terebinthus for active food packaging. *RSC Adv.* **2018**, *8*, 3941–3950. [CrossRef]
35. Sobral, P.d.A.; Menegalli, F.; Hubinger, M.; Roques, M. Mechanical, water vapor barrier and thermal properties of gelatin based edible films. *Food Hydrocoll.* **2001**, *15*, 423–432. [CrossRef]
36. Hosseini, S.F.; Rezaei, M.; Zandi, M.; Farahmandghavi, F. Bio-based composite edible films containing Origanum vulgare L. essential oil. *Ind. Crops Prod.* **2015**, *67*, 403–413. [CrossRef]
37. Aliheidari, N.; Fazaeli, M.; Ahmadi, R.; Ghasemlou, M.; Emam-Djomeh, Z. Comparative evaluation on fatty acid and Matricaria recutita essential oil incorporated into casein-based film. *Int. J. Biol. Macromol.* **2013**, *56*, 69–75. [CrossRef]
38. Scartazzini, L.; Tosati, J.; Cortez, D.; Rossi, M.; Flôres, S.; Hubinger, M.; Di Luccio, M.; Monteiro, A. Gelatin edible coatings with mint essential oil (*Mentha arvensis*): Film characterization and antifungal properties. *J. Food Sci. Technol.* **2019**, *56*, 4045–4056. [CrossRef]
39. Qin, Y.; Li, W.; Liu, D.; Yuan, M.; Li, L. Development of active packaging film made from poly (lactic acid) incorporated essential oil. *Prog. Org. Coat.* **2017**, *103*, 76–82. [CrossRef]
40. Magalhaes, M.; Manadas, B.; Efferth, T.; Cabral, C. Chemoprevention and therapeutic role of essential oils and phenolic compounds: Modeling tumor microenvironment in glioblastoma. *Pharmacol. Res.* **2021**, *169*, 105638. [CrossRef]
41. Suryanti, V.; Wibowo, F.; Khotijah, S.; Andalucki, N. Antioxidant activities of cinnamaldehyde derivatives. In Proceedings of the IOP Conference Series: Materials Science and Engineering, Kuala Lumpur, Malaysia, 13–14 August 2018; p. 012077.
42. Xu, T.; Gao, C.; Feng, X.; Yang, Y.; Shen, X.; Tang, X. Structure, physical and antioxidant properties of chitosan-gum arabic edible films incorporated with cinnamon essential oil. *Int. J. Biol. Macromol.* **2019**, *134*, 230–236. [CrossRef]
43. Tongnuanchan, P.; Benjakul, S.; Prodpran, T. Comparative studies on properties and antioxidative activity of fish skin gelatin films incorporated with essential oils from various sources. *Int. Aquat. Res.* **2014**, *6*, 62. [CrossRef]
44. Tongnuanchan, P.; Benjakul, S.; Prodpran, T. Properties and antioxidant activity of fish skin gelatin film incorporated with citrus essential oils. *Food Chem.* **2012**, *134*, 1571–1579. [CrossRef] [PubMed]
45. Pires, C.; Ramos, C.; Teixeira, B.; Batista, I.; Nunes, M.; Marques, A. Hake proteins edible films incorporated with essential oils: Physical, mechanical, antioxidant and antibacterial properties. *Food Hydrocoll.* **2013**, *30*, 224–231. [CrossRef]
46. Properties, A.S.D.o.M. Standard Test Method for Tensile Properties of Thin Plastic Sheeting. Available online: https://www.wewontech.com/testing-standards/190125019.pdf (accessed on 12 March 2023).
47. Erdem, B.G.; Dıblan, S.; Kaya, S. Development and structural assessment of whey protein isolate/sunflower seed oil biocomposite film. *Food Bioprod. Process.* **2019**, *118*, 270–280. [CrossRef]
48. Zhao, J.; Wang, Y.; Liu, C. Film Transparency and Opacity Measurements. *Food Anal. Methods* **2022**, *15*, 2840–2846. [CrossRef]
49. Brand-Williams, W.; Cuvelier, M.; Berset, C. Antioxidative activity of phenolic composition of commercial extracts of sage and rosemary. *LWT* **1995**, *28*, 25–30. [CrossRef]
50. Re, R.; Pellegrini, N.; Proteggente, A.; Pannala, A.; Yang, M.; Rice-Evans, C. Antioxidant activity applying an improved ABTS radical cation decolorization assay. *Free Radic. Biol. Med.* **1999**, *26*, 1231–1237. [CrossRef] [PubMed]

MDPI

Article

Chitosan-Based Green Pea (*Pisum sativum* L.) Pod Extract Gel Film: Characterization and Application in Food Packaging

Essam Mohamed Elsebaie [1,*], Mona Metwally Mousa [2], Samah Amin Abulmeaty [2], Heba Ali Yousef Shaat [2], Soher Abd-Elfttah Elmeslamy [2], Galila Ali Asker [2], Asmaa Antar Faramawy [3], Hala Ali Yousef Shaat [2], Wesam Mohammed Abd Elrahman [3], Hanan Salah Eldeen Eldamaty [3], Amira Lotfy Abd Allah [3] and Mohamed Reda Badr [4]

1 Food Technology Department, Faculty of Agriculture, Kafrelsheikh University, Kafr El-Sheikh 33516, Egypt
2 Food Science &Technology Department, Faculty of Home Economics, Al-Azhar University, Tanta 31732, Egypt
3 Nutrition & Food Science Department, Faculty of Home Economics, Al-Azhar University, Tanta 31512, Egypt
4 Food Science and Technology Department, Agriculture Faculty, Tanta University, Tanta 31512, Egypt
* Correspondence: essam.ahmed@kfs.edu.eg; Tel.: +201556126015

Abstract: This work focuses on studying the preparation, characterization (physical, mechanical, optical, and morphological properties as well as antioxidant and antimicrobial activities) and packaging application of chitosan (CH)-based gel films containing varying empty green pea pod extract (EPPE) concentrations (0, 1, 3, and 5% *w/w*). The experiments revealed that adding EPPE to CH increased the thickness (from 0.132 ± 0.08 to 0.216 ± 0.08 mm), density (from 1.13 ± 0.02 to 1.94 ± 0.02 g/cm^3), and opacity (from 0.71 ± 0.02 to 1.23 ± 0.04), while decreasing the water vapour permeability, water solubility, oil absorption ratio, and whiteness index from 2.34 to 1.08 × 10^{-10} g^{-1} s^{-1} pa^{-1}, from 29.40 ± 1.23 to 18.75 ± 1.94%, from 0.31 ± 0.006 to 0.08 ± 0.001%, and from 88.10 ± 0.43 to 77.53 ± 0.48, respectively. The EPPE films had better tensile strength (maximum of 26.87 ± 1.38 MPa), elongation percentage (maximum of 58.64 ± 3.00%), biodegradability (maximum of 48.61% after 3 weeks), and migration percentages than the pure CH-gel film. With the addition of EPPE, the antioxidant and antibacterial activity of the film improved. SEM revealed that as EPPE concentration increased, agglomerates formed within the films. Moreover, compared to control samples, packing corn oil in CH-based EPPE gel films slowed the rise of thiobarbituric acid and peroxide values. As an industrial application, CH-based EPPE films have the potential to be beneficial in food packaging.

Keywords: green pea pods; corn oil; chitosan; films; gel

Citation: Elsebaie, E.M.; Mousa, M.M.; Abulmeaty, S.A.; Shaat, H.A.Y.; Elmeslamy, S.A.-E.; Asker, G.A.; Faramawy, A.A.; Shaat, H.A.Y.; Abd Elrahman, W.M.; Eldamaty, H.S.E.; et al. Chitosan-Based Green Pea (*Pisum sativum* L.) Pod Extract Gel Film: Characterization and Application in Food Packaging. *Gels* **2023**, *9*, 77. https://doi.org/10.3390/gels9020077

Academic Editors: Aris Giannakas, Constantinos Salmas and Charalampos Proestos

Received: 29 December 2022
Revised: 13 January 2023
Accepted: 13 January 2023
Published: 18 January 2023

1. Introduction

Packaging is a crucial part of the food industry because it helps with food handling, storage, transportation, and preservation, as well as protecting against external contaminants, preventing substances inside the food from escaping into the surrounding environment, and reducing food waste [1,2]. Plastic polymers are commonly used for packaging because they are easy to manufacture, inexpensive, printable, and highly resistant to various mechanical and environmental variables [3]. Nevertheless, these packaging substances are harmful to the environment since they take a long time to disintegrate and present the danger of releasing chemicals that might affect food quality [4].

Therefore, the use of biodegradable packaging rather than plastic packaging is encouraged due to environmental and health concerns. Nowadays, the basic substance in biodegradable packaging films is usually derived from natural bio-polymers, such as polysaccharides, lipids, and proteins [5]. These may be recycled, degrade quickly, are non-toxic, and are environment-friendly [6,7].

Chitosan (CH) is the deacetylated chitin derivative and the second most frequent polysaccharide found naturally after cellulose. It is a linear polysaccharide composed of (1,4)-linked 2-amino-deoxy-β-D-glucan [8]. CH has demonstrated benefits in the creation of biodegradable films because of its unique characteristics, including ease of film formation, nontoxicity, good mechanical strength, excellent barrier capacity, antioxidant activity, biodegradability, and antimicrobial activity [9]. Despite the benefits mentioned above, a pure CH film is frequently fragile, has low force resistance, and is relatively susceptible to moisture [10]. Furthermore, continued development of its antioxidant and antibacterial characteristics is required for active packaging, and therefore, that becomes the focus of attention over time. Many modification procedures have been presented so far, such as the addition of antioxidants, antibacterial compounds, and reinforcing agents to either transmit or enhance certain characteristics of CH-based films [11]. The use of antioxidants in packaging films has grown in popularity since oxidation is a key issue impacting the quality of food. Butylated hydroxyl-toluene (BHT) and butylated hydroxy-anisole (BHA) are the most commonly added antioxidants to active packaging films nowadays [12]. Even though the great stability, low cost, and efficiency of these artificial antioxidants make them an effective choice for active food packaging, there are serious concerns about their toxicological properties [13].

Furthermore, due to the possible health risks posed by these substances, the consumption of artificial antioxidants is strictly regulated. In order to replace artificial antioxidants with natural ones, such as polyphenolic compounds, extensive investigations have been done in this area [14,15]. Several plant extracts have been incorporated into the CH film as a source of antioxidants, such as murta leaf extract [16], sweet potato extract [17], grape seed extract [18], banana peels extract [19], soybean seed coat extract [20], pomegranate peel extract [21], chestnut extract [22], etc., resulting in a boost in the film's antioxidant properties.

Pea *(Pisum sativum)*, sometimes referred to as "Besela" in Egypt, is an annual crop grown during the winter [23]. It is one of Egypt's most significant vegetable crops, and cooked green seeds are one of the most popular foods consumed. Empty pea pods (EPP), which make up between 30 and 67% of the entire weight of the pod, are a by-product of the pea processing sector [24]. Several high-value compounds are abundant in these EPP. They include a high amount of proteins, dietary fibre (over 50%, mostly water-insoluble), iron, potassium, and phenolic components [25]. The latter has significant promise and may be used in the feed, food, pharmaceutical, and cosmetic sectors if it is made from a cheap resource such as EPP [26]. Although the chemical composition of EPP has been extensively studied, little is currently known about its polyphenol compounds. Catechin, epicatechin, gallic acid, and other phenolic components found in abundance in empty pea pod extract (EPPE) provide EPP with a high antioxidant capacity [25,26].

To the best of our knowledge, no studies have been performed on the use of EPP or EPPE as a possible natural addition to boost the antioxidant activity of biodegradable film. Scientific research with the aim of developing active food packaging sheets incorporating EPPE might provide the pea processing sector with a new functionality and turn EPP into a very precious resource. Thus, the current study intended to develop an innovative CH-based active packaging gel film for the first time by incorporating EPPE and to assess the physical, optical, mechanical, morphological, and biological characteristics of the CH films produced. The gel films produced were also assessed as a package for oil storage because of their potential use in slowing the oxidation of corn oil.

2. Results and Discussion

2.1. Physical Properties

2.1.1. Film Thickness and Density

The thickness and density of a formulation are signs of its reliability and the quality preparation process. The film thickness and density values rose as the concentration of EPPE in the gel films increased, as indicated in Table 1. The film density and thickness both changed significantly ($p < 0.05$). Chitosan gel films that had been supplemented with 5% EPPE (Ch-5% EPPE) had the maximum film thickness (0.216 ± 0.08 mm) and density (1.94 ± 0.02 g cm^{-3}), whereas control films had the lowest values of thickness (0.132 ± 0.08 mm) and density (1.13 ± 0.02 g cm^{-3}) (Table 1). Riaz et al. [27] found a similar result, reporting that increasing the amount of apple peel polyphenolic extract in the chitosan gel matrix enhances film thickness and density. Peng et al. [28] discovered that with an increased amount of extract added to the film, the interaction between polyphenol components and CH increases. This increasing interaction causes stronger composite binding as the space between both the interacting molecules becomes smaller, enhancing film thickness and density with a rise in the applied EPPE's concentration in the CH-gel matrix.

Table 1. Thickness (T), density (D), water vapour permeability (WVP), solubility in water (S), and oil absorption ratio (OAR %) of the chitosan films modified with different percentages of EPPE.

Film Samples	Film Properties				
	T (mm)	D (g/cm^3)	WVP ($\times 10^{-10}$ g^{-1} s^{-1} pa^{-1})	S (%)	ORA (%)
Control (Ch-0% EPPE)	0.132 ± 0.08 ^{d}	1.13 ± 0.02 ^{d}	2.34 ± 0.04 ^{a}	29.40 ± 1.23 ^{a}	0.31 ± 0.006 ^{a}
Ch-1% EPPE	0.167 ± 0.09 ^{c}	1.52 ± 0.03 ^{c}	1.58 ± 0.09 ^{b}	28.17 ± 1.78 ^{a}	0.26 ± 0.004 ^{b}
Ch-3% EPPE	0.189 ± 0.06 ^{b}	1.75 ± 0.02 ^{b}	1.32 ± 0.07 ^{b}	22.43 ± 2.11 ^{b}	0.15 ± 0.003 ^{c}
Ch-5% EPPE	0.216 ± 0.08 ^{a}	1.94 ± 0.02 ^{a}	1.08 ± 0.06 ^{c}	18.75 ± 1.94 ^{c}	0.08 ± 0.001 ^{d}

Data are presented as mean ± SD. Means with different superscripts ($^{a–d}$) in lowercase letters in a column are significantly different at $p < 0.05$.

2.1.2. Film Water Vapour Permeability (WVP)

A film's WVP coefficient is a constant value for water vapour permeability at a specific temperature. A film's permeability is influenced by its chemical composition, morphology, type of permanence, and ambient temperature. WVP is a metric that measures how much moisture may pass through a film. It is more crucial in food preservation to protect a substance from moisture. The results of an investigation into the water vapour permeability of CH-gel films containing various EPPE concentrations are shown in Table 1. The results revealed that raising the EPPE concentration in the CH-gel matrix lowered the film's WVP. Significant changes ($p < 0.05$) were seen between all prepared films. The control film (Ch-0% EPPE) had the maximum WVP ($2.34 \pm 0.14 \times 10^{-10}$ g^{-1} s^{-1} pa^{-1}), followed by Ch-1% EPPE ($1.58 \pm 0.09 \times 10^{-10}$ g^{-1} s^{-1} pa^{-1}), Ch-3% EPPE ($1.32 \pm 0.07 \times 10^{-10}$ g^{-1} s^{-1} pa^{-1}), and Ch-5% EPPE ($1.08 \pm 0.06 \times 10^{-10}$ g^{-1} s^{-1} pa^{-1}). The lower WVP might be attributed to the restricted interaction between water molecules in the film as a result of the cross-linking action of chitosan, glycerol, and EPPE, culminating in less available free water [29]. The low WVP value for food packaging sheets is extremely desirable to reduce moisture transfer between the food and its immediate environment. The films' thickness is an important factor in defining their water barrier characteristics. WVP values are lower in thicker films, as water molecules take longer to flow through the film. Kurek et al. [30] observed comparable behaviour after incorporating blueberry and blackberry extracts into chitosan-based gel films.

2.1.3. Film Water Solubility (WS)

Greater solubility in water can increase film biodegradability by decreasing its degradation period in the environment. Furthermore, it will lessen the application of film on foods with high water content [29]. An increase in EPPE's addition percentage resulted in a decrease in EPPE films' WS values. Ch-1% EPPE film did not vary significantly ($p < 0.05$), while Ch-3% EPPE and Ch-5% EPPE films demonstrated a substantial reduction ($p < 0.05$) in WS in comparison to the control (Ch-0% EPPE). The WS of the EPPE films decreased, as demonstrated in Table 1, as a result of the extract's polyphenolic components forming strong hydrogen bonds with the CH matrix. Both Uranga et al. [31] and Riaz et al. [32] observed similar findings.

2.1.4. Film Oil Resistance Ability

The oil absorption ratio was used to assess Ch-EPPE gel films' capacity for oil resistance in possible food packaging use (OAR). The findings revealed that the Ch-5% EPPE film had a lower OAR percentage (0.08 ± 0.001%) than the control film (Ch-0%EPPE), which had an OAR percentage of 0.31 ± 0.006% (Table 1). Ch-1% EPPE, Ch-3% EPPE, and Ch-5% EPPE films all showed a substantial reduction in OAR percentages, from 0.26 ± 0.004 to 0.08 ± 0.001%. This reduction may be due to the fact that CH contains a hydrophilic hydroxyl group in its structure, which acts with EPPE to increase the film thickness, making it more difficult for oil molecules to cross through them. As a result, low OAR values demonstrated greater oil resistance capacity, which is a favourable attribute for packaging materials for oil goods. Riaz et al. [32] observed similar findings.

2.2. *Film Colour and Opacity*

The nutritional and flavour protection of food once exposed to visible and ultraviolet light are both determined by the optical properties of food packaging films [33,34]. These properties also impact how well the film is tolerated by a user. Colour parameters, whiteness index (WI), and opacity measurements were used to assess the prepared gel films' optical properties. Compared with the control gel film (Ch-0% EPPE), Ch-EPPE films showed significantly ($p < 0.05$) lower levels of L*. This demonstrated that the darkness intensity of Ch-EPPE gel films increases as their EPPE content increases. The films were darker as a result of light scattering and refraction produced by phenolic EPPE components [35]. Ch-5% EPPE film showed a maximum 11.94% reduction in lightness compared to the control film (Ch-0% EPPE).

Chitosan gel films' b* (yellowness/blueness) and a (redness/greenness) values were considerably ($p < 0.05$) impacted by the addition of EPPE. As the EPPE concentrations climbed from 0 to 5%, the gel films' b* values increased from (1.03 ± 0.01) to (1.98 ± 0.04) (an indication of the trend towards yellowness) and their a* values from 0.84 ± 0.01 to 1.12 ± 0.03 (an indication of the trend towards redness).

Based on these results, the existence of phenolic components and coloured compounds inside the integrated EPPE as well as the inner structure formed during film drying might be responsible for the addition of EPPE's remarkable effect on the colour of the resultant Ch-EPPE gel films [35,36]. Similar colour measurement findings were achieved when green tea extract and CH were combined, producing films that were a deeper shade of greenish yellow [29]. The films' level of whiteness and opacity were determined using opacity tests and a WI estimate. The WI of the Ch-EPPE was substantially different ($p < 0.05$) from that of the pure CH- film, as indicated in Table 2.

Table 2. Colour parameters and optical index of chitosan films modified with different percentages of EPPE.

Film Samples	L*	b*	a*	Whiteness Index	Opacity
Control (Ch-0% EPPE)	88.17 ± 0.55 ^{a}	1.03 ± 0.01 ^{a}	0.84 ± 0.01 ^{a}	88.10 ± 0.43 ^{a}	0.71 ± 0.02 ^{a}
Ch-1% EPPE	85.40 ± 0.72 ^{b}	1.19 ± 0.07 ^{b}	0.87 ± 0.01 ^{a}	85.33 ± 0.51 ^{b}	0.80 ± 0.03 ^{b}
Ch-3% EPPE	80.35 ± 0.61 ^{c}	1.47 ± 0.05 ^{c}	0.94 ± 0.03 ^{a}	80.27 ± 0.60 ^{c}	0.97 ± 0.03 ^{c}
Ch-5% EPPE	77.64 ± 0.66 ^{d}	1.98 ± 0.04 ^{d}	1.12 ± 0.03 ^{b}	77.53 ± 0.48 ^{d}	1.23 ± 0.04 ^{d}

Data are presented as mean ± SD. Means with different superscripts ($^{a–d}$) in lowercase letters in a column are significantly different at $p < 0.05$.

The reduced WI for Ch-EPPE films suggested that the film was beginning to become darker, allowing for the preservation of light-sensitive food items [36]. Opacity measurements revealed that the Ch-EPPE films were significantly ($p < 0.05$) impenetrable, with the Ch-5% EPPE film having a 73.24% increase in opacity over the control (Ch-0% EPPE). Because the chitosan polymer backbone has a mostly linear structure and offers the least resistance to light penetration compared to globular plasticizers, a comparison of the WI and opacity values of all the Ch-EPPE films with the control reveals a small quantitative drop in whiteness and a small rise in opacity [37]. Siripatrawan and Harte [29] obtained similar results.

2.3. *Mechanical Properties*

A film's greatest ability to endure applied tensile stress is measured by its tensile strength (TS), while its capacity to stretch is shown by its elongation percentage (EB%) [38]. Films with good mechanical characteristics are advantageous in industrial manufacturing, packing, transporting, and end-use applications. A comparative statistical analysis was done on the mechanical characteristics of chitosan gel films that contained various EPPE concentrations (Table 3). When the EPPE percentage rose from 0 to 5%, the TS and EB% both improved considerably, from 21.30 ± 1.191 to 26.87 ± 1.38 MPa and from 53.42 ± 3.02 to 58.64 ± 3.00%, respectively. The interaction between both the CH matrix and the polyphenolic components from EPPE may be responsible for the impact of EPPE inclusion on the enhancement in the mechanical characteristics of the related films. These interactions may result in tighter polymer chain-to-chain connections and better interfacial bonding between the CH monomers and the EPPE in the gel film layer, both of which increase the resistance against mechanical stress [39]. Similar findings were made by Balti et al. [40] and Siripatrawan and Harte [29], who noted that TS and EB increased as *spirulina* extract or green tea content increased from 0 to 5%.

Table 3. Mechanical properties of chitosan films modified with different percentages of EPPE.

Film Samples	Tensile Strength (MPa)	Elongation at Break (EB) %
Control (Ch-0% EPPE)	21.30 ± 1.19 ^{d}	53.42 ± 3.02 ^{d}
Ch-1% EPPE	23.16 ± 1.23 ^{c}	54.17 ± 2.98 ^{c}
Ch-3% EPPE	25.92 ± 1.40 ^{b}	56.83 ± 2.77 ^{b}
Ch-5% EPPE	26.87 ± 1.38 ^{a}	58.64 ± 3.00 ^{a}

Data are presented as mean ± SD. Means with different superscripts ($^{a–d}$) in lowercase letters in a column are significantly different at $p < 0.05$.

2.4. *Bioactivities of CH-EPPE Film*

2.4.1. Biodegradation Evaluation

The high biodegradability of a film by reducing its time of degradation in the environment is an important property that the new packaging must have because it reduces environmental problems [28]. The biodegradation of CH and Ch-EPPE gel films is depicted in Figure 1A. The weight loss of the Ch-EPPE and control (ch-0% EPPE) gel films increased ($p < 0.05$). As the duration for ground dumping was extended to 3 weeks, the weight loss rapidly accelerated for all the films examined. After three weeks, the Ch-5% EPPE film had the maximum weight reduction of 48.61%, compared to 27.35% for the control film (Ch-0%EPPE). A similar tendency towards biodegradation was observed when Chinese chive root extract was added to the CH-based sheet [32]. The main finding of the biodegradation assessment in this study was that adding EPPE to CH films enhance biodegradability by creating new polymer composites and reducing the environmental load through rapid breakdown.

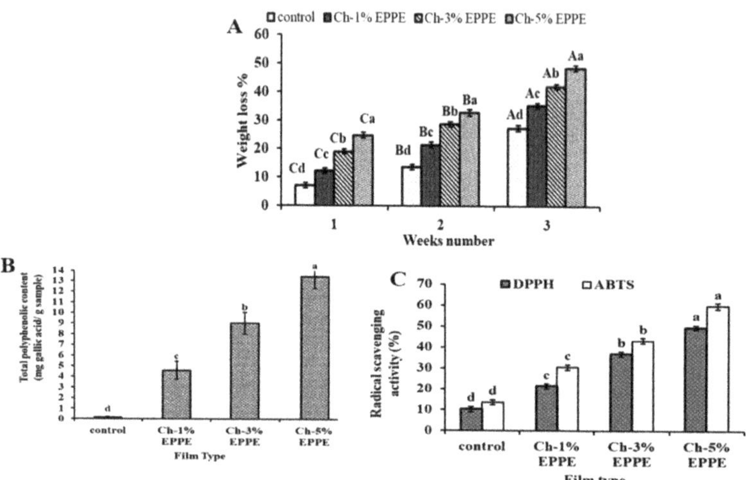

Figure 1. Biodegradability evaluation of chitosan films incorporated with EPPE (**A**), total phenolic content (**B**), and scavenging activities of the EPPE films on DPPH and ABTS radicals (**C**). Different letters indicate significant differences ($p < 0.05$). Values are given as mean ($n = 3$) ± SD.

2.4.2. Antioxidant Properties

Total phenolic content (TPC) and two other types of tests were used to assess the synthetic films' antioxidant activity (DPPH and ABTS). Figure 1 displays the results of the antioxidant activity. For active packaging sheets, TPC was utilized as a preliminary antioxidant evaluation [41]. Due to their capacity for free electron delocalization and H+ ion (of the hydroxyl group) donation, phenolic substances have been shown to have antioxidant action [42]. Figure 1B displays the TPC of CH edible films combined with EPPE. The findings revealed that as EPPE concentration increased, the total phenolic content of the CH films increased significantly ($p < 0.05$) (Figure 1B). In addition, CH films that did not have EPPE had a low TPC of 0.18 ± 0.003 mg GAE/g film. This finding may be explained by the generation of chromogens as a result of the Folin and Ciocalteu reagents reacting with non-phenolic reducing agents that may be identified spectrophotometrically [43]. Along with its phenolic components, EPPE's antioxidant action may also be attributed to other possible constituents, including the flavonoids that are present [25]. The DPPH and ABTS tests are therefore critically important in evaluating these components' antioxidant activity. As can be seen in Figure 1C, the CH- EPPE films' %DPPH radical scavenging was significantly ($p < 0.05$) greater than that of the pure CH films. The DPPH radical scavenging of the Ch-5% EPPE film was 49.71%, which was 4.83 times higher than that of the control (Ch-0% EPPE) film. The conclusions of the DPPH investigation followed a trend similar to the ABTS radical scavenging findings (Figure 1D). Films containing EPPE revealed significantly higher ABTS scavenging activity than the control ($p < 0.05$). The highest ABTS cation elimination (59.89%) was seen in the Ch-5% EPPE film as a result of its improved availability of antioxidant content. This increment was 4.40 times greater than the control group. When the antioxidant evaluations for the Ch-EPPE films were compared to one another, each antioxidant assay indicated a significant ($p < 0.05$) rise. This demonstrated how little the CH polysaccharide chain contributed to the antioxidant action [44].

2.4.3. Migration Test

Active packaging emits the active component via migration through the headspace or direct contact with the food's surface. Hence, it is crucial to assess the active packaging's capacity for releasing the active component using a migration experiment. For the migration experiment, the Food and Drug Administration of the United States has advised using food mimics, such as water (to simulate an aqueous medium) and 95% ethanol (to represent a fatty medium) [45]. The polyphenol data revealed that there is a movement of polyphenolic compounds that is greater in ethanol than in water (Table 4). Kurek et al. [46] maintain that if the structural integrity of the film is preserved, the active ingredient will continue to be present. This also has an impact on the increased GAE observed in ethanol compared with water. The Ch-5% EPPE film has the greatest migration value (5.09 ± 0.06 mg gallic acid/mL water and 9.12 ± 0.08 mg gallic acid/mL ethanol), which differs substantially ($p < 0.05$) from all other measures of polyphenol migration. It is preferable for active ingredients to migrate, such as polyphenols and antioxidants, since they can enhance the qualities of packed food items, prevent oxidation, increase their shelf life, and serve as a package having active characteristics [47].

Table 4. Migration test of chitosan films modified with 0 to 5% EPPE.

Film Samples	Simulant Type	
	Total Phenolic Content (mg gallic acid/mL Water)	Total Phenolic Content (mg gallic acid/mL Ethanol)
Control (Ch-0% EPPE)	0.002 ± 0.000 [a]	0.004 ± 0.001 [a]
Ch-1% EPPE	1.04 ± 0.05 [c]	1.92 ± 0.09 [c]
Ch-3% EPPE	3.19 ± 0.03 [b]	5.06 ± 0.06 [b]
Ch-5% EPPE	5.09 ± 0.06 [a]	9.12 ± 0.08 [a]

Data are presented as mean ± SD. Means with different superscripts (a–d) in lowercase letters in a column are significantly different at $p < 0.05$.

2.4.4. Antimicrobial Activity

For the food to be shielded from microbial development and kept fresh for a long time, the active food packaging sheet must have strong antimicrobial activity [48]. Table 5 displays the antibacterial activity of four Gram-positive and Gram-negative bacteria against CH edible gel films combined with EPPE at various percentages.

Table 5. Antibacterial activity of chitosan films modified with different percentages of EPPE.

Film Samples	Inhibition Zone Diameter (mm)			
	Salmonella typhimurium	*E. coli*	*Bacillus subtilis*	*Pseudomonas aeruginosa*
Control (Ch-0% EPPE)	NA	NA	NA	NA
Ch-1% EPPE	7.89 ± 0.10 [c]	8.12 ± 0.15 [c]	10.67 ± 0.13 [c]	10.35 ± 0.16 [c]
Ch-3% EPPE	11.38 ± 0.17 [b]	12.63 ± 0.12 [b]	15.94 ± 0.14 [b]	15.41 ± 0.11 [b]
Ch-5% EPPE	15.66 ± 0.14 [a]	16.25 ± 0.10 [a]	19.42 ± 0.20 [a]	18.98 ± 0.18 [a]

Data are presented as mean ± SD. Means with different superscripts (a–d) in lowercase letters in a column are significantly different at $p < 0.05$. NA means non-active.

As shown in Table 5, the control films (Ch-0%-EPPE) were ineffective against either of the four bacterial strains, but with the addition of EPPE, all the tested films showed antimicrobial activity on the contact area beneath the film discs. Our results were in accordance with those of Wang et al. [49], who found no significant inhibitory zone for the pure CH-gel film towards both Gram-positive and Gram-negative bacteria. This impact of CH may be connected to the fact that in the agar diffusion test method, chitosan does not diffuse across the neighbouring agar medium, meaning that only bacteria in direct contact with the active discs of CH are inhibited. The positively charged amino groups of the CH molecules, which may interact with the anionic groups on the microbial cell membrane, are also necessary for the antibacterial efficacy of chitosan [49]. In such instances, CH has been shown to exhibit intrinsic antibacterial activity against both Gram-positive and Gram-negative bacteria [22]. The Ch-EPPE films, in general, showed inhibitory effects ($p < 0.05$) on both Gram-positive and Gram-negative bacteria, and inhibition zones grew larger as the EPPE percentage rose in the film. According to the data reported in the same table, *Salmonella typhimurium* is the most sensitive to the films, followed by *E. coli, Pseudomonas aeruginosa,* and *Bacillus subtilis*. According to this study, EPPE worked better against Gram-positive bacteria than it did against Gram-negative bacteria. This could be due to variations in cell wall structure, as the cell walls of Gram-negative bacteria have lipopolysaccharides, which may prevent active constituents from entering the cytoplasmic membrane [50,51]. The primary location of contact for polyphenols with bacteria is the outer cell membrane [52]. In the polyphenols, the hydroxyl groups, conjugated double bonds, and galloyl groups are the active groups in charge of this interaction. The bacteria may die if the membrane, which protects the cell's integrity, is damaged as a result of this contact.

2.5. Films SEM Photographs

Figure 2A–D depict the results from a prepared film SEM study. The control gel film (ch-0%-EPPE) has a smooth, uniform surface. There were no impurities, delamination, or precipitates (Figure 2A). The studied films' surface morphology was unaffected by the addition of 1% EPPE, it appears (Figure 2B). Small agglomerates were seen when CH was added with 3% EPPE (Figure 2C). A more heterogeneous surface was produced as a result of the structure being disturbed by the rise in EPPE concentration from 3% to 5%, which led to the appearance of more and more white patches (Figure 2D). This could be because EPPE contains hydrophilic polyphenolic components.

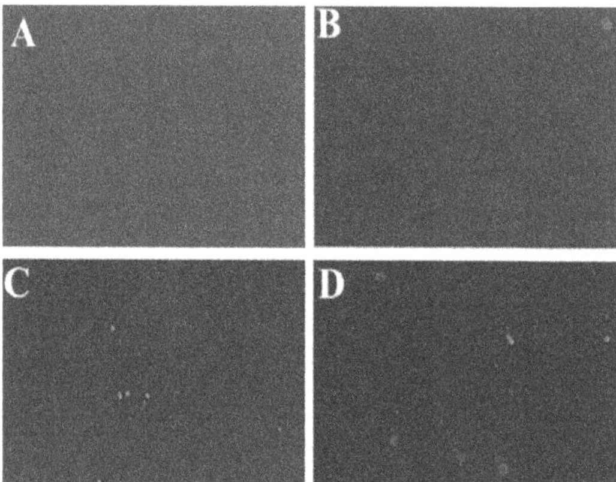

Figure 2. Active films' SEM images, where (**A**) Control (Ch-0% EPPE), (**B**) Ch-1% EPPE, (**C**) Ch-3% EPPE, and (**D**) Ch-5%EPPE.

2.6. Application of Films in Packaging Edible Oil

Edible oil oxidation occurs as a result of the effects of temperature, oxygen, and light [53]. In this context, oxidation stability is regarded as a crucial indicator of edible oil quality. Corn oil packed in four CH-based EPPE films (0, 1, 3, and 5%) and unpackaged corn oil (open control) were both evaluated for oxidation stability over a 10-day period at 50 °C. During oil storage, the values of peroxide value (PV) and thiobarbutic acid (TBA) steadily rose (Figure 3).

The oil's PV and TBARS levels were significantly reduced in the CH-based EPPE films. All treatments produced higher PV and TBARS values independent of duration; however, this increase was slower for CH-based EPPE films than with open control and Ch-0% EPPE films. The open control had the greatest values of PV (5.26 ± 0.41 mil-equivalent O2/kg oil) and TBARS (4.13 ± 0.38 mg malondialdhyde/kg oil) on the 10th day of storage, whereas Ch-5% EPPE had the lowest values of PV (2.73 ± 0.24 mil-equivalent O2/kg oil) and TBA (2.10 ± 0.07 mg malondialdhyde/kg oil). The bio-composite film's tight structure has an important role in reducing oxidation by making oxygen harder to get through [54]. Additionally, the inclusion of phenolic compounds in the film boosted the antioxidant potential of the sheet, which also aided in slowing the oxidation of the oil. Therefore, it may be assumed that CH-based EPPE films could be utilized as packaging films for foodstuffs that are extremely vulnerable to oxidation.

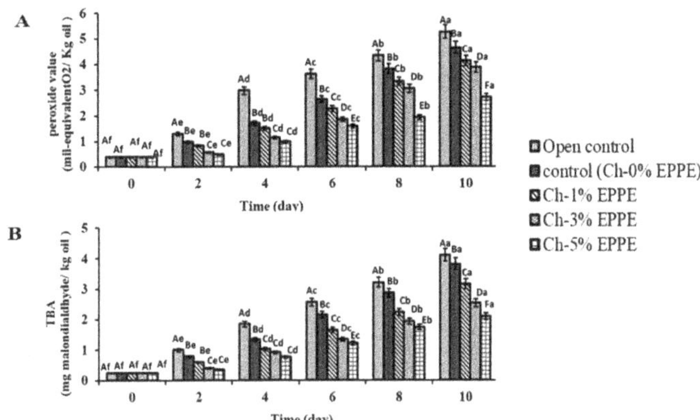

Figure 3. Changes in the PV (**A**) and TBARS (**B**) levels in corn oil stored in Ch-EPPE films at 50 °C for 10 days. Data are presented as means ± SD of triplicates. Different lowercase letters indicate the statistically significant difference ($p < 0.05$) within the same treatment group at different storage times. Different uppercase letters indicate the statistically significant difference ($p < 0.05$) among different treatment groups at the same storage time.

3. Conclusions

The empty pea pods resulting as residue from food factories are one of the sources that can be used in the separation of many important biological compounds, including phenolic compounds. The phenolic compounds present in EPPE have antioxidant and antimicrobial properties. In this study, active food films were prepared from chitosan-containing EPPE. The properties of these films were evaluated, and they were used to extend the shelf life of corn oil and protect it from oxidation. The results obtained showed that the addition of EPPE increased the physical parameters of the CH-gel film in terms of T and D. Furthermore, the overall colour characteristics improved from transparency to impenetrability. These films had an additional amount of EPPE in them, which resulted in a substantial ($p < 0.05$) improvement in TS. Increases in EPPE levels, on the other hand, resulted in substantial ($p < 0.05$) decreases in WVP, S, OAR%, and EB%. The SEM analysis confirmed the interactions between EPPE and CH by revealing a consistent structure for all Ch-EPPE films. The EPPE films demonstrated a significant ($p < 0.05$) enhancement in antioxidant and antimicrobial activity. The corn oil PV and TBA values were much lower in the CH-based EPPE gel films throughout the storage experiment. These findings indicate that EPPE films offer an environment-friendly active packaging alternative to synthetic polymers for use in the food industry.

4. Materials and Methods

4.1. Materials

The empty pea pods were bought from the Kaha Food Canning Company in Kaha, Egypt. They were cleaned and disinfected with sodium hypochlorite at a 50 ppm concentration after they arrived at the lab. They were then cut into small pieces (strips) that were about 2 cm long, placed in a single layer on stainless steel trays, and dried for 12 h at 55 °C in a hot-air oven (Memmert, UF). Then, the dried materials were ground in a FX1000 electrical grinder (Black & Decker, London, England) to pass through a 150 m sieve [25]. Each sample's dried powder was stored at 4 °C and sealed inside an airtight Kilner jar.

4.2. *Empty Pea Pods Extract (EPPE)*

Following the method outlined by Pinchao-Pinchao et al. [55], ultrasonic-assisted extraction (Elmasonic, Singen, Germany) was used to extract polyphenol components from powdered EPP. The best conditions (30 °C, 20 min, 50% ethanol concentration, and a liquid solid ratio of 1:40) were utilized to prepare EPPE with ethanol as the solvent. The extracted material was wrapped in aluminium foil and kept at −18 °C. Under the influence of a nitrogen gas stream, the solvent was eliminated.

4.3. *Gel Film Preparation*

The CH-gel film-forming solution was made in the manner described by Siripatrawan and Harte [29], with minor modifications. A number of initial experiments were carried out to determine the best type and amount of acid solvent and plasticizer to be utilized in the preparation of CH-based gel films. The findings showed that 2% CH in 1% acetic acid might be used to create the best CH-gel films. The findings also showed that the addition of glycerol as a plasticizer at a percentage of 30% w/w of CH powder enhanced the mechanical characteristics of the films. Hundred millilitres of glacial acetic acid solution (1%) and CH powder (2 g) (deacetylation degree: 75%, Sigma Aldrich Company, Darmstadt, Germany) were combined to create a film-forming solution. The film-forming solution was supplemented with glycerol (El-Gomhoria Chemical Co., Tanta, Egypt), a plasticizer, at a fixed percentage of 30% weight/weight of CH. The resulting solution was then heated in a water bath shaking incubator for 30 min at 60 °C and 100 rpm. To get rid of undissolved contaminants, the chitosan solution was then filtered using a coarse sintered glass filter. After being cooled to ambient temperature, the EPPE was mixed into the film-forming solution to produce percentages of 0, 1, 3, and 5% (w/v). The resultant solutions were homogenized with a Moulinex homogenizer (Courneuve, France). After that, a sonicator (Singen, Germany) was used to degas the film-forming solutions to get rid of air bubbles. A ceramic plate was used to cast each film-forming solution, and it was allowed to dry there. Prior to testing, the films were conditioned for 48 h at 25 °C and 50% RH in a chamber.

4.4. *Physical Properties*

4.4.1. Film Thickness (T)

A digital Mitutoyo Absolute Micrometer (Tokyo, Japan) was used to measure the T of the film. For every film sample, 10 replications were carried out. The mean values were obtained after the measures were conducted at random locations all over the film sample.

4.4.2. Film Density (D)

The film's weight and volume were used to calculate the D value of each film. The film's thickness and area were used to compute its volume.

4.4.3. Film WVP

The WVP was determined using the Zhang et al. [56] technique with the necessary changes. In total, 10 g of anhydrous $CaCl_2$ was placed in a glass cup with a diameter of 4 cm, and a film measuring 10 × 10 cm was placed on top, supported by rubber strands. The cups were then piled inside the desiccator and filled with sodium chloride saturation solution (75 RH). For up to seven hours, the cup's weight was registered every hour. Using linear regression, the slope of each line (K) (g/h) was determined. Equation (1) was used to calculate the WVP values.

$$\text{WVP} = \frac{(\text{Film thickness}) \times \text{K}}{(\text{Vapor pressure difference}) \times (\text{Film area})} \tag{1}$$

4.4.4. Film Water Solubility(S)

The technique of Rambabu et al. [37] was used to determine the film S percentage. The samples were cut into thin strips (20 × 20 mm) and dried in a hot-air oven for 24 h at 105 °C to a consistent weight. Each dried film sample was submerged for 24 h in 70 mL of distilled water. The film samples were then taken out of the solution and allowed to dry again at 100 °C for 24 h. Final weights were noted, and solubility was determined as follows:

$$S\ (\%) = \frac{(\text{Initial dry wieght} - \text{Final dry wieght})}{\text{Initial dry wieght!}} \times 100 \tag{2}$$

4.4.5. Films' ORA

The Wang et al. [49] technique was slightly modified to determine the ORA of the films. In a hot-air oven set at 50 °C, the filter paper (6 cm in diameter) was dried to a consistent weight. The film samples (4 cm × 4 cm) were placed upside down on the filter paper for 48 h while being secured with ropes on the top of glass test tubes containing 5 mL of oil. After 48 h, the filter paper was weighed, and the oil absorption rate (OAR) was determined using the equation

$$\text{OAR}\ (\%) = \frac{W - W1}{W1} \times 100 \tag{3}$$

where W is the weight of filter paper after 48 h and W1 is the weight of dried filter paper.

4.5. Optical Characteristics

Colour measurement and opacity estimates were used to conduct an optical investigation of the Ch-EPPE films. Using a Miniscan EZ colorimeter (HunterLab), the values of the Hunter colour (L*, a*, and b*) were calculated. Equation (2) was used to calculate the film's whiteness index using these values [57]. Six readings were taken at various locations on each film. Five film samples were utilized in each replication, which involved five replications for each treatment.

$$\text{Whiteness index} = 100 - \sqrt{a^2 + (100 - L)2 + b^2} \tag{4}$$

Utilizing the technique developed by Rambabu et al. [37], the opacity of the film was determined by measuring the absorbance of a rectangular film sample (1 × 4 cm in size) using a UV-Vis spectrophotometer (UV-3600, Shimadzu, Baltimore, MD, USA) with an absorbed wave length of 600 nm. The film's opacity was calculated using Equation (5).

$$\text{Opacity} = \frac{\text{Absorbance at 600 nm}}{\text{Film thickness (mm)}} \tag{5}$$

High transparency values, based on this equation, indicate poorer transparency and a greater level of opacity.

4.6. Mechanical Properties

The percentages of EB and TS, among other mechanical parameters according to ASTM [58], were assessed using a TA.XT Plus texture analyzer (Stable Micro-Systems, British). A dimension of 1.5 × 10 cm was made from films that had the same thickness. Forced paper was positioned between the two metal grips to support the ends of the film samples. Following machine calibration, the speed was set at 50 mm/min for initial grip separation and 100 mm/min for detection speed. Following the creation of the stress-strain curves, the TS and EB were determined using the following equations:

$$\text{TS(MPa)} = \frac{\text{Maximum extension force}}{\text{Initial crosssectional area}} \times 100 \tag{6}$$

$$\mathrm{EB}(\%) = \frac{\text{Film extension ratio}}{\text{Initial length}} \times 100 \quad (7)$$

4.7. Bioactivities of CH-EPPE Film

4.7.1. Biodegradation Test

The composting experiment described by Riaz et al. [32] was used to measure the biodegradation capability of the films produced. The topsoil was obtained from Kafrelshiekh University's experimental field (Kafr El-Shiekh, Egypt). The topsoil was put in a plastic container, and samples of each sheet (2 × 2 cm) were buried at a depth of 2 cm for three weeks. Twice a day, water was sprayed on the soil. The film samples were removed at the end of each week, and the weight loss of each film was recorded.

4.7.2. Antioxidant Properties

TPC Measurement

The Folin Ciocalteu reagent was used to estimate the EPPE films' TPC, as mentioned by Ruiz-Navajas et al. [59]. In order to partially dissolve the films and release the EPPE for the next tests, the film samples (25 mg) were soaked in ethanol (3 mL). The gel film's total phenolic content was determined.

Antioxidant Activity Measurement

The antioxidant activity % of the CH film was measured using the DPPH and ABTS tests outlined by Liu et al. [60]. In brief, 1.5 mL of the film extract was combined with 0.5 mL of a 0.1 mM ethanolic DPPH solution and stored in the dark for half an hour. The DPPH scavenging capacity % was determined from Equation (8) based on the absorbance at 515 nm.

$$\text{DPPH scavenging } (\%) = \left[1 - \frac{\text{film extract absorbation}}{\text{control absorbation}}\right] \times 100 \quad (8)$$

For the test of ABTS scavenging %, the film samples (10 mg) and 3 mL of ABTS stock solutions were combined, absorbance at 734 nm was measured, and scavenging activity was estimated using the following equation:

$$\text{ABTS scavenging } (\%) = \left[1 - \frac{\text{film extract absorbation}}{\text{control absorbation}}\right] \times 100 \quad (9)$$

Migration Test

The migration assessment was carried out in accordance with the method of Oliveira et al. [61]. CH-gel film specimens were chopped into small (2 × 2 cm) pieces and soaked in 5 mL water (to imitate an aqueous medium) and 95% ethanol (to simulate a fatty medium). Seven days were spent shaking it at 125 rpm at 25 °C. The TPC approach described in Section 4.7.2 was used to determine the migration of active molecules into food simulants.

Antimicrobial Activity

According to the method outlined by Elsebaie and Essa [62], the antimicrobial activity of the prepared gel films was assessed against two Gram-positive, *Bacillus subtilis* ATCC21331 and *Pseudomonas aeruginosa* (CGMCC1.8721), and two Gram-negative, *Escherichia coli* ATCC25921 and *Salmonella typhimurium* (CGMCC 1.10754), bacteria. A movable calliper was used to measure the inhibitory zone.

4.8. SEM Scanning

A Carl Zeiss SEM (EVO-LS-10, Hamburg, Germany) was used to examine the films created at 10 kV. Film sections with dimensions of 1 × 1 cm were cut out, fastened on aluminium stubs with carbon tape, and coated with gold on the outside. Additionally, the image was adjusted to ×5000.

4.9. Film Application in Corn Oil Packaging

The film specimens were stored at 50 °C for 10 days after being wrapped in line rope and placed on top of glass bottles containing corn oil. On days 0, 2, 4, 6, 8, and 10, a sample of 5 mL was removed from each glass tube in order to calculate the PV and TBA values using the technique outlined by Elsebaie et al. [63]. The analysis was carried out three times, and the average values were provided. A glass container with no film was applied as a control.

4.10. Statistical Analysis

To analyze differences between values, an ANOVA was performed using SPSS (Ver. 16.0, 2007)'s general linear regression model. For statistical testing, the probability degrees of $p < 0.05$ were substantially considered. Triplicates for each measurement and experiment were carried out.

Author Contributions: Conceptualization, S.A.A., G.A.A., H.A.Y.S.(Heba Ali Yousef Shaat) and H.S.E.E.; methodology, E.M.E.; software, M.M.M. and A.A.F.; validation, M.M.M., G.A.A. and H.A.Y.S. (Heba Ali Yousef Shaat); formal analysis, S.A.-E.E. and A.L.A.A.; investigation, E.M.E. and M.R.B.; resources, S.A.A., S.A.-E.E., H.S.E.E. and A.L.A.A.; data curation, H.A.Y.S. (Hala Ali Yousef Shaat) and W.M.A.E.; writing—original draft preparation, E.M.E.; writing—review and editing, E.M.E.; visualization, H.A.Y.S. (Hala Ali Yousef Shaat), M.R.B. and W.M.A.E. All authors have read and agreed to the published version of the manuscript.

Funding: This research received no external funding.

Institutional Review Board Statement: Not applicable.

Informed Consent Statement: Not applicable.

Data Availability Statement: The authors confirm that the data supporting the findings of this study are available within the article.

Conflicts of Interest: The authors declare no conflict of interest.

References

1. Pauer, E.; Wohner, B.; Heinrich, V.; Tacker, M. Assessing the environmental sustainability of food packaging: An extended life cycle assessment including packaging-related food losses and waste and circularity assessment. *Sustainability* **2019**, *11*, 925. [CrossRef]
2. Wohner, B.; Pauer, E.; Heinrich, V.; Tacker, M. Packaging-related food losses and waste: An overview of drivers and issues. *Sustainability* **2019**, *11*, 264. [CrossRef]
3. Bumbudsanpharoke, N.; Ko, S. Nanoclays in food and beverage packaging. *J. Nanomater.* **2019**, *2019*, 8927167. [CrossRef]
4. Horodytska, O.; Valdés, F.J.; Fullana, A. Plastic flexible films waste management—A state of art review. *Waste Manag.* **2018**, *77*, 413–425. [CrossRef] [PubMed]
5. Talukdar, M.; Nath, O.; Deb, P. Enhancing barrier properties of biodegradable film by reinforcing with 2D heterostructure. *Appl. Surf. Sci.* **2021**, *541*, 148464. [CrossRef]
6. Yao, Q.; Song, Z.; Li, J.; Zhang, L. Micromorphology, mechanical, crystallization and permeability properties analysis of HA/PBAT/PLA (HA, hydroxyapatite; PBAT, poly (butylene adipate-co-butylene terephthalate); PLA, polylactide) degradability packaging films. *Polym. Int.* **2020**, *69*, 301–307. [CrossRef]
7. Zhao, L.; Duan, G.; Zhang, G.; Yang, H.; He, S.; Jiang, S. Electrospun functional materials toward food packaging applications: A review. *Nanomaterials* **2020**, *10*, 150. [CrossRef]
8. Dehghani, S.; Hosseini, S.V.; Regenstein, J.M. Edible films and coatings in seafood preservation: A review. *Food Chem.* **2018**, *240*, 505–513. [CrossRef]
9. Manigandan, V.; Karthik, R.; Ramachandran, S.; Rajagopal, S. Chitosan applications in food industry. In *Biopolymers for Food Design*; Elsevier: Amsterdam, The Netherlands, 2018; pp. 469–491.
10. Bourtoom, T. Edible films and coatings: Characteristics and properties. *Int. Food Res. J.* **2008**, *15*, 237–248.
11. Nura, A. Advances in food packaging technology-a review. *J. Postharvest Technol.* **2018**, *6*, 55–64.
12. Quezada-Gallo, J. Delivery of food additives and antimicrobials using edible films and coatings. In *Edible Films and Coatings for Food Applications*; Embuscado, M.E., Huber, K.C., Eds.; Springer: New York, NY, USA, 2009; pp. 315–333.

13. Pokorný, J. Are natural antioxidants better–and safer–than synthetic antioxidants? *Eur. J. Lipid Sci. Technol.* **2007**, *109*, 629–642. [CrossRef]
14. Jongjareonrak, A.; Benjakul, S.; Visessanguan, W.; Tanaka, M. Antioxidative activity and properties of fish skin gelatin films incorporated with BHT and α-tocopherol. *Food Hydrocoll.* **2008**, *22*, 449–458. [CrossRef]
15. Yen, M.-T.; Yang, J.-H.; Mau, J.-L. Antioxidant properties of chitosan from crab shells. *Carbohydr. Polym.* **2008**, *74*, 840–844. [CrossRef]
16. Nguyen, T.T.; Dao, U.T.T.; Bui, Q.P.T.; Bach, G.L.; Thuc, C.H.; Thuc, H.H. Enhanced antimicrobial activities and physiochemical properties of edible film based on chitosan incorporated with Sonneratia caseolaris (L.) Engl. leaf extract. *Prog. Org. Coat.* **2020**, *140*, 105487. [CrossRef]
17. Yong, H.; Wang, X.; Bai, R.; Miao, Z.; Zhang, X.; Liu, J. Development of antioxidant and intelligent pH-sensing packaging films by incorporating purple-fleshed sweet potato extract into chitosan matrix. *Food Hydrocoll.* **2019**, *90*, 216–224. [CrossRef]
18. Sogut, E.; Seydim, A.C. The effects of Chitosan and grape seed extract-based edible films on the quality of vacuum packaged chicken breast fillets. *Food Packag. Shelf Life* **2018**, *18*, 13–20. [CrossRef]
19. Zhang, W.; Li, X.; Jiang, W. Development of antioxidant chitosan film with banana peels extract and its application as coating in maintaining the storage quality of apple. *Int. J. Biol. Macromol.* **2020**, *154*, 1205–1214. [CrossRef]
20. Wang, X.; Yong, H.; Gao, L.; Li, L.; Jin, M.; Liu, J. Preparation and characterization of antioxidant and pH-sensitive films based on chitosan and black soybean seed coat extract. *Food Hydrocoll.* **2019**, *89*, 56–66. [CrossRef]
21. Cui, H.; Surendhiran, D.; Li, C.; Lin, L. Biodegradable zein active film containing chitosan nanoparticle encapsulated with pomegranate peel extract for food packaging. *Food Packag. Shelf Life* **2020**, *24*, 100511. [CrossRef]
22. Kõrge, K.; Bajić, M.; Likozar, B.; Novak, U. Active chitosan–chestnut extract films used for packaging and storage of fresh pasta. *Int. J. Food Sci. Technol.* **2020**, *55*, 3043–3052. [CrossRef]
23. El Shaer, S.; El-Sharkawy, H.; El-Salehein, A. Efficiency of transmission pea enation mosaic virus (PEMV) by the pea aphid in Dakhlia governorate, Egypt. *J. Product. Dev.* **2022**, *27*, 189–200. [CrossRef]
24. Mateos-Aparicio, I.; Redondo-Cuenca, A.; Villanueva-Suárez, M.-J.; Zapata-Revilla, M.-A.; Tenorio-Sanz, M.-D. Pea pod, broad bean pod and okara, potential sources of functional compounds. *LWT-Food Sci. Technol.* **2010**, *43*, 1467–1470. [CrossRef]
25. Awad, K.; Abdel-Nabey, A.; Awney, H. Phenolic Composition and Antioxidant Activity of Some Agro-industrial Wastes. *Alex. J. Fd. Sci. Technol* **2018**, *15*, 33–44.
26. Hadjout, L.; Dahmoune, F.; Hentabli, M.; Spigno, G.; Madani, K. Physicochemical, Functional and Bioactive Properties of Pea (*Pisum sativum* L.) Pods Microwave and Convective Dried Powders. 2021. Available online: https://assets.researchsquare.com/files/rs-951507/v1_covered.pdf?c=1635862935 (accessed on 3 March 2021).
27. Riaz, A.; Lei, S.; Akhtar, H.M.S.; Wan, P.; Chen, D.; Jabbar, S.; Abid, M.; Hashim, M.M.; Zeng, X. Preparation and characterization of chitosan-based antimicrobial active food packaging film incorporated with apple peel polyphenols. *Int. J. Biol. Macromol.* **2018**, *114*, 547–555. [CrossRef] [PubMed]
28. Peng, Y.; Wu, Y.; Li, Y. Development of tea extracts and chitosan composite films for active packaging materials. *Int. J. Biol. Macromol.* **2013**, *59*, 282–289. [CrossRef] [PubMed]
29. Siripatrawan, U.; Harte, B.R. Physical properties and antioxidant activity of an active film from chitosan incorporated with green tea extract. *Food Hydrocoll.* **2010**, *24*, 770–775. [CrossRef]
30. Kurek, M.; Garofulić, I.E.; Bakić, M.T.; Ščetar, M.; Uzelac, V.D. Development and evaluation of a novel antioxidant and pH indicator film based on chitosan and food waste sources of antioxidants. *Food Hydrocoll.* **2018**, *84*, 238–246. [CrossRef]
31. Uranga, J.; Puertas, A.; Etxabide, A.; Dueñas, M.; Guerrero, P.; De La Caba, K. Citric acid-incorporated fish gelatin/chitosan composite films. *Food Hydrocoll.* **2019**, *86*, 95–103. [CrossRef]
32. Riaz, A.; Lagnika, C.; Luo, H.; Dai, Z.; Nie, M.; Hashim, M.M.; Liu, C.; Song, J.; Li, D. Chitosan-based biodegradable active food packaging film containing Chinese chive (Allium tuberosum) root extract for food application. *Int. J. Biol. Macromol.* **2020**, *150*, 595–604. [CrossRef]
33. Zhang, W.; Chen, J.; Chen, Y.; Xia, W.; Xiong, Y.L.; Wang, H. Enhanced physicochemical properties of chitosan/whey protein isolate composite film by sodium laurate-modified TiO2 nanoparticles. *Carbohydr. Polym.* **2016**, *138*, 59–65. [CrossRef]
34. Martins, J.T.; Cerqueira, M.A.; Vicente, A.A. Influence of α-tocopherol on physicochemical properties of chitosan-based films. *Food Hydrocoll.* **2012**, *27*, 220–227. [CrossRef]
35. Hopkins, E.J.; Chang, C.; Lam, R.S.; Nickerson, M.T. Effects of flaxseed oil concentration on the performance of a soy protein isolate-based emulsion-type film. *Food Res. Int.* **2015**, *67*, 418–425. [CrossRef]
36. Yang, H.; Li, J.G.; Wu, N.F.; Fan, M.M.; Shen, X.L.; Chen, M.T.; Jiang, A.M.; Lai, L.-S. Effect of hsian-tsao gum (HG) content upon rheological properties of film-forming solutions (FFS) and physical properties of soy protein/hsian-tsao gum films. *Food Hydrocoll.* **2015**, *50*, 211–218. [CrossRef]
37. Rambabu, K.; Bharath, G.; Banat, F.; Show, P.L.; Cocoletzi, H.H. Mango leaf extract incorporated chitosan antioxidant film for active food packaging. *Int. J. Biol. Macromol.* **2019**, *126*, 1234–1243.
38. Laohakunjit, N.; Noomhorm, A. Effect of plasticizers on mechanical and barrier properties of rice starch film. *Starch-Stärke* **2004**, *56*, 348–356. [CrossRef]

39. Pastor, C.; Sánchez-González, L.; Cháfer, M.; Chiralt, A.; González-Martínez, C. Physical and antifungal properties of hydroxypropylmethylcellulose based films containing propolis as affected by moisture content. *Carbohydr. Polym.* **2010**, *82*, 1174–1183. [CrossRef]
40. Balti, R.; Mansour, M.B.; Sayari, N.; Yacoubi, L.; Rabaoui, L.; Brodu, N.; Massé, A. Development and characterization of bioactive edible films from spider crab (Maja crispata) chitosan incorporated with Spirulina extract. *Int. J. Biol. Macromol.* **2017**, *105*, 1464–1472. [CrossRef]
41. Genskowsky, E.; Puente, L.; Pérez-Álvarez, J.; Fernandez-Lopez, J.; Muñoz, L.; Viuda-Martos, M. Assessment of antibacterial and antioxidant properties of chitosan edible films incorporated with maqui berry (Aristotelia chilensis). *LWT-Food Sci. Technol.* **2015**, *64*, 1057–1062. [CrossRef]
42. Dey, T.B.; Chakraborty, S.; Jain, K.K.; Sharma, A.; Kuhad, R.C. Antioxidant phenolics and their microbial production by submerged and solid state fermentation process: A review. *Trends Food Sci. Technol.* **2016**, *53*, 60–74.
43. Moradi, M.; Tajik, H.; Rohani, S.M.R.; Oromiehie, A.R.; Malekinejad, H.; Aliakbarlu, J.; Hadian, M. Characterization of antioxidant chitosan film incorporated with Zataria multiflora Boiss essential oil and grape seed extract. *LWT-Food Sci. Technol.* **2012**, *46*, 477–484. [CrossRef]
44. Atarés, L.; Pérez-Masiá, R.; Chiralt, A. The role of some antioxidants in the HPMC film properties and lipid protection in coated toasted almonds. *J. Food Eng.* **2011**, *104*, 649–656. [CrossRef]
45. Colín-Chávez, C.; Soto-Valdez, H.; Peralta, E.; Lizardi-Mendoza, J.; Balandrán-Quintana, R. Diffusion of natural astaxanthin from polyethylene active packaging films into a fatty food simulant. *Food Res. Int.* **2013**, *54*, 873–880. [CrossRef]
46. Kurek, M.; Guinault, A.; Voilley, A.; Debeaufort, F. Effect of relative humidity on carvacrol release and permeation properties of chitosan based films and coatings. *Food Chem.* **2014**, *144*, 9–17. [CrossRef] [PubMed]
47. Adilah, Z.M.; Jamilah, B.; Hanani, Z.N. Functional and antioxidant properties of protein-based films incorporated with mango kernel extract for active packaging. *Food Hydrocoll.* **2018**, *74*, 207–218. [CrossRef]
48. Wu, Y.; Luo, X.; Li, W.; Song, R.; Li, J.; Li, Y.; Li, B.; Liu, S. Green and biodegradable composite films with novel antimicrobial performance based on cellulose. *Food Chem.* **2016**, *197*, 250–256. [CrossRef]
49. Wang, L.; Liu, F.; Jiang, Y.; Chai, Z.; Li, P.; Cheng, Y.; Jing, H.; Leng, X. Synergistic antimicrobial activities of natural essential oils with chitosan films. *J. Agric. Food Chem.* **2011**, *59*, 12411–12419. [CrossRef] [PubMed]
50. Ouattara, B.; Simard, R.E.; Holley, R.A.; Piette, G.J.-P.; Bégin, A. Antibacterial activity of selected fatty acids and essential oils against six meat spoilage organisms. *Int. J. Food Microbiol.* **1997**, *37*, 155–162. [CrossRef] [PubMed]
51. Abdollahi, M.; Rezaei, M.; Farzi, G. Improvement of active chitosan film properties with rosemary essential oil for food packaging. *Int. J. Food Sci. Technol.* **2012**, *47*, 847–853. [CrossRef]
52. Ultee, A.; Bennik, M.; Moezelaar, R. The phenolic hydroxyl group of carvacrol is essential for action against the food-borne pathogen Bacillus cereus. *Appl. Environ. Microbiol.* **2002**, *68*, 1561–1568. [CrossRef]
53. Choe, E.; Min, D.B. Mechanisms and factors for edible oil oxidation. *Compr. Rev. Food Sci. Food Saf.* **2006**, *5*, 169–186. [CrossRef]
54. Mariniello, L.; Giosafatto, C.; Di Pierro, P.; Sorrentino, A.; Porta, R. Swelling, mechanical, and barrier properties of albedo-based films prepared in the presence of phaseolin cross-linked or not by transglutaminase. *Biomacromolecules* **2010**, *11*, 2394–2398. [CrossRef] [PubMed]
55. Pinchao-Pinchao, Y.A.; Ordoñez-Santos, L.E.; Osorio-Mora, O. Evaluation of the effect of different factors on the ultrasound assisted extraction of phenolic compounds of the pea pod. *Dyna* **2019**, *86*, 211–215.
56. Zhang, P.; Zhao, Y.; Shi, Q. Characterization of a novel edible film based on gum ghatti: Effect of plasticizer type and concentration. *Carbohydr. Polym.* **2016**, *153*, 345–355. [CrossRef] [PubMed]
57. Jouki, M.; Yazdi, F.T.; Mortazavi, S.A.; Koocheki, A. Physical, barrier and antioxidant properties of a novel plasticized edible film from quince seed mucilage. *Int. J. Biol. Macromol.* **2013**, *62*, 500–507. [CrossRef] [PubMed]
58. ASTM. *Annual Book of ASTM Standards*; American Society for Testing and Materials: West Conshohocken, PA, USA, 2003.
59. Ruiz-Navajas, Y.; Viuda-Martos, M.; Sendra, E.; Perez-Alvarez, J.; Fernández-López, J. In vitro antibacterial and antioxidant properties of chitosan edible films incorporated with Thymus moroderi or Thymus piperella essential oils. *Food Control* **2013**, *30*, 386–392. [CrossRef]
60. Liu, J.; Meng, C.-g.; Liu, S.; Kan, J.; Jin, C.-h. Preparation and characterization of protocatechuic acid grafted chitosan films with antioxidant activity. *Food Hydrocoll.* **2017**, *63*, 457–466. [CrossRef]
61. Oliveira, V.; Monteiro, M.; Silva, K.; Leite, R.; Aroucha, E. Analysis of the Barrier and Thermogravimetric Properties of Cassava Starch Biopolymeric Films with Addition of Beeswax. *J. Food Process. Technol.* **2018**, *10*, 2.
62. Elsebaie, E.M.; Essa, R.Y. Application of barnûf (Pluchea dioscoridis) leaves extract as a natural antioxidant and antimicrobial agent for eggs quality and safety improvement during storage. *J. Food Process. Preserv.* **2022**, *46*, e16061. [CrossRef]
63. Elsebaie, E.M.; Kassem, M.M.; Mousa, M.M.; Basuony, M.A.M.; Zeima, N.M.; Essa, R.Y. Cod Liver Oil's Encapsulation into Sodium Alginate/Lupin Protein Beads and Its Application in Functional Meatballs' Preparation. *Foods* **2022**, *11*, 1328. [CrossRef]

Article

Kiwi Fruits Preservation Using Novel Edible Active Coatings Based on Rich Thymol Halloysite Nanostructures and Chitosan/Polyvinyl Alcohol Gels

Constantinos E. Salmas [1,*], Aris E. Giannakas [2,*], Dimitrios Moschovas [1], Eleni Kollia [3], Stavros Georgopoulos [2], Christina Gioti [1], Areti Leontiou [2], Apostolos Avgeropoulos [1], Anna Kopsacheili [3], Learda Avdylaj [3] and Charalampos Proestos [3,*]

[1] Department of Material Science and Engineering, University of Ioannina, 45110 Ioannina, Greece
[2] Department of Food Science and Technology, University of Patras, 30100 Agrinio, Greece
[3] Laboratory of Food Chemistry, Department of Chemistry, National and Kapodistrian University of Athens, Zografou, 15771 Athens, Greece
* Correspondence: ksalmas@uoi.gr (C.E.S.); agiannakas@upatras.gr (A.E.G.); harpro@chem.uoa.gr (C.P.)

Abstract: The concept of this study is the replacement of previous fossil-based techniques for food packaging and food shelf-life extension, with novel more green processes and materials following the spirit of circular economy and the global trend for environmentally positive fingerprints. A novel adsorption process to produce thymol-halloysite nanohybrids is presented in this work. The high dispersion of this thymol-halloysite nanostructure in chitosan biopolymer is one of the goals of this study. The incorporation of this biodegradable matrix with poly-vinyl-alcohol produced a very promising food-packaging film. Mechanical, water-oxygen barrier, antimicrobial, and antioxidant properties were measured. Transparency levels were also tested using a UV-vis instrument. Moreover, the developed films were tested in-vivo for the preservation and the extension of the shelf-life of kiwi fruits. In all cases, results indicated that the increased fraction of thymol from thyme oil significantly enhances the antimicrobial and antioxidant activity of the prepared chitosan-poly-vinyl- alcohol gel. The use of the halloysite increases the mechanical and water-oxygen barrier properties and leads to a control release process of thymol which extends the preservation and the shelf-life of kiwi fruits. Finally, the results indicated that the halloysite improves the properties of the chitosan/poly-vinyl-alcohol films, and the thymol makes them further advantageous.

Keywords: thyme oil; halloysite; chitosan; polyvinyl alcohol; gel; active coatings; nanostructures; kiwi fruits

Citation: Salmas, C.E.; Giannakas, A.E.; Moschovas, D.; Kollia, E.; Georgopoulos, S.; Gioti, C.; Leontiou, A.; Avgeropoulos, A.; Kopsacheili, A.; Avdylaj, L.; et al. Kiwi Fruits Preservation Using Novel Edible Active Coatings Based on Rich Thymol Halloysite Nanostructures and Chitosan/Polyvinyl Alcohol Gels. *Gels* **2022**, *8*, 823. https://doi.org/10.3390/gels8120823

Academic Editor: Annarosa Gugliuzza

Received: 25 October 2022
Accepted: 9 December 2022
Published: 13 December 2022

1. Introduction

Nowadays, the prevalence of the circular economy spirit requires the replacement of conventional packaging plastics, which are derived from fossil resources, with biopolymers which are produced via the valorization of food and/or agricultural by-products [1,2]. Moreover, the greenhouse effect imposes the turn to more environmentally friendly activities in all life sectors. Finally, the food shortage and rising prices could be handled via the extension of food shelf-life. Under this spirit, this study aimed at the valorization of some food byproducts and the use of natural biodegradable raw materials to improve the preservation of foods. Some of the most frequently used and promising biopolymers for packaging applications are cellulose, starch, gelatin, and chitosan [3–7]. Chitosan is produced by the deacetylation reaction of chitin. Chitin can be extracted via chemical or biotechnological processes from seafood waste such as shrimp, lobster, and crayfish shells [7–10]. Chitosan has been extensively studied as a promising biopolymer to be used in active packaging films, coatings, and other industrial applications due to its antioxidant and antimicrobial properties [11–16]. Its poor mechanical properties can be enhanced by blending with other polymers [17,18] biopolymers [19–21] and/or incorporating nano-reinforcements such as nanoclays to give promising chitosan-based biopolymeric nanocomposite gels ready for film preparation [22]. Due to CS's water swelling, it has lower water barrier properties than other packaging materials but is well known for its good gas barrier properties [23].

Moreover, modern food packaging technologies follow the incorporation of natural preservatives, antioxidants, and antimicrobials such as essential oils (EOs) into the polymer matrix of the biopolymer, targeting the development of gels exhibiting controlled release properties of EOs into the food, and sequentially to the increase of food self-life and food safety [24–29]. Various procedures have been developed to protect the antioxidant and antimicrobial activity of EOs. One of them suggests the encapsulation of EOs into microemulsion or nanoemulsion nanostructures [30–32]. Another one suggests the adsorption of such EOs into cheap and naturally abundant adsorbents such as nanoclays and zeolites [33–36]. The excellent gas barrier properties combined with the intermediate moisture barrier can reduce fruits respiration rate without interrupting the supply of moisture on them. This makes CS a promising coating material to extend the shelf-life of fruits [37].

The kiwifruit is cultivated in many places in Greece. However, Pieria remains the main area of production [38]. Kiwifruit is unique because of its high nutritional content [39]. In our days there is a major effort from researchers and farmers to find low cost ways to produce, keep in storage, and deliver onto the market, fruits of high quality [38].

During our previous work, we developed a chitosan/polyvinyl alcohol (CS/PVOH) gel which led to composite films and exhibited improved mechanical, gas barrier, and antimicrobial properties compared to the relevant properties of pure CS film [18,40]. Furthermore, we developed a procedure for the adsorption of EOs such as thyme, oregano, and basil oil in montmorillonite and organophilic montmorillonite. The incorporation of these nanohybrids in Low-Density Poly-Ethylene (LDPE), Polystyrene (PS), and Poly-Lactide-Acid (PLA) active packaging films was also studied previously [27,36,41–43]. A modified method for the adsorption of a fraction rich in thymol from thyme oil (TO) in halloysite nanotube nanoclay (HNT) was applied in this work. The obtained TO@HNT nanohybrids dispersed at 5, 10, and 15 wt.%. nominal content into CS/PVOH matrix via a solution blending method. Pure HNT was also dispersed into CS/PVOH matrix at the same nominal contents and used as reference material. The obtained CS/PVOH/HNT and CS/PVOH/TO@HNT films were characterized via XRD analysis, FTIR spectroscopy, and SEM microscopy. They were also tested for their tensile properties, water/oxygen barrier properties, antioxidant activity, and antimicrobial capacity against food pathogens. Finally, the most active films are applied as a coating to enhance the preservation of kiwi fruits.

2. Results and Discussion

The significance of the result of this work could be summarized as the development of a novel active food packaging film using natural raw materials and food industry byproducts and avoiding the use of chemicals and fossil fuel-originated polymers. Two specific goals were achieved in this study, the first is the greener packaging film development and the second is the extension of kiwi fruit shelf-life and beyond this the improved food preservation using such packaging films.

2.1. Physicochemical Structural Characterization of TO@HNT Hybrid Nanostructure

The obtained TO@HNT hybrid nanostructure as well as the pure HNT were characterized with TG experiments, FTIR spectra, XRD analysis, and DSC measurements. TG, FTIR, XRD, and DSC plots of pure HNT and TO@HNT hybrid are shown in Figure 1a–d respectively. TGA plots of both materials in Figure 1a indicate that in both cases exist two mass loss steps. The first one starts at around 50 °C and ends at around 400 °C. The second mass loss step which is attributed to the HNT dehydration process starts at around 450 °C and ends at around 600–700 °C [44–46]. In the case of pure HNT, the first mass loss step represents the mass loss of superficially adsorbed water and the second step represents the mass loss of structural/trapped water. In the case of the TO@HNT hybrid, the first step represents the loss of both water and TO molecules. Hence, by subtracting the mass lost from the first mass loss step of pure HNT from the mass lost from the first mass loss step of the TO@HNT hybrid, we calculated an average TO loading on HNT equal to 34.5 wt.%.

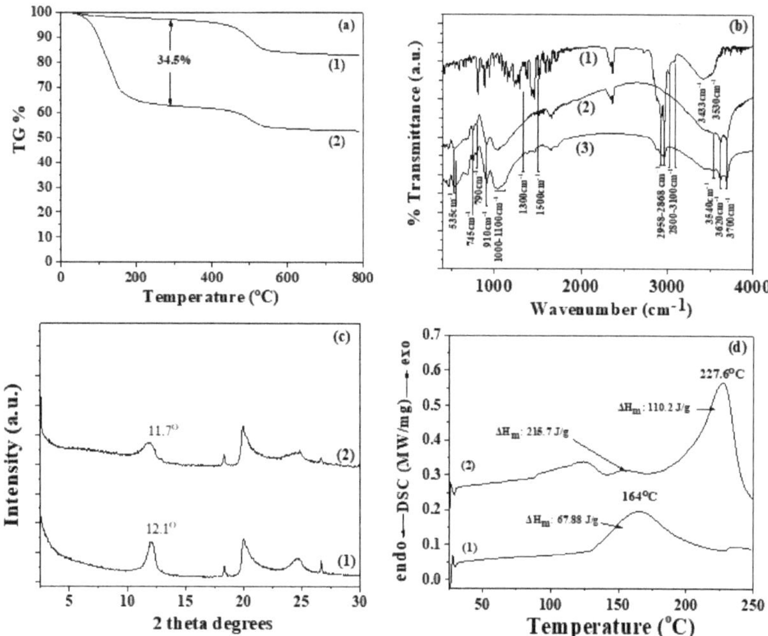

Figure 1. (**a**) TG plots of (1) pure HNT and (2) TO@HNT hybrid nanostructure, (**b**) FTIR plots of (1) TO, (2) pure HNT, and (3) TO@HNT hybrid nanostructure, (**c**) XRD plots of (1) pure HNT, and TO@HNT hybrid nanostructure, and (**d**) DSC plots of pure HNT (line 1) and modified TO@HNT (line 2).

In Figure 1b the FTIR spectra of both pure HNT and modified TO@HNT hybrid nanostructure are plotted. In the FTIR plot of TO, a broad peak in the range of 3530 to 3433 cm^{-1} was assigned to the stretching vibration of O-H groups [45]. The bands at ~3100–3000 cm^{-1} are corresponded to aromatic and alkenic -CH=CH- stretch vibrations [36]. The absorption bands in the range 2958 to 2868 cm^{-1} are assigned to the stretching mode of C-H groups [45]. The bands between 1500 cm^{-1} and 1300 cm^{-1} are assigned to the C-H bending of the C-O-H and aliphatic CH_2 groups bending [36].

In the FTIR plot of pure HNT, the bands at 3700 and 3620 cm^{-1} are assigned to hydroxyl groups in the internal HNT's surface. The weak band at 3540 cm^{-1} is assigned to the Si–O–Si (Al) groups. The intense absorption bands in the region of 1100–1000 cm^{-1} and at 790 cm^{-1} are assigned to Si–O group. The band at 910 cm^{-1} is assigned to the hydroxyl groups bending vibration. The band at 745 cm $^{-1}$ is assigned to the Si–O–Al bonds [44,45]. In the FTIR plot of hybrid TO@HNT are assigned the same bands with pure HNT and additionally the characteristic bands of TO mentioned hereabove. The characteristic bands of TO in the FTIR plot of TO@HNT imply the adsorption of TO on the HNT surface. No shift peak of HNT bands was obtained in the FTIR plot of the TO@HNT hybrid implying rather physisorbed than chemisorbed adsorption of TO on the HNT surface.

In Figure 1c the XRD plot of pure HNT and modified TO@HNT powders are shown. In both XRD plots which were obtained, the halloysite's distinct diffraction peaks at $2\theta = 12.0$, 20.1, and 24.6 are obvious, and correspond to (001), (100), and (002) planes respectively due to the crystalline property of the HNT [47]. In the case of pure HNT, the presence of the (001) peak at 2θ of 12.1° corresponds to a layer spacing of 0.73 nm. In the case of modified TO@HNT hybrid nanostructure, the peak at 2θ of 11.7o corresponds to a layer spacing of 0.76 nm. This difference of approx. 0.03 nm is too small and indicates probably the insertion of small water molecules in HNT's interlayer space. In the case of thymol molecules insertion in the HNT's interlayer space, it should be expected a larger opening of HNT's interlayer space as the thymol molecule size is bigger than that of phenol size (0.4 nm) [36]. So, XRD results indicated that adsorption of thymol took place on the external surface of HNT and no changes in the crystal structure of HNT are obtained due to the TO adsorption process. This result is in accordance with Shemesh et al. [48] where carvacrol a molecule similar to thymol loaded on the external surface of HNT.

In Figure 1d the DSC plots of pure HNT (line (1)) and modified TO@HNT (line (2)), nanohybrid are presented. In the DSC plot of pure HNT, the exothermic peak at 164 °C with a ΔH equal to 67.88 J/g is assigned to the desorption process of water molecules. In the DSC plot of TO@HNT nanohybrid, the exothermic peak at 227 °C with a ΔH equal to 227.6 J/g is assigned to the desorption of TO molecules in accordance with the previous report [49]. Thus, DSC analysis indicates the absorption of rich TO molecules on the HNT surface.

The overall characterization of TO@HNT nanohybrids concludes that a rich in TO fraction is physiosorbed on the external HNT's surface and validates the distillation/evaporation adsorption process followed.

2.2. XRD Analysis of CS/PVOH/HNT and CS/PVOH/TO@HNT Films

XRD patterns of all obtained CS/PVOH/HNT and CS/PVOH/TO@HNT films as well as of films from pure CS/PVOH gels are shown in Figure 2. The pattern of films from pure CS/PVOH gels exhibits three broad peaks at 8.5°, 11.5°, 18.5°, and 23° are observed. The peaks at 8.5° and 11.4° indicate the CS's hydrated crystallite structure due to the insertion of water molecules in the CS's crystal lattice [50,51] while the third peak at 18.5° is assigned to the CS's regular crystal lattice [51,52]. The later broaden peak around 23° assigned to the amorphous structure of CS [51,52].

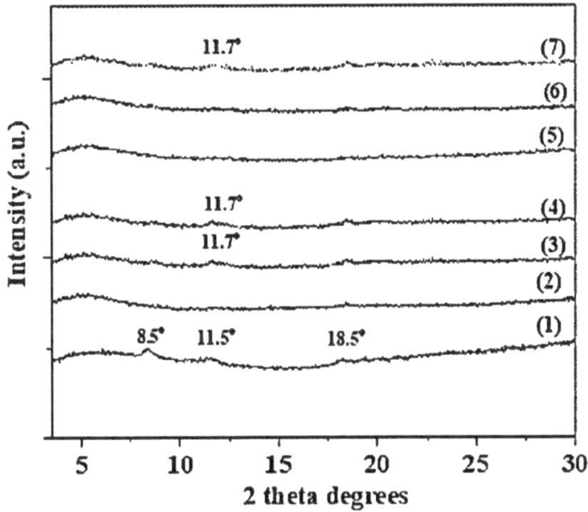

Figure 2. XRD plots of (1) CS/PVOH film, (2) CS/PVOH/5HNT, (3) CS/PVOH/10HNT, (4) CS/PVOH/15HNT, (5) CS/PVOH/5TO@HNT, (6) CS/PVOH/10TO@HNT, and (7) CS/PVOH/15TO@HNT nanocomposite films.

In the case of CS/PVOH/xTO@HNT film patterns, the diffraction peaks of HNT at 2θ = 11.7° do not observe for x = 5% wt. and 10% wt., i.e., line (5) and (6), and observed slightly for x = 15%wt., i.e., line (7). In the cases of CS/PVOH/xHNT, the HNT peak does not observe for x = 5%wt., i.e., line (2), and is observed slightly for x = 10%wt. and

15%wt. i.e., lines (3) and (4). These results indicate that in cases of TO@HNT the optimum dispersion of nanohybrid achieved for concentration x = 10%wt., while for concentration x = 15%wt. the aggregation started. In cases of HNT the optimum dispersion of nanohybrid achieved for concentration x = 5%wt. while for concentration x = 10%wt. or greater the aggregation started. The absence of shift of basal HNT's peak at $2\theta = 11.7^{\circ}$ indicates that CS/PVOH chains can not intercalate HNT's interlayer space [53]. The absence of HNT's peak in all CS/PVOH/TO@HNT films implied the higher dispersity of modified TO@HNT hybrid nanostructure than pure HNT in the CS matrix. The high dispersion of HNT and TO@HNT in the CS matrix is beneficial for such nanocomposite films.

2.3. FTIR Spectroscopy of CS/PVOH/HNT and CS/PVOH/TO@HNT Films

In Figure 3 representative spectra of pure CS/PVOH, CS/PVOH/HNT, and CS/PVOH/TO@HNT are observed.

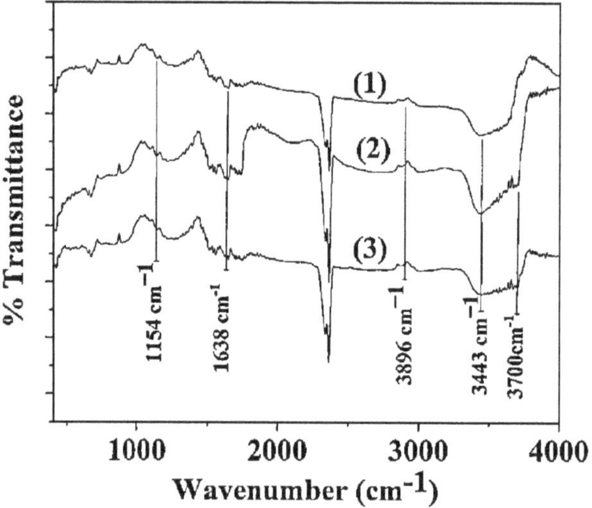

Figure 3. FTIR spectras of neat CS/PVOH film (1) as well as CS/PVOH/HNT (2), and CS/PVOH/TO@HNT (3) films for comparison.

FTIR spectra of pure CS/PVOH (see line (1) in Figure 3) are a combination of both CS and PVOH reflections. The large band at 3443 cm^{-1} is assigned to O–H groups stretching both presented in CS and PVOH and to N–H groups stretching of CS. The same band is also assigned to the intra- and inter-molecular hydrogen bonds of the CS/PVOH matrix.The band at2896 cm^{-1} is assigned to C–H asymmetric and symmetric stretching from CH_2 and CH groups. The band at 1637–1644 cm^{-1} is assigned to the associated water, C–OH from the glycosidic units of CS chains, and also the presence of residual N-acetyl groups (C=O stretching of amide I), and N-H in-plane deformation coupled with C–N stretching of amide II (secondary amide) from CS. The band at 1154 cm^{-1} is assigned to the glycosidic linkage (asymmetric bridge stretch) [54].

In the FTIR spectra of CS/PVOH/HNT and CS/PVOH/TO@HNT systems (see lines (2) and (3) in Figure 3) it additionally obtained the characteristic reflection band at 3620 cm^{-1} assigned to O-H groups in the internal HNT's surface. With a careful glance, it is obtained that the main difference between FTIR plots of CS/PVOH/HNT and CS/PVOH/TO@HNT is that in the case of CS/PVOH/HNT plot the band of O-H and N-H group stretching at 3443 cm^{-1} is more intense than the same band of CS/PVOH/TO@HNT plot. This implies a higher interaction between OH groups of the CS/PVOH matrix and OH groups of pure HNT than OH groups of the CS/PVOH matrix and modified TO@HNT.

2.4. SEM Images of CS/PVOH/HNT and CS/PVOH/TO@HNT Films

A Scanning Electron Microscopy (SEM) instrument was used for the surface/cross-section morphology investigation of the pure CS/PVOH film as well as of the CS/PVOH/HNT and CS/PVOH/TO@HNT nanocomposite films. The results confirmed that both HNT and the TO@HNT hybrid nanostructures were homogeneously dispersed in the polymer matrix. The chemical elements of the pure and the final nanocomposite active packaging films were identified by carrying out an Energy dispersive spectrometer (EDS) analysis on the surface of the materials.

The SEM images in Figure 4a,b show the expected smooth morphology of the neat polymer. The EDS spectra in Figure 4c certify the existence of carbon (C), and oxygen (O).

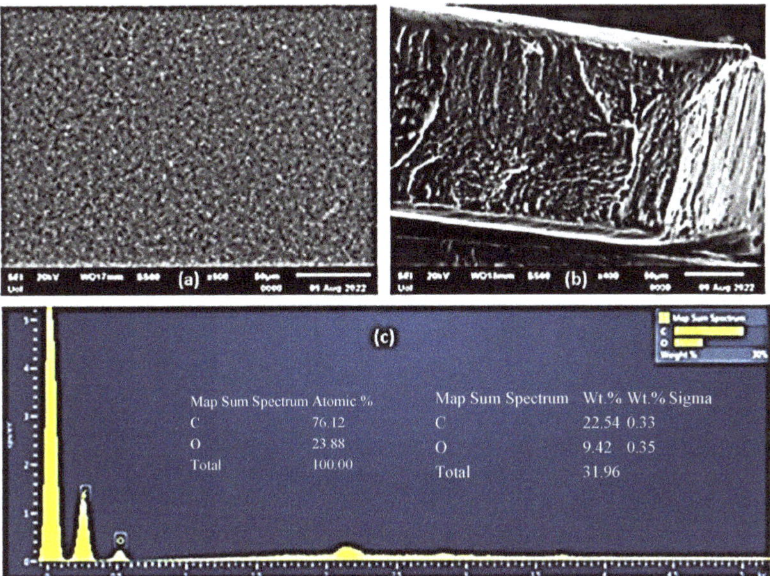

Figure 4. (**a**) SEM images of the surface and (**b**) cross-section for the pure CS/PVOH film. (**c**) Energy dispersive spectrometer (EDS) spectrum and relative elemental analysis of the surface (inset) from the SEM image (**a**).

Surface and relative cross-section images of CS/PVOH/HNT and CS/PVOH /TO@HNT with different ratios of HNT and TO@HNT are presented in Figures 5–7. Figures 5–7e,f show EDS chemical analysis of nanocomposite active packaging films with different concentrations of pure HNT and TO@HNT hybrid nanostructure i.e., 5, 10, and 15 wt.%. The presence of typical elements such as Si, Al, O, Fe, and Ca was confirmed.

As illustrated in Figures 5–7, the content of both HNT and TO@HNT increases the aggregation of the obtained CS/PVOH/HNT, and CS/PVOH/TO@HNT nanocomposites decrease further. Nevertheless, SEM images of the final nanocomposite films show that both pure HNT and TO@HNT nanohybrid were homogeneously dispersed and suggest enhanced compatibility with the polymer matrix. Moreover, SEM surface and cross-section images were shown more homogenous dispersion in the case of TO@HNT hybrid nanostructure in nanocomposite films compared to the relevant of pure HNT. This means that the TO@HNT hybrid nanostructure was incorporated significantly better in the polymer matrix compared to the incorporation of the respective pure HNT.

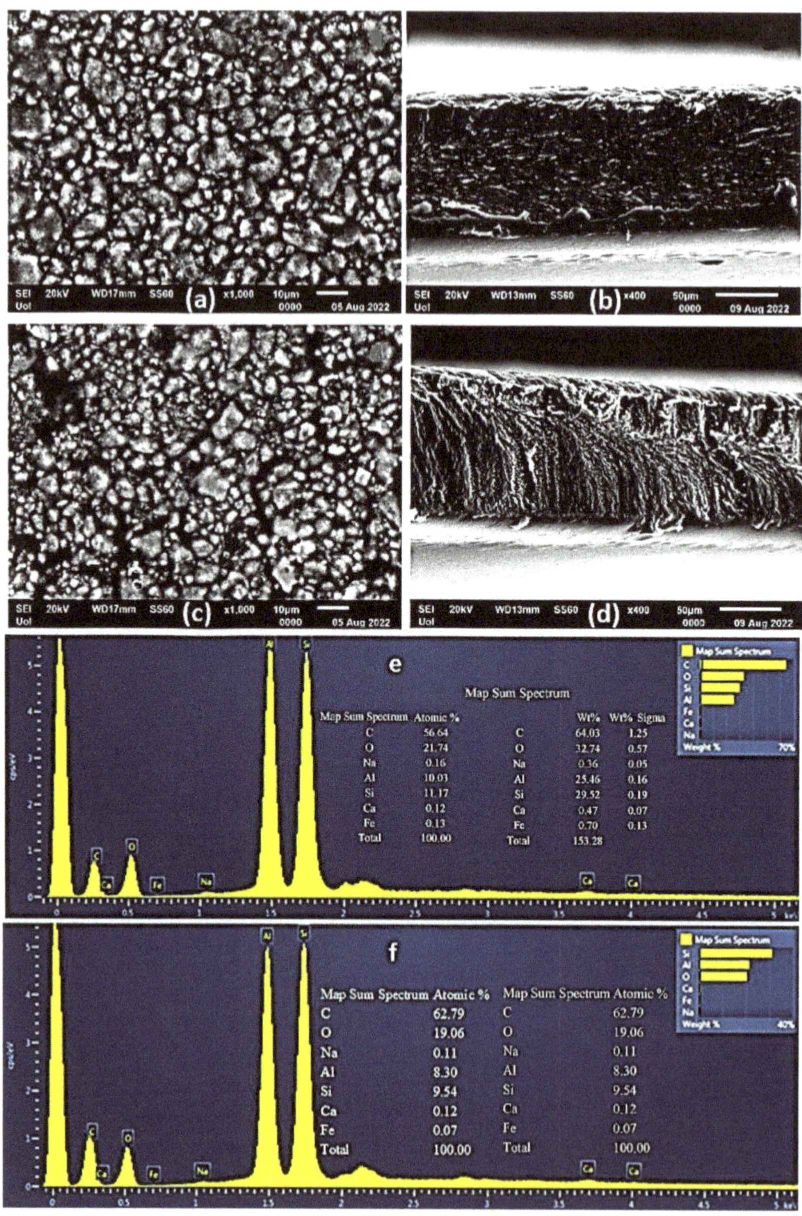

Figure 5. (**a**,**c**) SEM images of the surface and (**b**,**d**) cross-section for the nanocomposite films of ALG/G/5HNTNZ (**a**,**b**) and ALG/G/5TO@HNT (**c**,**d**) respectively. (**e**,**f**) Energy dispersive spectrometer (EDS) spectrum and relative elemental analysis of the surface (inset) from the SEM images (**a**,**c**).

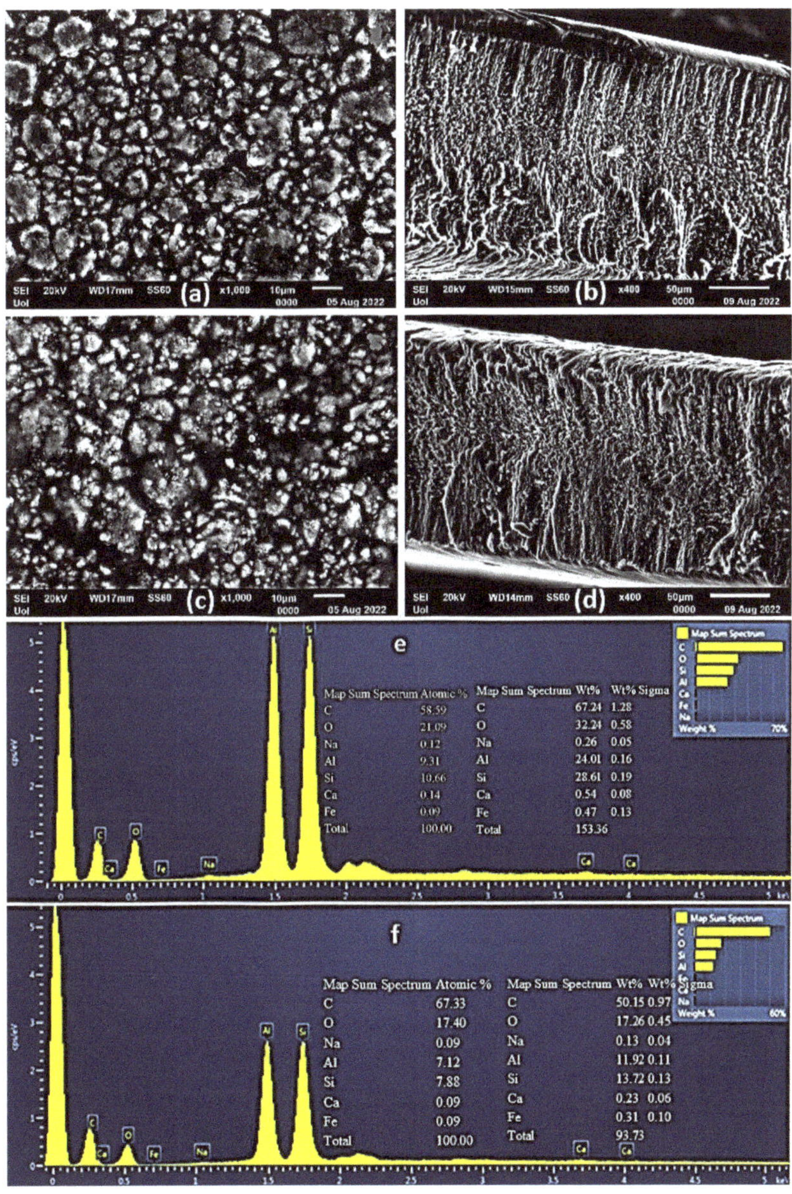

Figure 6. (**a**,**c**) SEM images of the surface and (**b**,**d**) cross-section for the nanocomposite films of ALG/G/10HNT (**a**,**b**) and ALG/G/10TO@HNT (**c**,**d**) respectively. (**e**,**f**) Energy dispersive spectrometer (EDS) spectrum and relative elemental analysis of the surface (inset) from the SEM images (**a**,**c**).

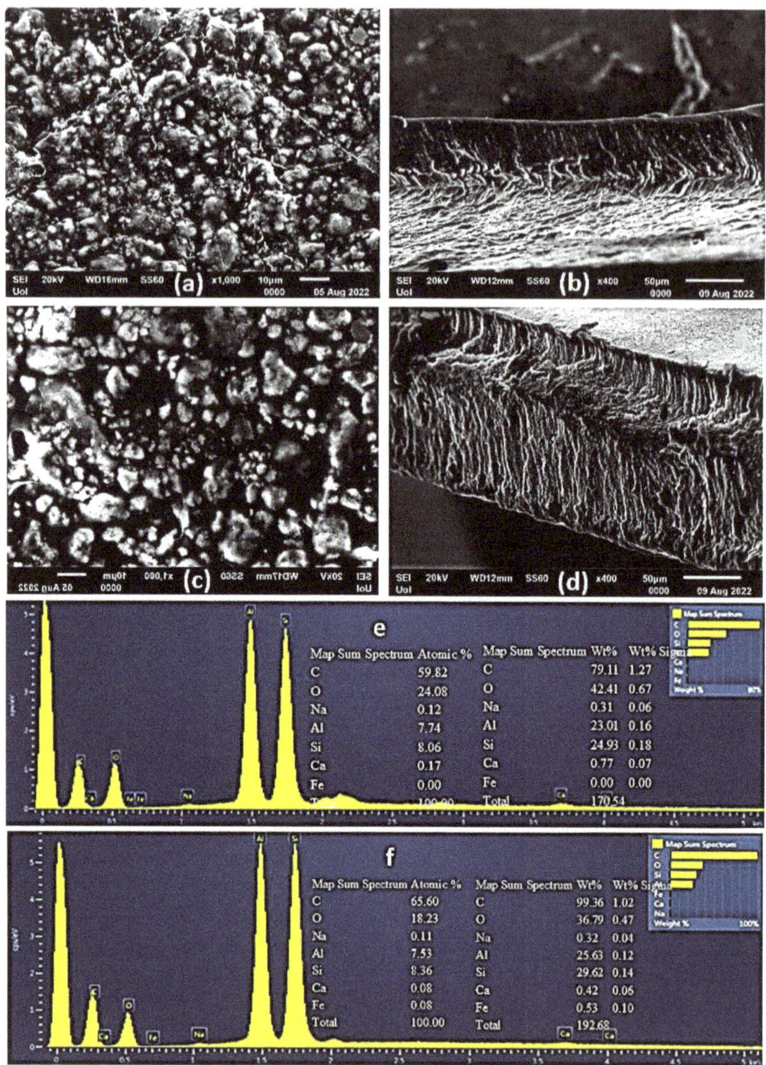

Figure 7. (**a**,**c**) SEM images of the surface and (**b**,**d**) cross-section for the nanocomposite films of ALG/G/15HNT (**a**,**b**) and ALG/G/15TO@HNT (**c**,**d**) respectively. (**e**,**f**) Energy dispersive spectrometer (EDS) spectrum and relative elemental analysis of the surface (inset) from the SEM images (**a**,**c**).

2.5. Tensile Properties of CS/PVOH/HNT and CS/PVOH/TO@HNT Films

Typical stress-strain curves of all CS/PVOH/HNT and CS/PVOH/TO@HNT films are shown in Figure 8.

From stress-strain curves in Figure 8 the Young's (E) Modulus, ultimate tensile strength (σ_{uts}), and % strain at break (ε_b) values were calculated and are listed in Table 1 for comparison.

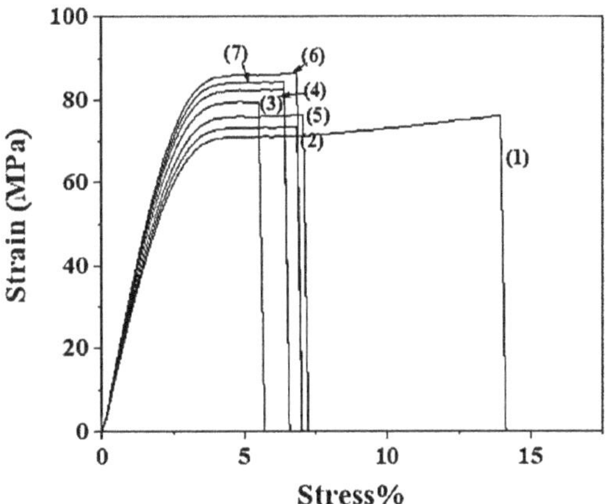

Figure 8. Stress-strain curves of (1) CS/PVOH, (2) CS/PVOH/5HNT, (3) CS/PVOH/10HNT, (4) CS/PVOH/15HNT, (5) CS/PVOH/5TO@HNT, (6) CS/PVOH/10TO@HNT and (7) CS/PVOH/15TO@HNT.

Table 1. Calculated values of Young's (E) Modulus, ultimate tensile strength (σuts), and % strain at break (ε_b).

Sample Name	E-Elastic Modulus (MPa)	σ_{uts} (MPa)	ε%-Elongation at Break
CS/PVOH	2249.3 ± 20.0	71.2 ± 1.8	14.8 ± 0.9
CS/PVOH/5HNT	2552.0 ± 21.3	74.1 ± 2.1	7.0 ± 1.6
CS/PVOH/10HNT	2766.0 ± 35.1	79.8 ± 1.7	5.7 ± 0.8
CS/PVOH/15HNT	2865.0 ± 27.4	96.0 ± 1.2	6.7 ± 0.6
CS/PVOH/5TO@HNT	2644.5 ± 13.4	74.9 ± 2.3	7.2 ± 2.1
CS/PVOH/10TO@HNT	2993.7 ± 27.6	103.7 ± 1.4	7.1 ± 0.2
CS/PVOH/15TO@HNT	2965.0 ± 29.4	98.5 ± 1.5	6.8 ± 0.7

From the Young's (E) Modulus, ultimate tensile strength (σ_{uts}), and % strain at break (ε_b) values, which are presented in Table 1, we could assume that CS/PVOH/HNT and CS/PVOH/TO@HNT nanocomposite films are stronger than pure CS/PVOH film. The higher the nominal content of the HNT and TO@HNT the higher ultimate strength and lower elongation at break values. In advance, TO@HNT-based nanocomposite films exhibited higher strength values than HNT-based nanocomposites. The increase of tensile strength with HNT and TO@HNT is in agreement with previous reports where HNT loaded in CS/PVOH with nominal content 0–5 wt.%. [53]. Here it is reported for the first time that this increment is existed also for higher HNT nominal loading up to 15 wt.%. It is also reported that a higher tensile increment is taking place for TO@HNT loading compared to the relevant HNT loading. This higher increase of tensile strength with TO@HNT addition is in agreement with the higher dispersion of the TO@HNT in the CS/PVOH matrix as was presented before by the XRD and SEM results.

2.6. UV-vis Transmittance of CS/PVOH/HNT and CS/PVOH/TO@HNT Films

In Figure 9 photo images (see Figure 9a) and UV-vis plots (see Figure 9b) of all prepared films are shown for comparison.

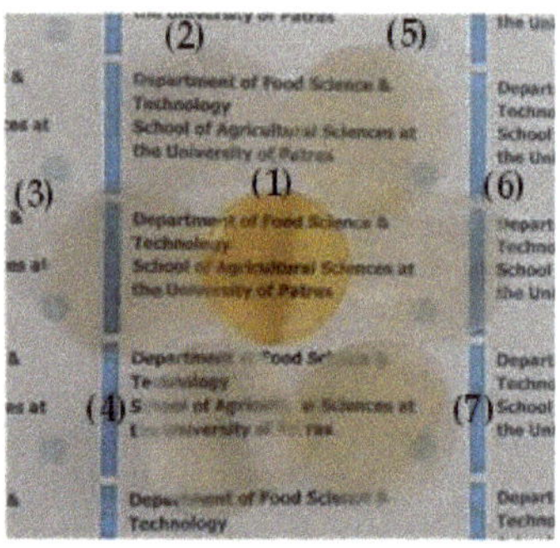

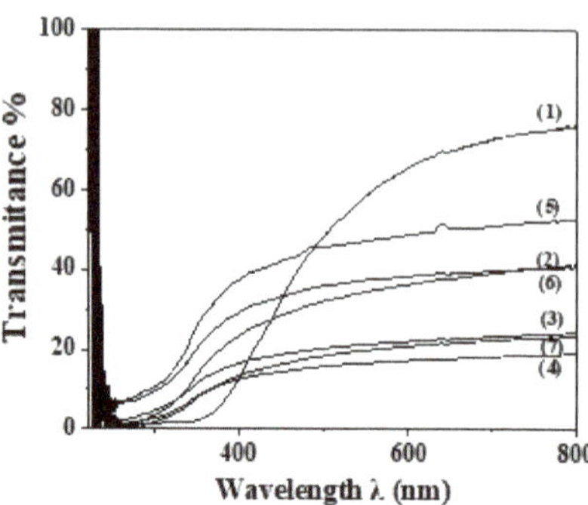

Figure 9. (**a**) photo images and (**b**) UV-vis transmittance plots of (1) pure CS/PVOH films, (2) CS/PVOH/5HNT, (3) CS/PVOH/10HNT, (4) CS/PVOH/15HNT, (5) CS/PVOH/5TO@HNT, (6) CS/PVOH/10TO@HNT, and (7) CS/PVOH/5TO@HNT films.

As it is obtained in UV-vis transmittance plots and illustrated in film images the addition of both HNT and TO@HNT decreases the transmittance of films. From UV-vis transmittance plots is obtained that TO@HNT-based films exhibited higher transmittance than HNT-based films. This result is in accordance with the higher dispersity of TO@HNT hybrid nanostructure in the CS/PVOH matrix which was discussed above in XRD and SEM results.

2.7. Water-Oxygen Barrier Properties of CS/PVOH/HNT and CS/PVOH/TO@HNT Films

In Table 2 the water-oxygen transmission rate values for all CS/PVOH/HNT and CS/PVOH/TO@HNT films as well as pure CS/PVOH films are listed. By multiplying water-oxygen transmission rate values with film thickness the water diffusivity (D) and the oxygen diffusivity (Pe_{O2}) values are obtained and are listed in Table 2 also for comparison.

Table 2. Film thickness, water vapor transmission rate (WVTR), water diffusivity (D), oxygen transmission rate (OTR), and oxygen diffusivity (Pe_{O2}) values of pure CS/PVOH film as well as CS/PVOH/HNT and CS/PVOH/TO@HNT films.

Sample Name	Film Thickness (mm)	WVTR $\times 10^{-6}$ (gr $\times$ cm^{-2} $\times$ s^{-1})	D $\times 10^{-4}$ (cm^2 $\times$ s^{-1})	OTR (mL $\times$ m^{-2} $\times$ day^{-1})	$Pe_{O2} \times 10^{-7}$ (cm^2 $\times$ s^{-1})
CS/PVOH	0.140 ± 0.010	1.15 ± 0.13	3.65 ± 0.31	49,577 ± 2478	8.03 ± 0.40
CS/PVOH/5HNT	0.113 ± 0.025	1.42 ± 0.18	3.67 ± 0.72	57,345 ± 2867	7.52 ± 0.38
CS/PVOH/10HNT	0.123 ± 0.015	1.02 ± 0.13	2.83 ± 0.62	32,785 ± 1639	4.68 ± 0.23
CS/PVOH/15HNT	0.117 ± 0.012	1.03 ± 0.20	2.75 ± 0.17	44,234 ± 2211	5.97 ± 0.30
CS/PVOH/5TO@HNT	0.140 ± 0.010	1.14 ± 0.13	3.59 ± 0.14	43,345 ± 2167	7.02 ± 0.35
CS/PVOH/10TO@HNT	0.117 ± 0.021	1.02 ± 0.21	2.62 ± 0.17	28,974 ± 1449	3.91 ± 0.20
CS/PVOH/15TO@HNT	0.170 ± 0.017	1.01 ± 0.67	3.64 ± 0.14	30,434 ± 1521	5.64 ± 0.28

For both water and oxygen diffusivity values, in general, could be stated that: (i) the addition of TO@HNT causes a higher increase of water-oxygen barrier than the addition of HNT, (ii) the highest water and oxygen barrier is achieved for films containing 10 wt.%. HNT and 10 wt.%. TO@HNT, and (iii) the film with the optimum water-oxygen barrier is CS/PVOH/10TO@HNT. So, it is concluded that around 10 wt.%. is the optimum content for both HNT and TO@HNT nanostructures to obtain the highest water/oxygen barrier. This conclusion is consistent with the results of the XRD measurements mentioned above herein. In other words, with 10 % wt. addition of both HNT and TO@HNT in the CS/PVOH matrix the optimum dispersions are achieved to obtain the highest water/oxygen barrier.

2.8. Antioxidant Activity of CS/PVOH/HNT and CS/PVOH/TO@HNT Films

Antioxidant activity in active food packaging has a key role to extend the shelf-life of food products. Especially for such edible coatings enhances nutritional and aesthetic quality aspects of food without affecting its integrity.

The calculated % antioxidant activity values of all CS/PVOH/HNT and CS/PVOH/TO@HNT films as well as pure CS/PVOH film are obtained in Table 3.

Table 3. Antioxidant activity values of pure CS/PVOH and all CS/PVOH/HNT, CS/PVOH/TO@HNT films.

Sample Name	Antioxidant Activity after 24 h [1] (%)
CS/PVOH	3.5 ± 1.6
CS/PVOH/5HNT	5.4 ± 2.6
CS/PVOH/10HNT	5.5 ± 2.2
CS/PVOH/15HNT	6.1 ± 3.1
CS/PVOH/5TO@HNT	18.1 ± 5.0
CS/PVOH/10TO@HNT	25.5 ± 3.6
CS/PVOH/15TO@HNT	32.2 ± 5.1

[1] DPPH assey.

As it is observed no significant antioxidant activity is obtained for pure CS/PVOH and CS/PVOH/HNT films. For CS/PVOH/TO@HNT films antioxidant activity is increased as the TO@HNT nominal content is increased. The highest antioxidant activity value is obtained for CS/PVOH/15TO@HNT film.

2.9. Antibacterial Activity of CS/PVOH/HNT and CS/PVOH/TO@HNT Films

Figure 10 depicts the petri dishes used for the antimicrobial activity measurements of CS, CS/PVOH, CS/PVOH/xHNT, and CS/PVOH/xTO@HNT films against *E. coli* and *Staphylococcus* bacteria. Table 4 presents the antibacterial activity of the developed films that were based on the CS/PVOH nano-reinforcement. Four foodborne pathogenic bacteria cultivations i.e., *Escherichia coli*, *Staphylococcus aureus*, *Salmonella enterica*, and *Listeria monocytogenes* were used to test the antibacterial capacity of all films. The clear zone's diameter around the tested films indicates the magnitude of the inhibition of the microorganisms' growth. The absence of a clear zone, which means zero value of diameter, entails the absence of an inhibitory zone. Moreover, in this work, the bacteria growth in the area of direct contact of film with the agar surface was also studied.

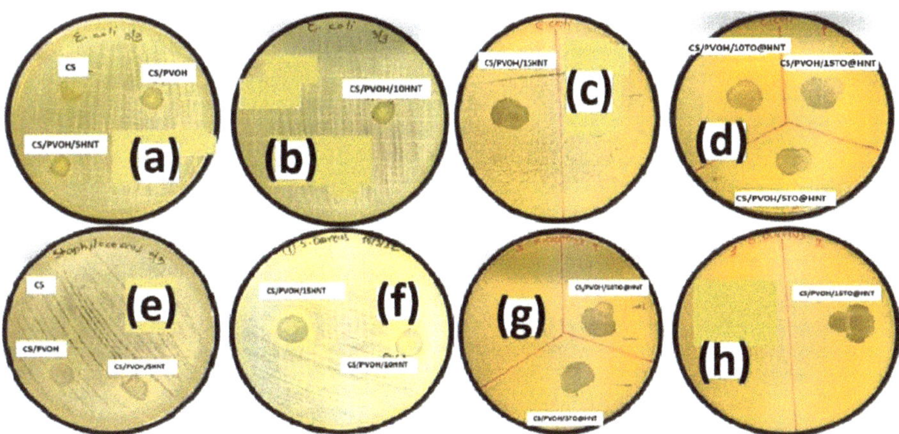

Figure 10. Petri dishes images of (**a–c**,**e**,**f**) CS, CS/PVOH, CS/PVOH/xHNT, and (**d**,**g**,**h**) CS/PVOH/xTO@HNT films against *E. coli*, *S. aureus*, *S. enterica*, and *L. monocytogenes*.

Table 4. Antibacterial activity of active films against food pathogenic bacteria *E. coli*, *S. enterica*, *S. aureus*, and *L. monocytogenes*.

Film Material	*E. coli*		*S. enterica*		*S. aureus*		*L. monocytogenes*	
	Inhibition [a]	**Contact [b]**	**Inhibition [a]**	**Contact [b]**	**Inhibition [a]**	**Contact [b]**	**Inhibition [a]**	**Contact [b]**
CS	0.00	-	0.00	-	0.00	-	0.00 [9]	+
CS/20PVOH	3.50 ± 0.87	-	3.83 ± 0.76 [5,6,8]	-	4.47 ± 0.81 [3]	-	0.00 [9]	+
CS/20PVOH/5HNT	4.83 ± 0.29 [1]	-	5.00 ± 0.00 [5,6,7,8]	-	5.40 ± 0.66 [3,4]	-	5.00 ± 0.00 [10]	-
CS/20PVOH/10HNT	5.00 ± 0.00 [1]	-	5.33 ± 0.76 [6,7,8]	-	5.77 ± 0.75 [3,4]	-	5.00 ± 0.00 [10]	-
CS/20PVOH/15HNT	7.00 ± 0.50 [2]	-	6.33 ± 0.29 [6,7,8]	-	6.00 ± 0.00 [4]	-	6.57 ± 0.51	-
CS/20PVOH/5TO@HNT	7.50 ± 0.50 [2]	-	6.83 ± 0.29 [8]	-	9.00± 0.00	-	8.00 ± 0.50	-
CS/20PVOH/10TO@HNT	7.80 ± 0.20	-	8.00 ± 0.00	-	9.50 ± 0.50	-	9.03 ± 0.45	-
CS/20PVOH/15TO@HNT	8.00 ± 0.50	-	9.00 ± 0.87	-	10.00 ± 0.00	-	9.00 ± 0.50	-

[a] Inhibitory zone surrounding film discs measured in mm after the subtraction of the disc diameter (6 mm); [b] Contact area of film discs with the agar surface; (+) indicates bacterial growth in the area, (-) indicates no bacterial growth in the area; Results expressed as mean ± standard deviation (n = 3); Means in the same column baring same superscript numbers are significantly equal ($p > 0.05$).

The chitosan films CS/PVOH/5HNT, CS/PVOH/10HNT, CS/PVOH/15HNT, CS/PVOH/5TO@HNT, CS/PVOH/10TO@HNT, CS/PVOH/15TO@HNT were compared to pure CS and CS/PVOH films. Pure CS films inhibited the growth of all tested bacteria but only by direct contact; except *L. monocytogenes* where no antibacterial activity was observed either in the contact area or by the formation of clear surroundings zones.

Furthermore, the CS/PVOH film showed antibacterial activity by formatting a clear zone of 3.50 mm for *E. coli*, 3.83 mm for *S. enterica*, and, 4.47 mm for *S. aureus* while no antibacterial effect against *L. monocytogenes* was observed.

All the incorporated CS films displayed antibacterial effectiveness. The noted inhibition of the bacteria growth seems to have a dependency on the HNT and thyme oil (TO) concentration.

By reviewing the results, it is obvious that the growth inhibition was amplified upon increasing the concentration of the nanostructures and the EO. The CS/PVOH/HNT (5%, 10%, 15%) films showed pronounced antibacterial activity against the tested bacteria.

Specifically, the CS/PVOH/15HNT film exhibited higher antibacterial activity against *E. coli*, *S. enterica*, *S. aureus*, and *L. monocytogenes* if compared to CS/PVOH/5HNT, and CS/PVOH/10HNT. The inhibitory clear zones were noticeably higher for CS/PVOH/HNT films when thyme oil (TO) was incorporated.

The CS/PVOH/5TO@HNT film inhibited all the tested bacteria by formatting clear zones of 7.50 mm for *E. coli*, 7.00 mm for *S. enterica*, 9.00 mm for *S. aureus*, and 8.00 mm for *L. monocytogenes*.

Increasing the thyme oil concentration, it was also enhanced the zone of inhibition of bacteria growth. Finally, CS/PVOH/HNT films containing 15% thyme oil displayed the highest antibacterial activity, resulting in a clear zone formation of 8.00 mm for *E. coli*, 9.00 mm for *S. enterica*, 10.00 mm for *S. aureus*, and 9.00 mm for *L. monocytogenes*.

In all cases, the nano-enforcement films showed significant antibacterial activity against Gram-negative bacteria and slightly stronger activity against Gram-positive bacteria.

It is known that chitosan possesses important antibacterial activity against a wide spectrum of bacteria. This activity is ascribed to its cationic nature (positively charged ammonium (NH_4^+)) that interacts with the negatively charged compounds of the bacteria cell wall [55]. However, CS does not show any migrated inhibitory activity [56]. Bacterial cell wall barry a negative charge, therefore electrostatic interaction between bacteria and positively-charged clays such as HNT, under specific conditions (pH, ionic force) is probable [57]. In order for clay to exhibit antibacterial activity, it is crucial to have the ability to maintain metal ions in solution and to have also sufficient interlayer cation exchange capacities [58]. Theoretically, HNTs do not meet these criteria, however, the literature refers to a wide range of possible modes of action of HNTs against bacteria. Abhinayaa et al., 2019 found that HNT at a concentration of 2.5 mg mL^{-1}, was able to inhibit the growth of the phytopathogenic bacteria *Agrobacterium tumifaciens* and *Xanthomonas oryzae*, while at lower concentrations it was observed decreased bacteria growth rate and damages on the cell membrane. These results were probably attributed to the effect of siloxane groups of HNTs surface in combination with the production of reactive oxygen species [59]. Moreover, increased antibacterial activity has been observed after the modification of the HNT surface. The functionalized HNT displayed strong antibacterial activity against the food-borne bacteria, *L. monocytogenes*, and *E. coli* [60,61].

HNT has been also found to be toxic to *E. coli* and *Salmonella typhimurium*, especially as a result of light-dependent oxidative stress [59,62]. Several antibacterial compounds such as antibiotics, essential oils, antibacterial peptides, etc. have been loaded into the HNTs in order to enhance the antibacterial activity [63,64]. The HNT structure seems to allow the sustained release of the incorporated antibacterial agents such as thyme oil, leading to bacteria inhibition. By this controlled release, the thyme oil could enter the lipid layer of the bacterial cell wall causing cell death. Moreover, the interaction of bacterial cell walls with the thyme oil-HNT matrix may produce an oxidation-reduction response leading to cell death due to the production of reactive oxygen species [65].

Concluding, it is once more noted that in the present work, the bacteria growth was inhibited in a dose-dependent manner. The antibacterial efficacy of the tested films could be due to the nanocomposite films themselves but also might be due to the controlled release/migration of thyme oil. Many parameters play a role in the final antibacterial effect of a nanostructure, such as bacterial strain, nanoparticle type/size, chitosan molecular weight, growth media type, assay type, and bacterial cell concentration. Consequently, the increased antibacterial activity of the CS/PVOH/HNT films reported in this work is attributed to the synergistic effect of chitosan, HNT, thyme oil concentration, etc.

2.10. Packaging Test-Application of CS/PVOH/HNT and CS/PVOH/TO@HNT Films as Coating on Fresh Kiwifruits

In Figure 11 of uncoated kiwifruits and coated kiwifruits with CS/PVOH, CS/PVOH/10HNT, and CS/PVOH/10TO@HNT solution after 21 days of storage at 25 °C and ambient humidity are shown. As it is obtained from the uncoated kiwifruits the weight loss change is visible from the 3rd day of storage. The visible changes for uncoated samples are more visible on the 6th day and the deterioration increased until the 15th day when the uncoated kiwi fruits were rejected. In the kiwifruits coated with pure CS/PVOH solution the deterioration is visible 3 days later in the 6th day and is increased until the 15th day when also these samples were rejected. For the kiwifruits coated with CS/PVOH/10HNT and CS/PVOH/10TO@HNT solution, the deterioration is much slower. The first visible weight loss change is starting for the kiwifruits coated with CS/PVOH/10HNT solution on the 12th day while for the kiwifruits coated with CS/PVOH/10TO@HNT solution three days later on the 15th day. From the 15th to 21st day of storage as it is obtained (see the last image in the down part of Figure 11) the deterioration of kiwifruits coated with CS/PVOH/10HNT was much more accelerated than the kiwifruits coated with CS/PVOH/10TO@HNT solution which are in a much better optical condition in the last day of the experiment. To conclude a significant deceleration of kiwifruits deterioration was obtained for all coated samples. The uncoated kiwifruits deterioration starts on the 3rd day, while for the kiwifruits coated with CS/PVOH, CS/PVOH/HNT, and CS/PVOH/10TO@HNT solution on the 9th, 12th, and 15th day correspondingly. At this point, it must be mentioned, that the delayed deterioration of the kiwi fruits could be attributed also to the antimicrobial activity of the tested films. Although the films were studied for certain food-borne bacteria, it is likely to have also antimicrobial activity against other spoilage microorganisms, leading to extended life.

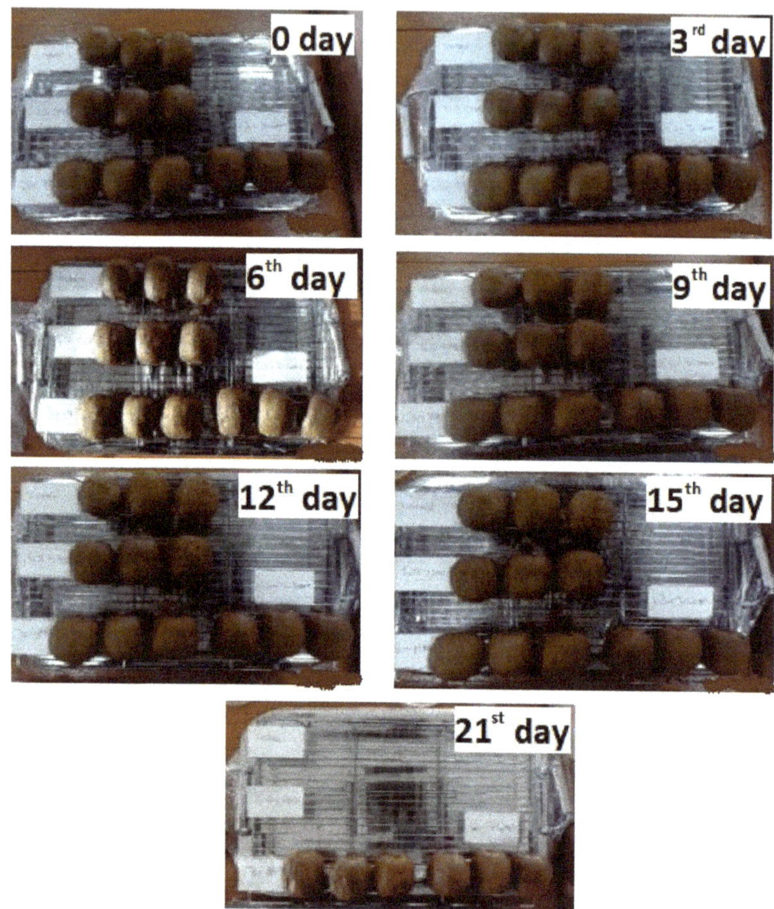

Figure 11. Images of uncoated kiwifruits, coated with CS/PVOH solution kiwifruits, coated with CS/PVOH/10HNT solution kiwifruits and coated with CS/PVOH/10TO@HNT solution kiwifruits stored at 25 °C and ambient humidity at zero, 3rd, 6th, 9th, 12th, 15th and 21st day.

3. Conclusions

Concluding this study, according to the TG and FTIR results the proposed modified adsorption process led to ~35 wt.%. of TO loading onto HNT nano-clay combined with very good incorporation. Furthermore, the FTIR results indicate a prevailing physisorption mechanism as it is compared to chemisorption. This led to a controlled release mechanism of the TO which was confirmed by the antibacterial tests. We are heading in the same direction if we interpret the XRD results shown, that the thymol was adsorbed on the external surface of the HNT in a high order and thus, such molecules could be easily released. The comparison of the HNT with the TO@HNT nanostructures behaviour via SEM, EDS, and XRD measurements implies that the dispersion of the second one in the CS matrix is higher than the relevance of the first one, which is beneficial for the final developed film. Even though CS/PVOH/15TO@HNT exhibits higher antioxidant and antibacterial activity, the CS/PVOH/10TO@HNT exhibits a higher water-oxygen barrier. According to the XRD measurements, this result originates from dispersion collapse for higher nanohybrid addition. Such results were verified by the in-vivo experiments of kiwi fruits preservation where the uncoated food started to decline on the 3rd day, the coated food with CS/PVOH/15HNT film started to decline on the 12th day, and the coated food with CS/PVOH/15TO@HNT film started to decline at the 15th day. The rejection day of the uncoated kiwi fruits was the 15th day while the rejection of the coated fruits was the 21st day. The final situation of the CS/PVOH/15TO@HNT kiwi samples was better than the relevant of the CS/PVOH/15HNT coated samples. Finally, UV-vis and SEM measurements show that the TO addition is beneficial for the film's transparency and to the HNT dispersion in the CS matrix.

As a final result of this study, we could say that a promising active packaging film was developed through a more environmentally friendly procedure and tested successfully for kiwi fruit preservation. As future work, we could say that this product should be also tested with other food kinds and develop an industrial process for bulk film production via a scale-up procedure.

4. Materials and Methods

4.1. Materials

Acros-Organics company was the supplier of Chitosan (CS) with a molecular weight of 100,000–300,000 (Zeel West Zone 2, Janssen Pharmaceuticalan 3a, B2440, Geel, Belgium). Poly(vinyl alcohol) (PVOH) (low molecular weight i.e., 13,000–23,000, hydrolysis degree 87–89%) was purchased by SIGMA-ALDRICH (Co., 3050 Spruce Street, St. Louis, MO, USA, 314-771-5765). Nanocor Inc. was provided the powder nanoclay (2870 Forbs Avenue, Hoffman Estates, IL, USA). Montmorillonite which contained 1.53% halloysite nanotubes ($Al_2Si_2O_5(OH)_4 \cdot 2H_2O$, 99.5% clay, 1.68% CaO, 3.35% Fe_2O_3, 62.9% SiO_2, 19.6% Al_2O_3, and 3.05% MgO) was obtained by Sigma-Aldrich (product 685445, Sigma-Aldrich, St. Louis, MO, USA). The used Thyme Oil (TO) was produced by the Chemco company (Via Achille Grandi, 13–13/A, 42030 Vezzanosul, Crostolo, Italy).

4.2. Preparation of TO@HNT Hybrid Nanostructure

The modification of the HNT nanoclay with TO EO was based on a modified evaporation/adsorption process [36]. Firstly, 20 mL of TO was placed in a glass distiller flask and heated at 200 °C to remove the D-limonene and L-cymene (see part (1) in Figure 12). The remaining rich in thymol TO was placed in a round bottom glass flask. Above the flask, 2 g of HNT bed was adapted and above the HNT bed, a reflux condenser was placed. With this modified reflux condenser, the remaining rich in thymol TO was heated at 300 °C and the evaporated thymol molecules adsorbed on HNT (see part (2) in Figure 12).

When the HNT bed color turned from white to brown the process stopped and the modified HNT was removed and weighted. The wt.%. TO loading to HNT was calculated at approx. 30%. The as-prepared rich in thymol-modified HNT was labeled as TO@HNT and used as active nano-reinforcement in the development of CS/PVOH/TO@HNT active films/coatings.

4.3. Preparation of CS/PVOH/HNT and CS/PVOH/TO@HNT Films/Coatings

For the preparation of each film/coating, 100 mL of an aqueous 2% *w*/*v* CS solution activated with 1% *v*/*v* acetic acid was used. In this 100 mL of as-prepared 2% *w*/*v* CS solution with 1% *v*/*v* acetic acid PVOH added to achieve final PVOH nominal content at 30% wt. The obtained mixture was refluxed overnight to solubilize PVOH and to achieve a homogeneous CS/PVOH solution. In this, 100 mL of CS/PVOH mixed solution amounts of TO@HNT hybrid nanostructure was added and homogenized for 5 min at 18,000 rpm to achieve final TO@HNT nominal content 5, 10, and 15 wt.% (see part (3) in Figure 12). For comparison amounts of pure HNT were also added into CS/PVOH mixed solution to achieve final HNT nominal content 5, 10, and 15 wt.%. The as-obtained CS/PVOH/xHNT and CS/PVOH/xTO@HNT (x takes values 5, 10, and 15) homogenized coatings were placed in 11 cm diameter petri dishes and dried at ambient temperature. The obtained films were peeled off and preserved inside PE plastic bags at 25 °C and 50% RH before the measurements.

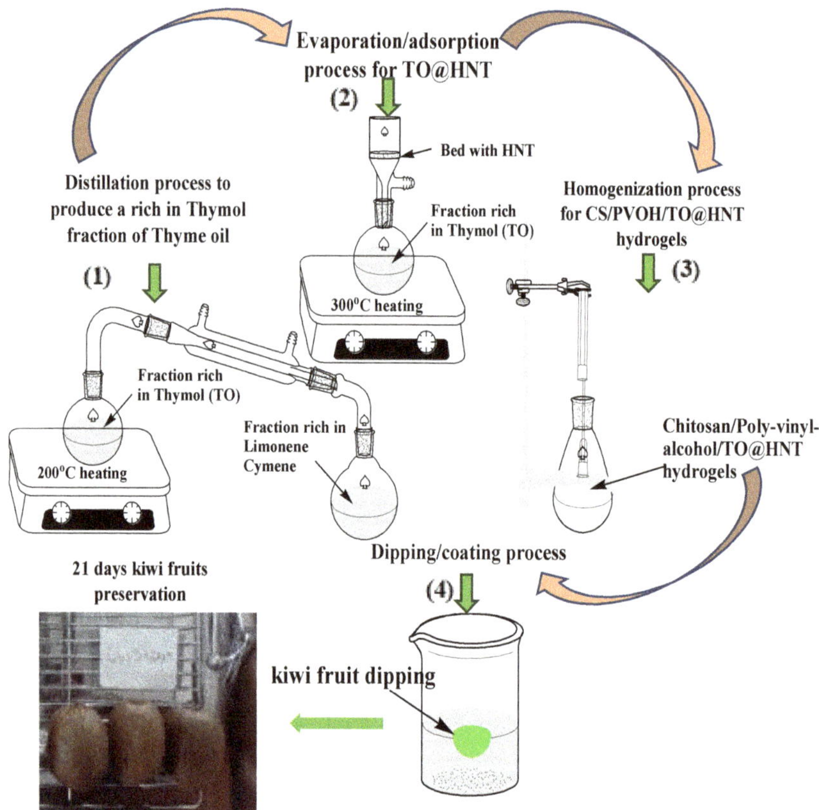

Figure 12. Schematic presentation of (**1**) distillation process to produce a rich in thymol fraction from pure thyme oil, (**2**) evaporation/adsorption process to modify pure HNT and develop TO@HNT nanohybrids, (**3**) homogenization process to develop CS/PVOH/TO@HNT gels and (**4**) kiwi fruit dipping/coating process.

4.4. Characterization of TO@HNT Hybrid Nanostructure

The obtained TO@HNT hybrid nanostructure was characterized with XRD analysis, FTIR spectroscopy, and DSC analysis. For XRD analysis a Brücker D8 advance instrument was employed (Brüker, Analytical Instruments, S.A., Athens, Greece) and the measurements were carried out in the range 2θ = 2–30°. For the FTIR spectroscopy measurements, an FT/IR-6000 JASCO Fourier transform spectrometer (JASCO, Interlab, S.A., Athens, Greece) was employed. The FTIR spectra were recorded in the range of 4000–400 cm^{-1} and the obtained spectra was was the average of 32 scans at 2 cm^{-1} resolution. For the DSC experiments, a DSC214 Polyma Differential Scanning Calorimeter (NETZSCH manufacturer, Selb, Germany) was employed. Samples with an average weight in the range of 1.2–3.3 mg were tested under a nitrogen atmosphere with a heating rate of 10 K/min from 50 to 300 °C.

4.5. XRD Analysis of CS/PVOH/HNT and CS/PVOH/TO@HNT Films

The prepared CS/PVOH/HNT and CS/PVOH/TO@HNT films were characterized with XRD analysis by using a Brücker D8 advance instrument was employed (Brüker, Analytical Instruments, S.A., Athens, Greece) in the range of 2θ = 2–30°.

4.6. FTIR Analysis of CS/PVOH/HNT and CS/PVOH/TO@HNT Films

The interaction of pure HNT with CS/PVOH matrix in CS/PVOH/HNT films and TO@HNT hybrid nanostructure with CS/PVOH matrix in CS/PVOH/TO@HNT films were tested with FTIR analysis by using an FT/IR-6000 JASCO Fourier transform spectrometer (JASCO, Interlab, S.A., Athens, Greece) in the range of 4000–400 cm^{-1}.

4.7. SEM Images

The surface and cross-section morphology of CS/PVOH/HNT and CS/PVOH/TO@HNT films was recorded by using a JEOL JSM-6510 LV SEM (Microscope Ltd., Tokyo, Japan) were used equipped with an X-Act EDS-detector by Oxford Instruments, Abingdon, Oxfordshire, UK (an acceleration voltage of 20 kV was applied).

4.8. Tensile Properties of CS/PVOH/HNT and CS/PVOH/TO@HNT Films

The influence of pure HNT and TO@HNT addition in the tensile characteristics of the CS/PVOH matrix was evaluated with tensile measurements by using a Simantzü AX-G 5kNt instrument (Simandzu, Asteriadis, S.A., Athens, Greece). According to the ASTM D638 method, three to five samples of type V ASTMD638 specimens of each film were tensioned at an across-head speed of 2 mm/min.

4.9. UV-vis Transparency of CS/PVOH/HNT and CS/PVOH/TO@HNT Films

UV–vis transmittance measurements for all obtained CS/PVOH/HNT and CS/PVOH/TO@HNT films were carried out with a Shimatzu 1900 spectrophotometer in the range of 200 to 800 nm.

4.10. Water and Oxygen Barrier Properties of CS/PVOH/HNT and CS/PVOH/TO@HNT Films

For the evaluation of the water vapor transmission rate (WVTR) of CS/PVOH/HNT and CS/PVOH/TO@HNT films, a hand-made apparatus was used according to the ASTM E96/E 96M-05 method [66]. According to the methodology described extensively in previous reports [17,40–42,66] experiments were carried out inside thermostated chamber at 38 °C and constant 95% RH. Water vapor transmission rate WVTR ($g \cdot cm^{-2} \cdot s^{-1}$) and water vapor diffusion coefficient D ($cm^2 \cdot s^{-1}$) were estimated according to Equations (2) and (3) respectively as they reported to number [50] referenced paper and based on the theory described in this paper [67].

4.11. Oxygen Permeability of CS/PVOH/HNT and CS/PVOH/TO@HNT Films

ASTM D 3985 method (23 °C and 0% RH) was followed to estimate the oxygen transmission rate (OTR). OTR values in cc $O_2 \cdot m^{-2} \cdot day^{-1}$ achieved using an (8001, Systech Illinois Instruments Co., Johnsburg, IL, USA) oxygen permeation analyzer. The oxygen permeability coefficient Pe_{gas} (cm^2/s), was estimated following Equations (4) and (5) of the number [50] referenced paper and according to the theory described in the literature [68].

4.12. Antioxidant Activity of CS/PVOH/HNT and CS/PVOH/TO@HNT Films

The antioxidant activity of CS/PVOH/HNT and CS/PVOH/TO@HNT films were evaluated with the diphenyl-1-picrylhydrazyl (DPPH) method. Briefly, 300 mg of each film was cut into small pieces and placed inside dark glass bottles with 10 mL of an ethanolic solution of diphenyl-1-picrylhydrazyl (DPPH) with 40 ppm concentration. A sample with 10 mL of ethanolic DPPH solution without the addition of any film was used as the blank sample. The absorbance of the DPPH solution at 517 nm at 0 h and after 24 h incubation was measured using a Jasco V-530 UV-vis spectrophotometer. For each film, three to five different samples were made and measured. After 24 h incubation of films, the % antioxidant activity was estimated using the Equation (7) reported in previous reports [27,40,41,49].

4.13. Antibacterial Activity of CS/PVOH/HNT and CS/PVOH/TO@HNT Films

Antibacterial properties of the films were investigated by utilizing the agar diffusion method against Gram-negative bacteria *Escherichia coli* (ATCC 25922), *Salmonella enterica* subsp. enterica (DSMZ 17420), Gram-positive bacteria *Staphylococcus aureus* (DSMZ 12463), and *Listeria monocytogenes* (DSMZ 27575). The tested foodborne microorganisms were obtained from the Institute of Technology of Agricultural Products, ELGO-DEMETER, Lykovryssi, Greece.

Fresh cultures of the bacterial strains were prepared in Mueller Hinton Broth. The cultures were inoculated at 37 °C for 24 h in order to achieve a range of 10^7–10^8 CFU mL^{-1}. After that, the bacteria were swabbed on Mueller-Hinton agar dishes by rotating the plate every 60° to ensure consistent growth.

The tested films were cut into 6 mm diameter discs by a circular knife and were placed on a Mueller-Hinton inoculated plate. The dishes were incubated at 37 °C overnight. The diameter of the inhibitory zones, and the contact area of the discs with agar surface, were measured. The experiment was performed thrice.

4.14. Packaging Test of CS/PVOH/HNT and CS/PVOH/TO@HNT Coatings in Preservation of Kiwifruits

15 kiwifruits of as close to the same shape as possible and the same ripeness were purchased from the local supermarket and they were divided into four groups with three fruits each. The first group of three kiwifruits was used as blank uncoated samples. The other three groups with three kiwifruits each were coated with pure CS/PVOH solution, CS/PVOH/10HNT solution, and CS/PVOH/10TO@HNT. CS/PVOH/10TO@HNT coating solution was selected as the optimum one according to each higher water/oxygen barrier properties shown hereabove which are critical for such active fruit coatings. Therefore, the uncoated and coated kiwifruits were put in a tray, stored under room humidity at 25 °C, and observed daily for any visible changes or fungal growth on their surfaces over 21 days. For the coating of kiwi fruits, dipping for 1 min in the selected coating solutions was followed (see part (4) in Figure 12).

4.15. Statistical Analysis

Three pieces of every film sample were tested to obtain the values presented in Tables 1–4. The final value of each property is the mean value of such measurements. All experimental data were processed with the SPSS vr. 20 statistical software and the mean values and standard deviation values which are tabulated above, resulted assumed a confidence interval of C.I. = 95%. Hypothesis tests ran assuming a statistical significance level of $p = 0.05$ to ensure that different mean values of a property for different samples are also statistically different. The non-positive normality tests implied the non-parametric Kruskal–Wallis method for such investigations and statistically unequal mean values of all properties were confirmed. The equality or inequality assurance was tested according to the empirical Equations (1) and (2) which were explained in detail in previous works [40]:

$$EA(\%) = \frac{Sig. - p}{1 - p} \cdot 100 \quad (1)$$

$$IA(\%) = \frac{p - Sig.}{p} \cdot 100 \quad (2)$$

All the mechanical and barrier properties of all kinds of films exhibit statistical inequal mean values. The significance Sig. value range resulting from the SPSS software is presented in the following table:

It is obvious from Table 5 that the values of properties E, σ_{uts}, ε%, WVTR, and OTR are different for different kinds of films while the values of the antioxidant activity are different but, in some cases, close to each other. Finally, it is obvious from Tables 4 and 5 that mean values of antimicrobial activity, in some cases are statistically equal while in other cases are statistically unequal.

Table 5. Significance level, equality, and inequality, assurance of mean values for Young Modulus (E), σ_{uts}, % elongation at break (ε%), WVTR, OTR, % Antioxidant.

	E	σ_{uts}	ε%	WVTR	OTR	Antiox.	*E. coli*	*Saur.*	*Senter.*	*L. monoc.*
Sig. < 0.05	0–0.0110	0–0.0185	0–0.014	0–0.0220	0–0.0120	0–0.0400	0–0.037	0–0.024	0–0.039	0–0.002
IA (%)	78–100	63–100	72–100	56–100	76–100	20–100	26–100	52–100	22–100	96–100
Sig. > 0.05	-	-	-	-	-	-	0.862–1	0.071–0.999	0.081–0.991	1
EA (%)	-	-	-	-	-	-	85–100	2–100	3–99	100

Significance level $p < 0.05$.

Author Contributions: Synthesis experiment design, A.E.G., C.P. and C.E.S.; characterization measurements and interpretation, A.E.G., D.M., A.A., C.G., E.K., A.K., L.A., A.L., S.G. and C.E.S.; paper writing, A.E.G., C.P. and C.E.S.; overall evaluation of this work, A.E.G. and C.E.S.; experimental data analysis and interpretation, A.E.G., C.E.S., C.G., S.G., E.K., A.K., L.A. and C.P.; XRD, FTIR, OTR, tensile measurements, UV-vis spectroscopy, antioxidant activity, WVTR experimental measurements, and kiwi fruits packaging test A.E.G., A.L., A.A., S.G. and C.E.S.; SEM images D.M. and A.A.; antibacterial activity tests, E.K., A.K., L.A. and C.P. All authors have read and agreed to the published version of the manuscript.

Funding: This research was funded by the funding program "MEDICUS", Project F.K. 81541, of the University of Patras, Greece.

Data Availability Statement: The datasets generated for this study are available on request to the corresponding author.

Conflicts of Interest: The authors declare no conflict of interest.

References

1. Hamam, M.; Chinnici, G.; Di Vita, G.; Pappalardo, G.; Pecorino, B.; Maesano, G.; D'Amico, M. Circular Economy Models in Agro-Food Systems: A Review. *Sustainability* **2021**, *13*, 3453. [CrossRef]
2. Guillard, V.; Gaucel, S.; Fornaciari, C.; Angellier-Coussy, H.; Buche, P.; Gontard, N. The Next Generation of Sustainable Food Packaging to Preserve Our Environment in a Circular Economy Context. *Front. Nutr.* **2018**, *5*, 121. [CrossRef] [PubMed]
3. Jabeen, N.; Majid, I.; Nayik, G.A. Bioplastics and Food Packaging: A Review. *Cogent Food Agric.* **2015**, *1*, 1117749. [CrossRef]
4. Taherimehr, M.; Yousefnia Pasha, H.; Tabatabaeekoloor, R.; Pesaranhajiabbas, E. Trends and Challenges of Biopolymer-Based Nanocomposites in Food Packaging. *Compr. Rev. Food Sci. Food Saf.* **2021**, *20*, 5321–5344. [CrossRef]
5. Sid, S.; Mor, R.S.; Kishore, A.; Sharanagat, V.S. Bio-Sourced Polymers as Alternatives to Conventional Food Packaging Materials: A Review. *Trends Food Sci. Technol.* **2021**, *115*, 87–104. [CrossRef]
6. Liu, Y.; Ahmed, S.; Sameen, D.E.; Wang, Y.; Lu, R.; Dai, J.; Li, S.; Qin, W. A Review of Cellulose and Its Derivatives in Biopolymer-Based for Food Packaging Application. *Trends Food Sci. Technol.* **2021**, *112*, 532–546. [CrossRef]
7. Flórez, M.; Guerra-Rodríguez, E.; Cazón, P.; Vázquez, M. Chitosan for Food Packaging: Recent Advances in Active and Intelligent Films. *Food Hydrocoll.* **2022**, *124*, 107328. [CrossRef]
8. Mahmud, N.; Islam, J.; Tahergorabi, R. Marine Biopolymers: Applications in Food Packaging. *Processes* **2021**, *9*, 2245. [CrossRef]
9. Ravi Kumar, M.N.V. A Review of Chitin and Chitosan Applications. *React. Funct. Polym.* **2000**, *46*, 1–27. [CrossRef]
10. Elsabee, M.Z.; Abdou, E.S. Chitosan Based Edible Films and Coatings: A Review. *Mater. Sci. Eng. C* **2013**, *33*, 1819–1841. [CrossRef]
11. Yang, Y.; Khan, H.; Gao, S.; Khalil, A.K.; Ali, N.; Khan, A.; Show, P.L.; Bilal, M.; Khan, H. Fabrication, Characterization, and Photocatalytic Degradation Potential of Chitosan-Conjugated Manganese Magnetic Nano-Biocomposite for Emerging Dye Pollutants. *Chemosphere* **2022**, *306*, 135647. [CrossRef]
12. Cai, D.-L.; Thanh, D.T.H.; Show, P.-L.; How, S.-C.; Chiu, C.-Y.; Hsu, M.; Chia, S.R.; Chen, K.-H.; Chang, Y.-K. Studies of Protein Wastes Adsorption by Chitosan-Modified Nanofibers Decorated with Dye Wastes in Batch and Continuous Flow Processes: Potential Environmental Applications. *Membranes* **2022**, *12*, 759. [CrossRef]
13. Chen, Y.-S.; Ooi, C.W.; Show, P.L.; Hoe, B.C.; Chai, W.S.; Chiu, C.-Y.; Wang, S.S.-S.; Chang, Y.-K. Removal of Ionic Dyes by Nanofiber Membrane Functionalized with Chitosan and Egg White Proteins: Membrane Preparation and Adsorption Efficiency. *Membranes* **2022**, *12*, 63. [CrossRef]
14. Ortiz-Duarte, G.; Pérez-Cabrera, L.E.; Artés-Hernández, F.; Martínez-Hernández, G.B. Ag-Chitosan Nanocomposites in Edible Coatings Affect the Quality of Fresh-Cut Melon. *Postharvest Biol. Technol.* **2019**, *147*, 174–184. [CrossRef]

15. Díaz-Montes, E.; Castro-Muñoz, R. Edible Films and Coatings as Food-Quality Preservers: An Overview. *Foods* **2021**, *10*, 249. [CrossRef]
16. Antimicrobial Chitosan and Chitosan Derivatives: A Review of the Structure—Activity Relationship | Biomacromolecules. Available online: https://pubs.acs.org/doi/10.1021/acs.biomac.7b01058 (accessed on 20 May 2022).
17. Giannakas, A.E.; Salmas, C.E.; Leontiou, A.; Baikousi, M.; Moschovas, D.; Asimakopoulos, G.; Zafeiropoulos, N.E.; Avgeropoulos, A. Synthesis of a Novel Chitosan/Basil Oil Blend and Development of Novel Low Density Poly Ethylene/Chitosan/Basil Oil Active Packaging Films Following a Melt-Extrusion Process for Enhancing Chicken Breast Fillets Shelf-Life. *Molecules* **2021**, *26*, 1585. [CrossRef]
18. Giannakas, A.; Vlacha, M.; Salmas, C.; Leontiou, A.; Katapodis, P.; Stamatis, H.; Barkoula, N.-M.; Ladavos, A. Preparation, Characterization, Mechanical, Barrier and Antimicrobial Properties of Chitosan/PVOH/Clay Nanocomposites. *Carbohydr. Polym.* **2016**, *140*, 408–415. [CrossRef]
19. Salmas, C.E.; Giannakas, A.E.; Baikousi, M.; Leontiou, A.; Siasou, Z.; Karakassides, M.A. Development of Poly(L-Lactic Acid)/Chitosan/Basil Oil Active Packaging Films via a Melt-Extrusion Process Using Novel Chitosan/Basil Oil Blends. *Processes* **2021**, *9*, 88. [CrossRef]
20. Lozano-Navarro, J.I.; Díaz-Zavala, N.P.; Velasco-Santos, C.; Martínez-Hernández, A.L.; Tijerina-Ramos, B.I.; García-Hernández, M.; Rivera-Armenta, J.L.; Páramo-García, U.; Reyes-de la Torre, A.I. Antimicrobial, Optical and Mechanical Properties of Chitosan–Starch Films with Natural Extracts. *Int. J. Mol. Sci.* **2017**, *18*, 997. [CrossRef]
21. Meng, W.; Shi, J.; Zhang, X.; Lian, H.; Wang, Q.; Peng, Y. Effects of Peanut Shell and Skin Extracts on the Antioxidant Ability, Physical and Structure Properties of Starch-Chitosan Active Packaging Films. *Int. J. Biol. Macromol.* **2020**, *152*, 137–146. [CrossRef]
22. Qu, B.; Luo, Y. A Review on the Preparation and Characterization of Chitosan-Clay Nanocomposite Films and Coatings for Food Packaging Applications. *Carbohydr. Polym. Technol. Appl.* **2021**, *2*, 100102. [CrossRef]
23. Wiles, J.L.; Vergano, P.J.; Barron, F.H.; Bunn, J.M.; Testin, R.F. Water Vapor Transmission Rates and Sorption Behavior of Chitosan Films. *J. Food Sci.* **2000**, *65*, 1175–1179. [CrossRef]
24. Bhargava, N.; Sharanagat, V.S.; Mor, R.S.; Kumar, K. Active and Intelligent Biodegradable Packaging Films Using Food and Food Waste-Derived Bioactive Compounds: A Review. *Trends Food Sci. Technol.* **2020**, *105*, 385–401. [CrossRef]
25. De Carvalho, A.P.A.; Conte Junior, C.A. Green Strategies for Active Food Packagings: A Systematic Review on Active Properties of Graphene-Based Nanomaterials and Biodegradable Polymers. *Trends Food Sci. Technol.* **2020**, *103*, 130–143. [CrossRef]
26. Domínguez, R.; Barba, F.J.; Gómez, B.; Putnik, P.; Bursać Kovačević, D.; Pateiro, M.; Santos, E.M.; Lorenzo, J.M. Active Packaging Films with Natural Antioxidants to Be Used in Meat Industry: A Review. *Food Res. Int.* **2018**, *113*, 93–101. [CrossRef]
27. Giannakas, A. Na-Montmorillonite vs. Organically Modified Montmorillonite as Essential Oil Nanocarriers for Melt-Extruded Low-Density Poly-Ethylene Nanocomposite Active Packaging Films with a Controllable and Long-Life Antioxidant Activity. *Nanomaterials* **2020**, *10*, 1027. [CrossRef]
28. Giannakas, A.E.; Leontiou, A.A. Montmorillonite Composite Materials and Food Packaging. In *Composites Materials for Food Packaging*; John Wiley & Sons, Ltd.: Hoboken, NJ, USA, 2018; pp. 1–71. ISBN 978-1-119-16024-3.
29. Sánchez-González, L.; Vargas, M.; González-Martínez, C.; Chiralt, A.; Cháfer, M. Use of Essential Oils in Bioactive Edible Coatings: A Review. *Food Eng. Rev.* **2011**, *3*, 1–16. [CrossRef]
30. Aswathanarayan, J.B.; Vittal, R.R. Nanoemulsions and Their Potential Applications in Food Industry. *Front. Sustain. Food Syst.* **2019**, *3*, 95. [CrossRef]
31. Maurya, A.; Singh, V.K.; Das, S.; Prasad, J.; Kedia, A.; Upadhyay, N.; Dubey, N.K.; Dwivedy, A.K. Essential Oil Nanoemulsion as Eco-Friendly and Safe Preservative: Bioefficacy against Microbial Food Deterioration and Toxin Secretion, Mode of Action, and Future Opportunities. *Front. Microbiol.* **2021**, *12*, 751062. [CrossRef]
32. Full Article: Microemulsions: A Potential Delivery System for Bioactives in Food. Available online: https://www.tandfonline.com/doi/full/10.1080/10408690590956710 (accessed on 20 May 2022).
33. De Oliveira, L.H.; Trigueiro, P.; Souza, J.S.N.; de Carvalho, M.S.; Osajima, J.A.; da Silva-Filho, E.C.; Fonseca, M.G. Montmorillonite with Essential Oils as Antimicrobial Agents, Packaging, Repellents, and Insecticides: An Overview. *Colloids Surf. B Biointerfaces* **2022**, *209*, 112186. [CrossRef]
34. Li, Q.; Ren, T.; Perkins, P.; Hu, X.; Wang, X. Applications of Halloysite Nanotubes in Food Packaging for Improving Film Performance and Food Preservation. *Food Control* **2021**, *124*, 107876. [CrossRef]
35. Villa, C.C.; Valencia, G.A.; López Córdoba, A.; Ortega-Toro, R.; Ahmed, S.; Gutiérrez, T.J. Zeolites for Food Applications: A Review. *Food Biosci.* **2022**, *46*, 101577. [CrossRef]
36. Giannakas, A.; Tsagkalias, I.; Achilias, D.S.; Ladavos, A. A Novel Method for the Preparation of Inorganic and Organo-Modified Montmorillonite Essential Oil Hybrids. *Appl. Clay Sci.* **2017**, *146*, 362–370. [CrossRef]
37. Shiekh, K.A.; Ngiwngam, K.; Tongdeesoontorn, W. Polysaccharide-Based Active Coatings Incorporated with Bioactive Compounds for Reducing Postharvest Losses of Fresh Fruits. *Coatings* **2022**, *12*, 8. [CrossRef]
38. Kiwifruit Production and Research in Greece | International Society for Horticultural Science. Available online: http://www.actahort.org/books/444/444_3.htm (accessed on 27 September 2022).
39. Göksel, Z.; Atak, A. Kiwifruit processing studies. *Acta Hortic.* **2015**, *1096*, 99–107. [CrossRef]

40. Salmas, C.E.; Giannakas, A.E.; Baikousi, M.; Kollia, E.; Tsigkou, V.; Proestos, C. Effect of Copper and Titanium-Exchanged Montmorillonite Nanostructures on the Packaging Performance of Chitosan/Poly-Vinyl-Alcohol-Based Active Packaging Nanocomposite Films. *Foods* **2021**, *10*, 3038. [CrossRef]
41. Giannakas, A.E.; Salmas, C.E.; Karydis-Messinis, A.; Moschovas, D.; Kollia, E.; Tsigkou, V.; Proestos, C.; Avgeropoulos, A.; Zafeiropoulos, N.E. Nanoclay and Polystyrene Type Efficiency on the Development of Polystyrene/Montmorillonite/Oregano Oil Antioxidant Active Packaging Nanocomposite Films. *Appl. Sci.* **2021**, *11*, 9364. [CrossRef]
42. Giannakas, A.E.; Salmas, C.E.; Leontiou, A.; Moschovas, D.; Baikousi, M.; Kollia, E.; Tsigkou, V.; Karakassides, A.; Avgeropoulos, A.; Proestos, C. Performance of Thyme Oil@Na-Montmorillonite and Thyme Oil@Organo-Modified Montmorillonite Nanostructures on the Development of Melt-Extruded Poly-L-Lactic Acid Antioxidant Active Packaging Films. *Molecules* **2022**, *27*, 1231. [CrossRef]
43. Tsagkalias, I.S.; Loukidi, A.; Chatzimichailidou, S.; Salmas, C.E.; Giannakas, A.E.; Achilias, D.S. Effect of Na- and Organo-Modified Montmorillonite/Essential Oil Nanohybrids on the Kinetics of the In Situ Radical Polymerization of Styrene. *Nanomaterials* **2021**, *11*, 474. [CrossRef]
44. Saucedo-Zuñiga, J.N.; Sánchez-Valdes, S.; Ramírez-Vargas, E.; Guillen, L.; Ramos-de Valle, L.F.; Graciano-Verdugo, A.; Uribe-Calderón, J.A.; Valera-Zaragoza, M.; Lozano-Ramírez, T.; Rodríguez-González, J.A.; et al. Controlled Release of Essential Oils Using Laminar Nanoclay and Porous Halloysite/Essential Oil Composites in a Multilayer Film Reservoir. *Microporous Mesoporous Mater.* **2021**, *316*, 110882. [CrossRef]
45. Jang, S.; Jang, S.; Lee, G.; Ryu, J.; Park, S.; Park, N. Halloysite Nanocapsules Containing Thyme Essential Oil: Preparation, Characterization, and Application in Packaging Materials. *J. Food Sci.* **2017**, *82*, 2113–2120. [CrossRef]
46. Lee, M.H.; Seo, H.-S.; Park, H.J. Thyme Oil Encapsulated in Halloysite Nanotubes for Antimicrobial Packaging System. *J. Food Sci.* **2017**, *82*, 922–932. [CrossRef]
47. Le Ba, T.; Alkurdi, A.Q.; Lukács, I.E.; Molnár, J.; Wongwises, S.; Gróf, G.; Szilágyi, I.M. A Novel Experimental Study on the Rheological Properties and Thermal Conductivity of Halloysite Nanofluids. *Nanomaterials* **2020**, *10*, 1834. [CrossRef]
48. Shemesh, R.; Krepker, M.; Natan, M.; Danin-Poleg, Y.; Banin, E.; Kashi, Y.; Nitzan, N.; Vaxman, A.; Segal, E. Novel LDPE/Halloysite Nanotube Films with Sustained Carvacrol Release for Broad-Spectrum Antimicrobial Activity. *RSC Adv.* **2015**, *5*, 87108–87117. [CrossRef]
49. Giannakas, A.E.; Salmas, C.E.; Moschovas, D.; Zaharioudakis, K.; Georgopoulos, S.; Asimakopoulos, G.; Aktypis, A.; Proestos, C.; Karakassides, A.; Avgeropoulos, A.; et al. The Increase of Soft Cheese Shelf-Life Packaged with Edible Films Based on Novel Hybrid Nanostructures. *Gels* **2022**, *8*, 539. [CrossRef]
50. Lavorgna, M.; Piscitelli, F.; Mangiacapra, P.; Buonocore, G.G. Study of the Combined Effect of Both Clay and Glycerol Plasticizer on the Properties of Chitosan Films. *Carbohydr. Polym.* **2010**, *82*, 291–298. [CrossRef]
51. Giannakas, A.; Grigoriadi, K.; Leontiou, A.; Barkoula, N.-M.; Ladavos, A. Preparation, Characterization, Mechanical and Barrier Properties Investigation of Chitosan–Clay Nanocomposites. *Carbohydr. Polym.* **2014**, *108*, 103–111. [CrossRef]
52. Grigoriadi, K.; Giannakas, A.; Ladavos, A.K.; Barkoula, N.-M. Interplay between Processing and Performance in Chitosan-Based Clay Nanocomposite Films. *Polym. Bull.* **2015**, *72*, 1145–1161. [CrossRef]
53. A Chitosan/Poly(Vinyl Alcohol) Nanocomposite Film Reinforced with Natural Halloysite Nanotubes—Huang—2012—Polymer Composites—Wiley Online Library. Available online: https://onlinelibrary.wiley.com/doi/epdf/10.1002/pc.22302 (accessed on 14 June 2022).
54. Suflet, D.M.; Popescu, I.; Pelin, I.M.; Ichim, D.L.; Daraba, O.M.; Constantin, M.; Fundueanu, G. Dual Cross-Linked Chitosan/PVA Hydrogels Containing Silver Nanoparticles with Antimicrobial Properties. *Pharmaceutics* **2021**, *13*, 1461. [CrossRef] [PubMed]
55. Bonilla, J.; Fortunati, E.; Atarés, L.; Chiralt, A.; Kenny, J.M. Physical, Structural and Antimicrobial Properties of Poly Vinyl Alcohol–Chitosan Biodegradable Films. *Food Hydrocoll.* **2014**, *35*, 463–470. [CrossRef]
56. Giannakas, A.E.; Salmas, C.E.; Moschovas, D.; Baikousi, M.; Kollia, E.; Tsigkou, V.; Karakassides, A.; Leontiou, A.; Kehayias, G.; Avgeropoulos, A.; et al. Nanocomposite Film Development Based on Chitosan/Polyvinyl Alcohol Using ZnO@Montmorillonite and ZnO@Halloysite Hybrid Nanostructures for Active Food Packaging Applications. *Nanomaterials* **2022**, *12*, 1843. [CrossRef]
57. Unuabonah, E.I.; Ugwuja, C.G.; Omorogie, M.O.; Adewuyi, A.; Oladoja, N.A. Clays for Efficient Disinfection of Bacteria in Water. *Appl. Clay Sci.* **2018**, *151*, 211–223. [CrossRef]
58. What Makes a Natural Clay Antibacterial? Environmental Science & Technology. Available online: https://pubs.acs.org/doi/10.1021/es1040688 (accessed on 27 September 2022).
59. Abhinayaa, R.; Jeevitha, G.; Mangalaraj, D.; Ponpandian, N.; Meena, P. Toxic Influence of Pristine and Surfactant Modified Halloysite Nanotubes on Phytopathogenic Bacteria. *Appl. Clay Sci.* **2019**, *174*, 57–68. [CrossRef]
60. Duan, L.; Zhao, Q.; Liu, J.; Zhang, Y. Antibacterial Behavior of Halloysite Nanotubes Decorated with Copper Nanoparticles in a Novel Mixed Matrix Membrane for Water Purification. *Environ. Sci. Water Res. Technol.* **2015**, *1*, 874–881. [CrossRef]
61. Wang, L.-F.; Rhim, J.-W. Functionalization of Halloysite Nanotubes for the Preparation of Carboxymethyl Cellulose-Based Nanocomposite Films. *Appl. Clay Sci.* **2017**, *150*, 138–146. [CrossRef]
62. Taylor, A.A.; Aron, G.M.; Beall, G.W.; Dharmasiri, N.; Zhang, Y.; McLean, R.J.C. Carbon and Clay Nanoparticles Induce Minimal Stress Responses in Gram Negative Bacteria and Eukaryotic Fish Cells. *Environ. Toxicol.* **2014**, *29*, 961–968. [CrossRef]
63. Boelter, J.F.; Brandelli, A.; Meira, S.M.M.; Göethel, G.; Garcia, S.C. Toxicology Study of Nanoclays Adsorbed with the Antimicrobial Peptide Nisin on Caenorhabditis Elegans. *Appl. Clay Sci.* **2020**, *188*, 105490. [CrossRef]

64. Krepker, M.; Shemesh, R.; Danin Poleg, Y.; Kashi, Y.; Vaxman, A.; Segal, E. Active Food Packaging Films with Synergistic Antimicrobial Activity. *Food Control* **2017**, *76*, 117–126. [CrossRef]
65. Maruthupandy, M.; Seo, J. Allyl Isothiocyanate Encapsulated Halloysite Covered with Polyacrylate as a Potential Antibacterial Agent against Food Spoilage Bacteria. *Mater. Sci. Eng. C* **2019**, *105*, 110016. [CrossRef]
66. Giannakas, A.; Giannakas, A.; Ladavos, A. Preparation and Characterization of Polystyrene/Organolaponite Nanocomposites. *Polym. Plast. Technol. Eng.* **2012**, *51*, 1411–1415. [CrossRef]
67. Bastarrachea, L.; Dhawan, S.; Sablani, S.S. Engineering Properties of Polymeric-Based Antimicrobial Films for Food Packaging: A Review. *Food Eng. Rev.* **2011**, *3*, 79–93. [CrossRef]
68. Units of Gas Permeability Constants—Yasuda—1975—Journal of Applied Polymer Science—Wiley Online Library. Available online: https://onlinelibrary.wiley.com/doi/abs/10.1002/app.1975.070190915 (accessed on 11 November 2021).

 gels

MDPI

Article

The Increase of Soft Cheese Shelf-Life Packaged with Edible Films Based on Novel Hybrid Nanostructures

Aris E. Giannakas [1,*], Constantinos E. Salmas [2,*], Dimitrios Moschovas [2], Konstantinos Zaharioudakis [1,3], Stavros Georgopoulos [1], Georgios Asimakopoulos [2], Anastasios Aktypis [3], Charalampos Proestos [4], Anastasios Karakassides [2], Apostolos Avgeropoulos [2], Nikolaos E. Zafeiropoulos [2] and George-John Nychas [3]

[1] Department of Food Science and Technology, University of Patras, 30100 Agrinio, Greece
[2] Department of Material Science and Engineering, University of Ioannina, 45110 Ioannina, Greece
[3] School of Food and Nutritional Sciences, Department of Food Science and Human Nutrition, Laboratory of Microbiology and Biotechnology of Foods, Agricultural University of Athens, Iera Odos 75, 11855 Athens, Greece
[4] Laboratory of Food Chemistry, Department of Chemistry, National and Kapodistrian University of Athens, 15771 Athens, Greece
* Correspondence: agiannakas@upatras.gr (A.E.G.); ksalmas@uoi.gr (C.E.S.)

Abstract: This study presents, the development of a green method to produce rich in thymol natural zeolite (TO@NZ) nanostructures. This material was used to prepare sodium-alginate/glycerol/xTO@NZ (ALG/G/TO@NZ) nanocomposite active films for the packaging of soft cheese to extend its shelf-life. Differential scanning calorimetry (DSC), X-ray analysis (XRD), scanning electron microscopy (SEM), and Fourier-transform infrared spectroscopy (FTIR) instruments were used for the characterization of such nanostructures and films, to identify the thymol adsorbed amount, to investigate the thermal behaviour, and to confirm the dispersion of nanostructure powder into the polymer matrix. Water vapor transmission rate, oxygen permeation analyzer, tensile measurements, antioxidant measurements, and antimicrobial measurements were used to estimate the film's water and oxygen barrier, mechanical properties, nanostructure's nanoreinforcement activity, antioxidant and antimicrobial activity. The findings from the study revealed that ALG/G/TO@NZ nanocomposite film could be used as an active packaging film for foods with enhanced, mechanical properties, oxygen and water barrier, antioxidant and antimicrobial activity, and it is capable of extending food shelf-life.

Keywords: active packaging; soft cheese preservation; sodium alginate; thyme oil; natural zeolite; shelf-life extension

Citation: Giannakas, A.E.; Salmas, C.E.; Moschovas, D.; Zaharioudakis, K.; Georgopoulos, S.; Asimakopoulos, G.; Aktypis, A.; Proestos, C.; Karakassides, A.; Avgeropoulos, A.; et al. The Increase of Soft Cheese Shelf-Life Packaged with Edible Films Based on Novel Hybrid Nanostructures. *Gels* **2022**, *8*, 539. https://doi.org/10.3390/gels8090539

Academic Editor: Bjørn Torger Stokke

Received: 30 July 2022
Accepted: 23 August 2022
Published: 26 August 2022

1. Introduction

In the last few years, petroleum-based food packaging materials have been replaced with biodegradable biopolymers as a result of circular economy and sustainability [1–5]. Thus, the utilization of protein and polysaccharide-based biopolymer hybrid nanostructure materials in the food industry has been increased due to their non-toxicity, biodegradability, ability to form gels, encapsulate and deliver bioactive compounds such as essential oils [6–8]. Such novel biopolymer-based food packaging systems are suitable for active food packaging applications. Active food packaging is defined as "packaging in which subsidiary constituents have been deliberately included in or on either the packaging material or the package headspace to enhance the performance of the package system" [4].

Alginate (ALG) has become one of the most popular natural polysaccharides extensively used in the development of delivery systems for food bioactive ingredients. The suitability of this material for such purposes is due to its ionic crosslinking ability, pH responsiveness, excellent biocompatibility, biodegradability, and low price [5]. Alginate (ALG) is an unbranched anionic polysaccharide consisting of β-D-mannuronic acid (M) and α-L-guluronic acid (G) linked by glycosidic bonds. The structure of alginate depends primarily on the monomer composition, the M/G ratio, the polymer sequence, and the molecular weight of the linear chain. The structure of alginate and the M/G ratio are crucial for its capability to deliver bioactive compounds. For example, a higher concentration of G blocks generates more rigid hydrogels with larger pores [6], which leads to the easier release of immobilized bioactive components from the polymer matrix. On the contrary, higher M block content is more suitable for the formulation of softer edible films and coatings with lower gas permeability [7,8].

Under the same spirit, there is a trend to replace the "commonly" used antioxidant and/or antimicrobial chemical agents such as Butylated hydroxytoluene (BHT) and Butylated Hydroxyanisole (BHA) which are added directly to the food. The use of such chemicals was replaced with the use of essential oils [9,10] or other bioactive phytochemicals [11] in active packaging film gels, and coating is taking place. Essential oil loss due to evaporation phenomena was reduced using various nanomaterials as nanocarriers in food delivery systems and active food packaging applications. Such materials were developed based on the food nanotechnology concept [12–14]. Nanoclays, such as montmorillonite [15–17] and halloysite [18–20] were used both as reinforcements and as essential oil nanocarriers for controlled released applications [21]. These nanoclay based essential oil nanocarriers were incorporated into polymer [21–23] or biopolymer [16] networks. The new composite materials were promising for active packaging film applications with antioxidant and/or antimicrobial activity. Natural zeolite (NZ) is another, abundant nanomaterial promising for food preservation and food packaging applications [24] as is reported in the literature [25–27]. Reščеk et al., 2018 [28] developed double-layered polyethylene/caprolactone packaging films modified with zeolite and magnetite. It was shown that the addition of zeolite improved the mechanical and barrier properties of obtained films. Youssef et al., 2019 [29] prepared carboxymethyl cellulose/polyvinyl alcohol films modified with zeolite which was firstly doped with Ag and Au ions. It was shown that the addition of this modified zeolite enhanced the mechanical, barrier and antimicrobial properties of the obtained packaging films. Recently, Nascimento Souza et al., 2020 [30] prepared chitosan packaging film and used zeolite as an ethylene scavenger. To the best of our knowledge, there is no study on the use of natural zeolite (NZ) as nanoreinforcement and/or essential oil nanocarrier in ALG based film preparation.

Cottage cheese is a highly consumed type of cheese which, however, is easily acidified. Because of its high moisture content i.e., about 75%, and pH values over 4.5, the shelf-life of this product is restricted to 15 days [31]. It is well documented that various types of spoilage bacteria, yeasts, and molds that may develop on the cheese surface during storage can influence the shelf life of cheese, particularly in the case of soft and spread cheese. To prevent damage and spoilage, soft and fresh cheese are currently packaged with active cheese technology [32].

In this study natural zeolite (NZ) was firstly modified with thyme essential oil (TO) and produced a novel TO@NZ hybrid nanostructure. These nanostructures were characterized with XRD analysis and FTIR spectrometry. They were directly added to sodium alginate (ALG) plasticized with glycerol (G) hydrogels and produced novel ALG/G/TO@NZ active packaging films. The TO@NZ hybrid nanostructure content was fixed to 5, 10, and 15% wt. The properties of these films were compared with the properties of films prepared with pure NZ. The obtained ALG/G/NZ and ALG/G/TO@NZ films were also characterized with XRD analysis and FTIR spectrometry. Moreover, they were tested for their mechanical and water/oxygen barrier properties. The antioxidant and antimicrobial capacity of the obtained films was also evaluated. Finally, the most active films were used as packaging films to extend the shelf-life of soft cheese. The innovation of the current study can be summarized in the following three points: (1) modification/preparation of a rich in thymol content NZ nanostructure via a green evaporation/adsorption method, (2) development of edible active packaging films based on an ALG/G biopolymer matrix by using the novel TO@NZ nanostructure, and (3) use of such edible active packaging films to extend the self–life of a soft cottage-cheese.

2. Results and Discussion

2.1. GC-MS Results

The results of the GC-MS analysis of the TO as received and of the remaining after the first stage distillation process TO are summarized in Tables S1 and S2. The main substances of the TO as received are p-cymene 12.3%, D-limonene 16.5%, and thymol 56.7% (see Table S1). At the end of the first stage distillation process, the remaining TO does not contain p-cymene and D-limonene and contains 86.7% thymol (see Table S2).

2.2. DSC Results

Figure 1 presents DSC plots of pure NZ, TO@NZ, and TO_NZ hybrid nanostructures in the range of 50–250 °C.

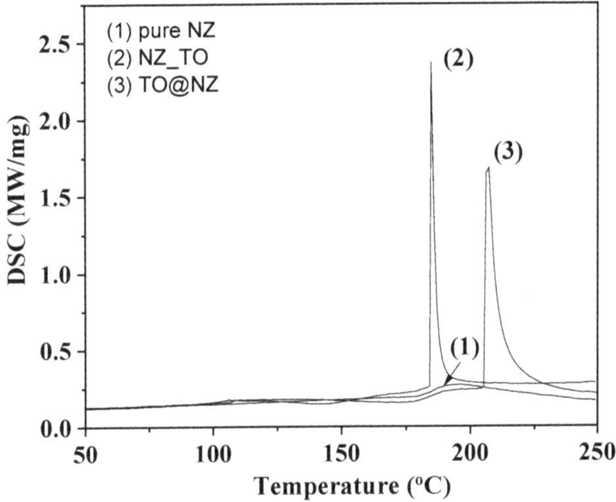

Figure 1. DSC plots of (1) pure NZ, (2) TO_NZ hybrid nanostructure and (3) TO@NZ hybrid nanostructure.

All curves exhibit a small, wide, and broad peak starting at around above 100 °C. This peak is attributed to the exothermic water evaporation process. In the case of TO_NZ and TO@NZ hybrid nanostructures two sharp exothermic peaks at approximately 185 °C and 230 °C are observed correspondingly. Additionally, in the case of TO_NZ hybrid nanostructure, there is also a wide broad peak starting above 200 °C. Precisely in the case of TO_NZ hybrid nanostructure the sharp peak at 185 °C does not ends and continues above 200 °C. This peak at 185 °C corresponds to p-Cymene and D-Limonene molecules' evaporation while the peak at 230 °C corresponds to the thymol molecules' evaporation [33–35]. This observation indicates that molecules that existed in the TO_NZ hybrid nanostructure are of different kinds of molecules that existed in the TO@NZ hybrid nanostructure. In the case of TO_NZ hybrid nanostructures, Limonene, Cymene, and Thymol molecules were adsorbed while in the case of TO@NZ hybrid nanostructures the Limonene and Cymene were removed from TO during the distillation process, and thymol was mainly adsorbed. Additionally, in the case of TO_NZ hybrid nanostructures, higher amounts of Limonene and Cymene molecules were adsorbed.

The DSC results indicate that the TO evaporation process led to the adsorption of higher quantities of D-Limonene and p-Cymene than thymol on the NZ surface (TO_NZ hybrid nanostructure). This happens probably because of their lower evaporation temperature. On the contrary, when D-Limonene and p-Cymene molecules were removed from TO via the distillation process, thymol is the main substance which was adsorbed on the NZ surface (TO@NZ hybrid nanostructure). In other words, from DSC plots, it is clear that the TO_NZ hybrid nanostructure was rich in D-Limonene and p-Cymene molecules while TO@NZ was rich in thymol molecules.

2.3. XRD Analysis

The XRD plots of the as received NZ and of the modified TO@NZ hybrid nanostructure are presented in Figure 2.

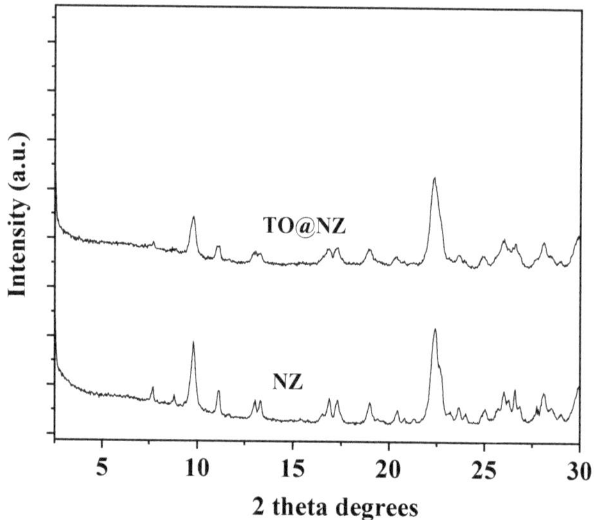

Figure 2. XRD plots of as received NZ and TO@NZ hybrid nanostructure.

The observed reflections in patterns of both NZ and TO@NZ materials are attributed to Heulandite $Ca(Si_7Al_2)O_{16} \times 6H_2O$ monoclinic crystal phase (PDF-41-1357). This means that the adsorption of TO into NZ did not affect the crystal phase.

The XRD plots of pure ALG/G, ALG/G/xNZ, and ALG/G/xTO@NZ nanocomposite films are presented in Figure 3 (where x is the nanostructure composition).

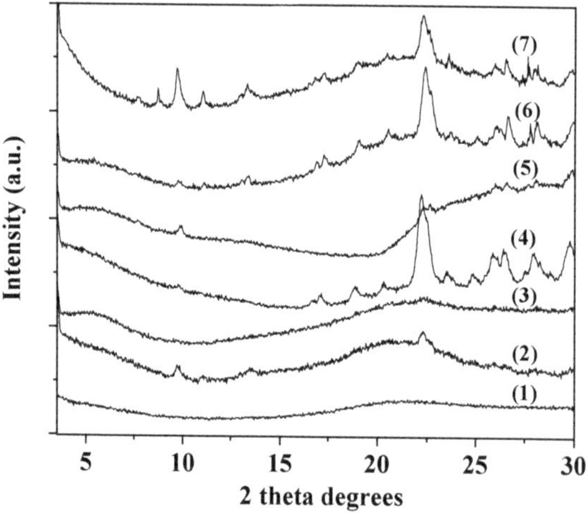

Figure 3. XRD plots of: (1) ALG/G, (2) ALG/G/5NZ, (3) ALG/G/10NZ, (4) ALG/G/15NZ, (5) ALG/G/5TO@NZ, (6) ALG/G/10TO@NZ AND (7) ALG/G/15TO@NZ obtained films.

Unplasticized alginate films exhibited two broad peaks with central positions at $2\theta = 13.5°$ and 21.6° [36,37]. As it is observed in Figure 3, in the case of such ALG/G films the peak at $2\theta = 13.5°$ disappeared indicating a lower proportion of the amorphous structure with larger chain distances. This is a result of water and glycerol de-structuration [36]. No changes of the ALG crystallinity were observed with the addition of either NZ or TO@NZ hybrid nanostructures. Moreover, after an initial additive loading into the polymeric matrix, as the % wt. content of NZ or TO@NZ hybrid nanostructure increases the reflections of zeolite's crystal phase are increase. This indicates that the higher dispersion of such materials in the ALG/G film is obtained only for low % wt. loadings i.e., <10% wt.

2.4. FTIR Spectroscopy

Line (1) in Figure 4 represents the FTIR spectra of the as received TO. In the same figure, Line (2) is assigned to the as received natural zeolite FTIR spectra, and Line (3) to the modified rich in thymol natural zeolite TO@NZ.

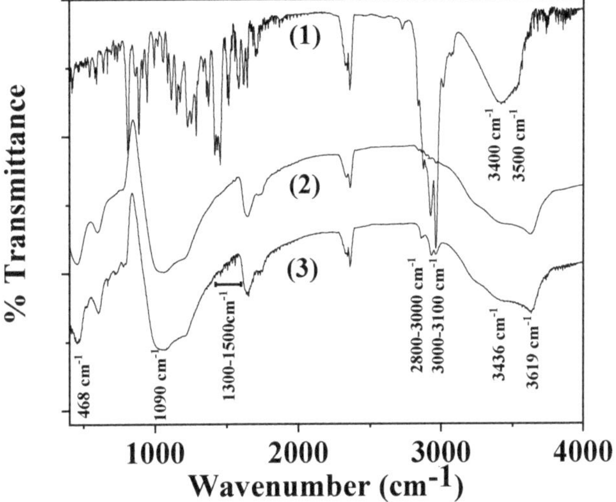

Figure 4. FTIR plots of: (1) TO as received, (2) NZ as received, (3) modified TO@NZ hybrid nanostructure.

In the FTIR plot of the TO material, i.e., Line (1) in Figure 4, the bands at ~3400 cm^{-1} and at ~3500 cm^{-1} are assigned to hydrogen-bonded OH stretching, and the bands at ~3100–3000 cm^{-1} are devoted to aromatic and alkenic C-H=C-H stretch vibrations. Three more bands are observed between 2800 and 3000 cm^{-1} which are attributed to the C-H stretch vibration of aliphatic CH_2 bonds. These bands are the strongest evidence that TO was adsorbed on NZ substrate because they are not covered by pristine NZ bands. For wavenumbers between 1500 cm^{-1} and 1300 cm^{-1} as well as for lower than 1000 cm^{-1}, there are several bands assigned to TO which are attributed to the C-H bending of the aliphatic CH_2 groups and C-O-H bending. These bands do not be overlapped with those of NZ. spectra and thus they could be visible in an NZ spectrum with adsorbed TO [15,38].

In the FTIR plots of NZ and TO@NZ powders ((see line (2) and line (3) in Figure 4), the bands at 3619 and 3436 cm^{-1} are assigned to the OH group stretching mode. The band at 1650 cm^{-1} corresponds to the OH group bending mode. The band at 1090 cm^{-1} to the Si-O stretching vibration and at 468 cm^{-1} to the -SiO_4- bending mode [39–41]. It is obvious from the FTIR plot of TO@NZ that characteristic bands of TO exist in the range of 2800–3100 cm^{-1}, 1300–1500 cm^{-1}, and 500–1000 cm^{-1}. This indicates that the adsorption of TO molecules in the NZ occurs. The absence of band shift between NZ and TO@NZ plots means that the adsorption process is rather a physisorption than chemisorption.

Line (1) in Figure 5 depicts the FTIR plots of ALG/G while Line (2) shows the FTIR spectra of ALG/G/NZ material, and Line (3) of ALG/G/TO@NZ films.

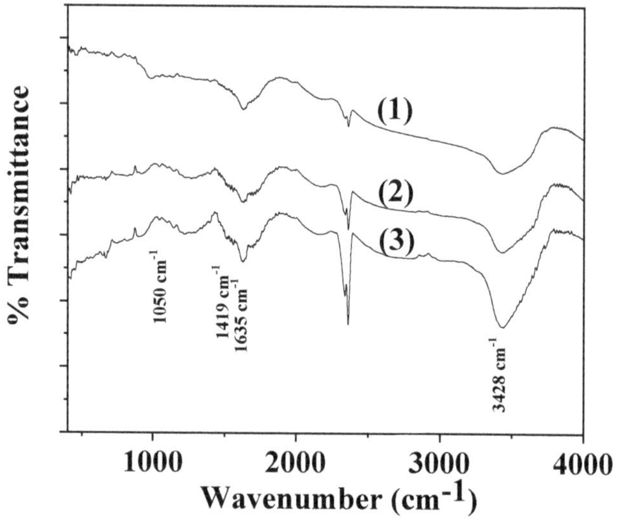

Figure 5. Representative FTIR spectra of (1) ALG/G, (2) ALG/G/NZ, (3) ALG/G/TO@NZ obtained films.

Line 1 of Figure 5 corresponds to the FTIR plot of pure ALG/G film while line 2 is a representative FTIR plot of ALG/G/10NZ. Finally, line 3 is assigned to FTIR measurements of ALG/G/10TO@NZ nanocomposite films. The characteristic sodium-alginate peaks are observed in all plots. A broad band at 3.428 cm^{-1} is assigned to hydrogen-bonded O–H stretching vibrations [42]. The band at 1635 cm^{-1} is attributed to the asymmetric stretching vibration of COO groups, the band at 1419 cm^{-1} to the symmetric stretching vibration of COO groups, and the band at 1050 cm^{-1} to the elongation of C-O groups [43]. It is obvious from lines (2) and (3) that the addition of NZ and TO@NZ hybrid nanostructures causes an increase to the bands at 3.428 cm^{-1} and 1635 cm^{-1} which could be attributed to strong interactions of ALG chains with NZ and TO@NZ hybrid nanostructures. This interaction is higher in the case of the TO@NZ hybrid nanostructure. Thus, it is revealed that modified TO@NZ hybrid structure interacts better with ALG/G matrix compared to the relevant of the pure NZ material. Furthermore, the absence of TO peaks in ALG/G/10TO@NZ spectra indicates that the TO molecules are not in the surface but in the inner area of the ALG/G matrix and supports the relaxation between the NZ material and the ALG/G matrix.

2.5. SEM Images

A SEM instrument equipped with an EDS detector was used to investigate the surface/cross-section morphology of the pure ALG/G film as well as of the ALG/G/xNZ and ALG/G/xTO@NZ hybrid nanocomposite films. The results confirmed that the NZ and the TO@NZ hybrid nanostructures were homogeneously dispersed in the ALG/G polymeric matrix. The chemical elements contained in the pure and final nanocomposite active packaging films were identified by carrying out EDS analysis on the surface of the materials.

The SEM images in Figure 6a,b show the expected smooth morphology inside and outside of the neat ALG/G polymer matrix. The EDS spectra in Figure 6c certify the existence of carbon (C), oxygen (O), and sodium (Na) on the surface of such films which is expected because of the sodium alginate. Figures 7e, 8e and 9e show EDS chemical analysis of nanocomposite active packaging films with different concentrations of pure NZ and TO@NZ hybrid nanostructure i.e., 5, 10, and 15% wt. In addition to the above mentioned presence of (C), (O), and (Na), the presence of typical elements, such as Si, Al, Fe, K, and Ca, confirm the existence of NZ and TO@NZ in such nanocomposite films. Moreover, the increase of (Na) content of the ALG/G/xNZ films i.e., ~10% compared to the relevant content of the pure ALG/G films i.e., ~2% indicates the incorporation of the natural zeolite into the polymer matrix. Surface and relative cross-section images of ALG/G/xNZ and ALG/G/xTO@NZ with different ratios x of NZ and TO@NZ are presented in Figures 7–9.

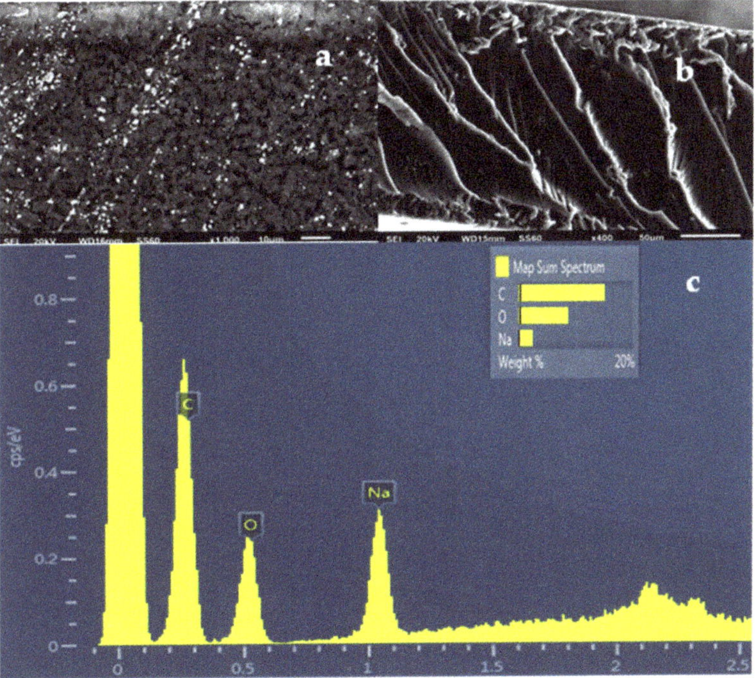

Figure 6. (**a**) SEM images of the surface and (**b**) cross-section for the pure film of ALG/G. (**c**) Energy dispersive spectrometer (EDS) spectrum and relative elemental analysis of the surface (inset) from the SEM image (**a**).

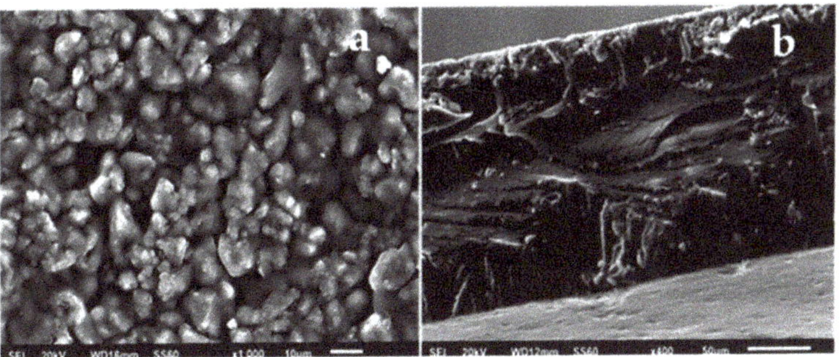

Figure 7. *Cont.*

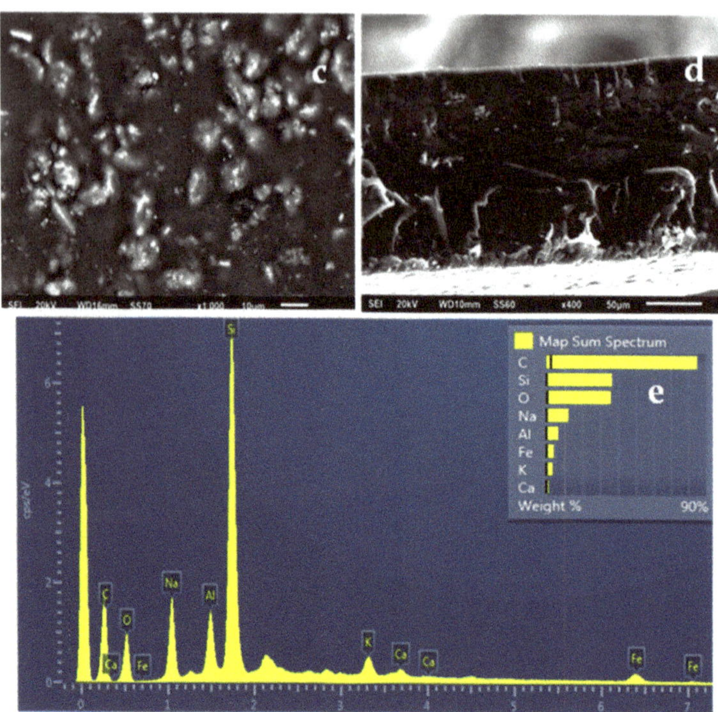

Figure 7. (**a**,**c**) SEM images of the surface and (**b**,**d**) cross-section for the nanocomposite films of ALG/G/5NZ (**a**,**b**) and ALG/G/5TO@NZ (**c**,**d**) respectively. (**e**) Energy dispersive spectrometer (EDS) spectrum and relative elemental analysis of the surface (inset) from the SEM image (**a**).

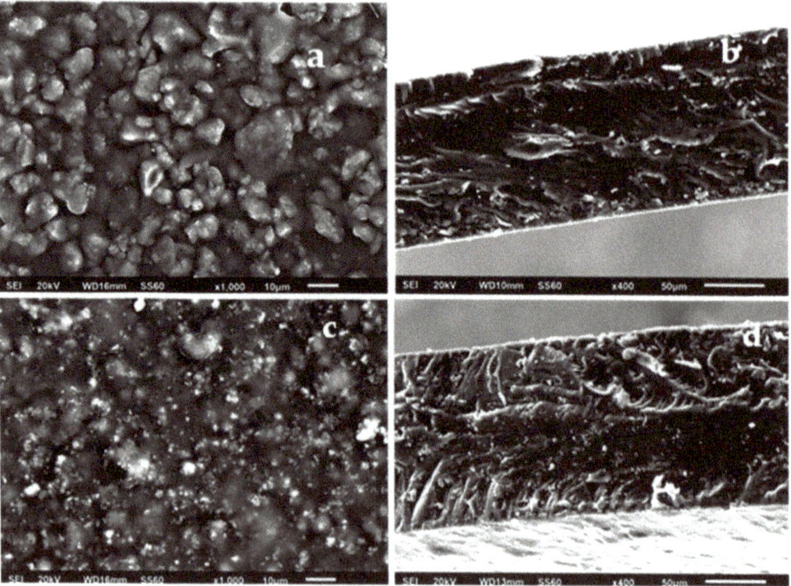

Figure 8. *Cont.*

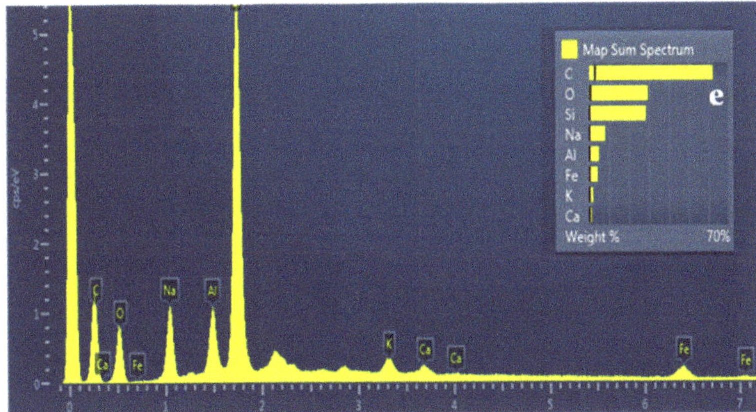

Figure 8. (**a**,**c**) SEM images of surface and (**b**,**d**) cross-section for the nanocomposite films of ALG/G/10NZ (**a**,**b**) and (**c**,**d**) ALG/G/10TO@NZ (**c**,**d**) respectively. (**e**) Energy dispersive spectrometer (EDS) spectrum and relative elemental analysis of the surface (inset) from the SEM image (**a**).

It is obvious from Figures 7–9 that after the incorporation into the polymer matrix, the increase of the content of NZ or TO@NZ nanocomposite material caused an increase to the aggregation degree. Nevertheless, SEM images of the final nanocomposite films show that the nanohybrids were homogeneously dispersed, which indicates their enhanced compatibility with the polymer matrix. Moreover, SEM surface and cross-section images were shown more homogenous dispersion in the case of TO@NZ hybrid nanostructure in nanocomposite films compared to the relevant of pure NZ. This means that the TO@NZ hybrid nanostructure was incorporated significantly better in the polymer matrix compared to the incorporation of the respective pure NZ.

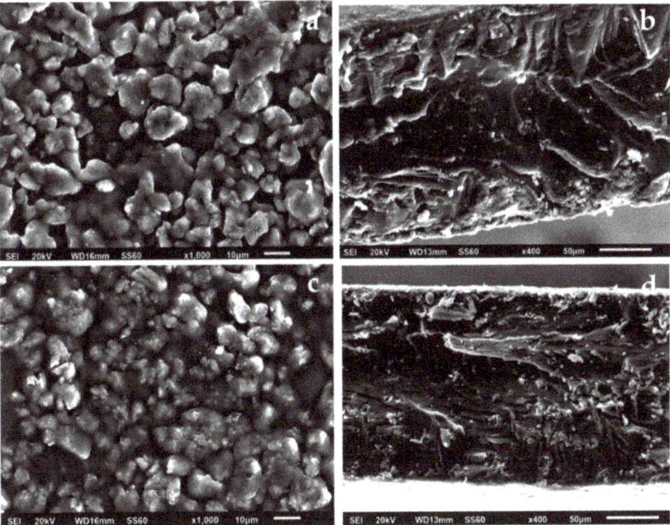

Figure 9. *Cont.*

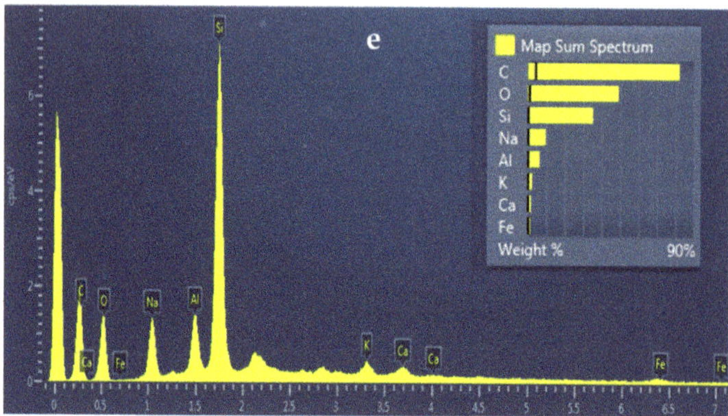

Figure 9. (**a**,**c**) SEM images of surface and (**b**,**d**) cross-section for the nanocomposite films of ALG/G/15NZ (**a**,**c**) and ALG/G/15TO@NZ (**c**,**d**) respectively. (**e**) Energy dispersive spectrometer (EDS) spectrum and relative elemental analysis of the surface (inset) from the SEM image (**a**).

2.6. Tensile Properties

The calculated values of elastic modulus (E), ultimate strength (σ_{uts}), and elongation at break (%ε) for all ALG/G/xNZ and ALG/G/xTO@NZ nanocomposite films are listed in Table 1.

Table 1. Calculated values of Young's (E) Modulus, ultimate tensile strength (σ_{uts}) and % strain at break (ε_b).

Code Name	E-Elastic Modulus (MPa)	σ_{uts} (MPa)	%ε
ALG/G	445.5 (63.8)	15.2 (2.4)	40.2 (4.7)
ALG/G/5NZ	755.6 (67.3)	22.7 (0.9)	24.7 (12.4)
ALG/G/10NZ	669.3 (24.3)	21.1 (5.9)	20.3 (2.7)
ALG/G/15NZ	785.3 (146.6)	23.1 (5.5)	23.1 (2.5)
ALG/G/5TO@NZ	739.4 (20.3)	20.9 (3.5)	28.4 (8.2)
ALG/G/10TO@NZ	651.5 (76.2)	18.5 (2.9)	28.3 (6.6)
ALG/G/15TO@NZ	798.5 (177.5)	22.6 (1.4)	25.3 (2.5)

It is obvious from Table 1 that the addition of both NZ and TO@NZ hybrid nanostructure increases stiffness and strength and decreases %elongation at break values. The nanocomposite film with the higher strength was the ALG/G/15NZ and ALG/G/15TO@NZ. This result is in accordance with previous reports where zeolite was successfully incorporated into polyethylene/caprolactone [28], cellulose [29,44], and chitosan [30] films as nano-reinforcement. The result also agrees with the FTIR morphological evaluation of such films where an interplay between NZ, TO@NZ hybrid nanostructures and ALG/G matrix was obtained. In general, ALG/G/TO@NZ based nanocomposite films exhibited higher elongation at break values than the ALG/G/NZ due to the presence of TO molecules which acted as plasticizers [22,45].

2.7. UV-vis Transmittance of Films

Figure 10a presents the images of as prepared pure ALG/G film as well as of ALG/G/xNZ and ALG/G/xTO@NZ nanocomposite films. Figure 8b presents the UV-vis transmittance plots of all obtained films. It is obvious from both images and UV-vis plots that the TO@NZ based films are more transparent than NZ based counterparts. Higher transparency indicates higher nanofiller dispersion and integration inside the polymer matrix. Thus, it seems that the TO enhances the dispersion of the NZ in ALG/G/xTO@NZ nanocomposite films. Moreover, the lowest transparency is obtained for ALG/G/15NZ and ALG/G/15TO@NZ films.

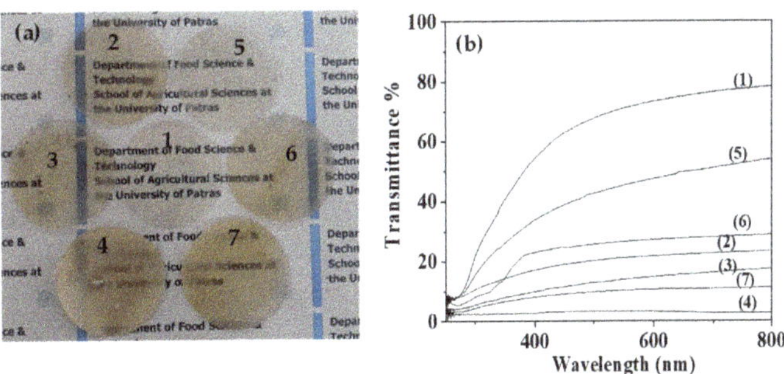

Figure 10. (**a**) photo images of all prepared films, (**b**) UV-vis transmittance of all prepared films. (1) pure ALG/G, (2) ALG/G/5NZ, (3) ALG/G/10NZ, (4) ALG/G/15NZ, (5) ALG/G/5TO@NZ, (6) ALG/G/10TO@NZ and (7) ALG/G/15TO@NZ films.

2.8. Water-Oxygen Barrier Properties

Water vapor transmission rate (WVTR) and oxygen transmission rate (OTR) values for all ALG/G/xNZ and ALG/G/xTO@NZ films are listed in Table 2. Using these values, water vapor diffusivity (D_{WV}) and oxygen permeability (Pe_{O2}) values were calculated and listed in the same table.

Table 2. Measured values of water-vapor transmission rate WVTR and oxygen transmission rate OTR. Calculated values of water diffusion coefficient D_{WV} and oxygen permeability coefficient Pe_{O2} for all obtained films.

Code Name	Film Thickness (mm)	WVTR (10^{-6}) ($gr \cdot cm^{-2} \cdot s^{-1}$)	D_{WV} (10^{-4}) ($cm^2 \cdot s^{-1}$)	OTR (10^{-4}) ($ml \cdot cm^{-2} \cdot day^{-1}$)	Pe_{O2} (10^{-7}) ($cm^2 \cdot s^{-1}$)
ALG/G	0.11 (0.01)	2.45 (0.14)	5.96 (0.45)	111,939 (234)	15.70 (0.21)
ALG/G/5NZ	0.07 (0.01)	3.00 (0.45)	5.40 (0.11)	70,476 (124)	5.71 (0.10)
ALG/G/10NZ	0.08 (0.01)	2.31 (0.21)	3.97 (0.42)	85,456 (174)	9.23 (0.19)
ALG/G/15NZ	0.09 (0.01)	2.47 (0.25)	5.55 (0.67)	127,556 (435)	15.3 (0.52)
ALG/G/5TO@NZ	0.08 (0.01)	2.36 (0.21)	4.06 (0.11)	93,984 (205)	9.06 (0.20)
ALG/G/10TO@NZ	0.10 (0.01)	2.26 (0.16)	5.16 (0.23)	90,549 (345)	6.99 (0.27)
ALG/G/15TO@NZ	0.13 (0.01)	2.28 (0.27)	6.89 (0.74)	78,476 (234)	11.8 (0.35)

Observing the water diffusivity values, we can conclude that the addition of NZ initially caused a reduction to water diffusivity i.e., a minimum value of 3.97 cm^2/s to the content of 10% wt. Beyond that, the water vapor diffusivity starts to increase and, for NZ content 15% wt., its value became almost equal to the relevant of the initial raw material. In the case of TO@NZ hybrid nanostructure the minimum water diffusivity value observed for 5% wt. TO@NZ content and for 15% wt. content the water vapor diffusivity was higher than the relevant of the initial raw material. In general, 10% wt. NZ content in the polymer matrix exhibits a water vapor barrier almost equal to the relevant of 5% wt. TO@NZ content and lower enough compared to the relevant of the initial raw material. Beyond these concentrations, the extra addition of such nanostructured materials to the polymer matrix caused an increase to the water vapor diffusivity. Concerning the oxygen permeability, the minimum coefficient values compared to the relevant of the initial raw material were observed for 5% wt. NZ content and for 10 %wt. TO@NZ content. Increasing these concentrations, the oxygen permeability starts to increase. Considering Table 2 and according to the previous observations, we could say that the optimum NZ or TO@NZ additive concentration for the highest water vapor or oxygen barrier lies in the range of 5–10% wt.

2.9. Antioxidant Activity of Films

Antioxidant activity values of all the tested films were measured following the diphenyl-1-picrylhydrazyl (DPPH) assay method and are listed in Table 3.

Table 3. Antioxidant activity of all obtained films.

Code Name	% Antioxidant Activity 24 h
ALG/G	8.7 (0.6)
ALG/G/5NZ	7.7 (2.7)
ALG/G/10NZ	10.4 (1.9)
ALG/G/15NZ	15.1 (2.0)
ALG/G/5TO@NZ	17.3 (1.3)
ALG/G/10TO@NZ	25.3 (3.9)
ALG/G/15TO@NZ	46.4 (4.3)

It is already known that Sodium alginate exhibits antioxidant activity [46,47]. According to our measurements, the antioxidant activity of the pure ALG/G film was 8.7%. In the case of NZ incorporation into the polymer matrix, this activity was increased by the addition of 10 %wt. material and over. According to literature, NZ acts as an antioxidant agent due to the entrapment of free radicals in its porous [48]. The microporosity presence in the NZ pore structure enhances the adsorption properties of this material and consequently the antioxidant activity [49]. Thus, the increased antioxidant activity of the NZ containing films measured in this study could be attributed to the adsorption of DPPH ions and free radicals in its structure. According to Table 3 values, even the lower TO@NZ load exhibits stronger antioxidant activity compared to the relevant value of the film with the higher NZ content. This happens because essential oils and especially TO exhibit significant antioxidant activity [50]. As an overall conclusion, we could say that, based on Table 3 values, ALG/G/15TO@NZ films show the highest antioxidant activity.

2.10. Antimicrobial Tests

2.10.1. MICs and MBCs Determination of TO@NZ Hybrid Nanostructure against LAB and Pathogen Bacteria

The MICs and MBCs of TO@NZ against different bacteria are presented in Table 4.

Table 4. Minimum inhibitory concentrations (MIC) and minimum bactericidal concentrations (MBC) of TO@NZ hybrid nanostructure against lactic acid bacteria and pathogens (% *w*/*v*).

Bacteria (10^6 cfu/mL)	MIC	MBC
Lactococcus lactis ssp. *lacts* ACA-DC127	0.025 (0.1)	0.025 (0.1)
S. thermophilus ACA-DC112	0.025 (0.1)	0.025 (0.1)
S. aureus ATCC1538	0.05 (0.4)	0.05 (0.4)
L. monocytogenes NCTC10527	0.1 (0.2)	0.1 (0.2)
E. faecalis EF1	0.1 (0.3)	0.1 (0.4)

TO@NZ hybrid nanostructure exerted antimicrobial activity against all tested bacteria for at least 2 days, although its effectiveness was dependent on the targeted strain. It was observed that both the lactic acid bacteria were inhibited at 0.025% concentration, which was the lowest MIC compared with that of the pathogenic strains. However, the MICs measured for the pathogenic strains were two times higher (0.05%) for *S. aureus* and 20 times higher (0.1) for *L. monocytogenes* and *Enterococcus faecalis* strains. These results indicate that the hybrid nanostructure can inactivate LAB and presumptive pathogenic microorganisms associated with spoilage and the safety of dairy products. Similar results have been reported [51] for the antimicrobial activity of thyme essential oil in Coahlo fresh cheese, where the MIC of thyme oil on pathogenic *S. aureus* and *L. monocytogenes* was two times higher (2.5 μL/mL) than that (1.25 μL/mL) for starters *L. lactis* ssp. *lactis* and *L. lactis* ssp. *cremoris*. These results suggest that the doses of EO thyme alone and in the form of TO@NZ hybrid nanostructure to control pathogenic bacteria should affect the growth and survival of starter cultures. On the other hand, carvacrol and thymol have been reported as the most inhibitory essential oils against non-starter lactic starter bacteria (NSLAB) with MICs of 0.1% (*w*/*v*) [52]. Reduction of 2-log CFU/mL against *L. buchneri* and

P. acidilactici was achieved for thymol 0.1% (w/v) while this concentration was bactericidal against L. citrovorum (>4-log reduction). These results indicate that thymol oil can inactivate the non-starter LAB. According to the common opinion, such bacteria cause spoilage of shelf-stable low-acid dairy products. In particular, antimicrobial activity of thyme oil on milk and dairy products was established by previous studies [53,54].

2.10.2. Antimicrobial Activity of Active Films Application on Cheese against *S. aureus*

The results for the antimicrobial activity of hybrid nanostructured films used for packaging of cottage cheese under 10 °C storage temperature are shown in Figure 11.

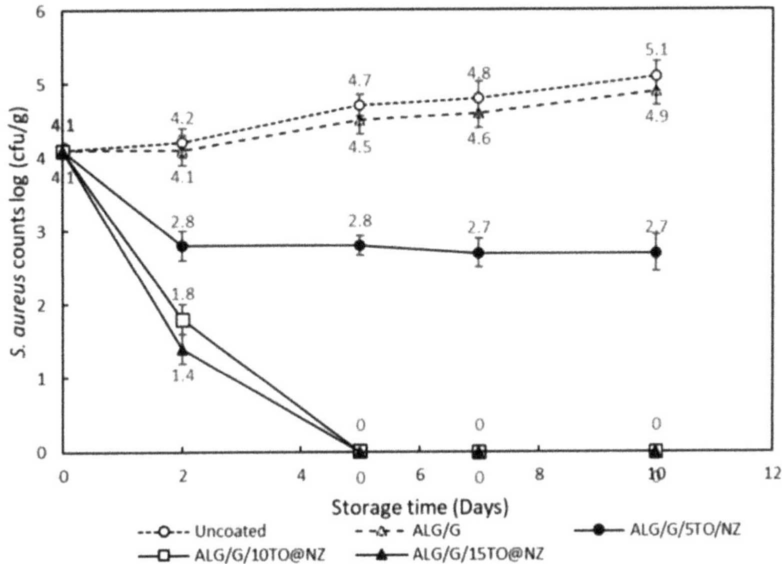

Figure 11. Changes in *Staphylococcus aureus* ATCC1534 counts in cottage cheese samples wrapped with alginate-based edible films (ALG/G/xTO@NZ) with the hybrid nanostructure TO@NZ in different concentrations.

Given the strong antibacterial activity of the TO@NZ bioactive hybrid nanostructure against the pathogen strains, the prepared ALG/G/TO@NZ active films were studied for the antimicrobial activity against the pathogenic bacteria *S. aureus* ATCC1538 in cottage fresh cheese, as a frequently associated toxinogenic bacteria with fresh or low-ripened cheeses. During 15 days of storage of cheese at an abuse temperature of 10 °C, it was found that *S. staphylococcus* ATCC1538 levels increased in the control uncoated sample and the sample coated with ALG/G film. However, a significant decrease of this bacterium concentration occurred in all cheese samples coated with ALG/G/xTO@NZ films. Furthermore, the antimicrobial activity was significantly related with films containing essential oils ($p < 0.05$). The initial count of *S. aureus* ATCC1538 in the uncoated control sample was 4.25 $\log_{10}$ cfu/g on the 1st day of storage and this value significantly increased ($p < 0.05$) to 4.75 (0.50 $\log_{10}$ cfu/g increase) by the 15th day of storage. However, in the case of the coated sample with the ALG/G film without the TO@NZ hybrid nanostructure, there was a slight decrease (not statically significant) of count to 4.1 $\log_{10}$ (cfu/g), which shows a bacteriostatic activity on *S. aureus* strain. The application of ALG/G/5TO@NZ active film on cheese resulted in a significant reduction of about 1-$\log_{10}$ (cfu/g) ($p < 0.05$) against the *S. aureus* ATCC1538 population from the 2nd day of storage, which remained at this level during the whole storage period (15 days), indicating a bacteriostatic activity. However, in the case of ALG/G/10TO@NZ and ALG/G/15TO@NZ active films application on cheese, a significant decrease of 2 $\log_{10}$ (cfu/g) of *S. aureus* population was counted from the

2nd day of storage and from the 5th day no counts detected up to 15th day, indicating a strong bactericidal effect of both active films against the pathogen. Similarly, a significant relationship has been detected between the thyme-fortified edible whey isolate-based films (WPIOF) and *S. aureus* inactivation in Kashar cheese [55]. Opposite of the studies in which the essential oils were added directly to foods, the number of studies in which the essential oils were used as a composite of packaging materials are fewer [56]. As the time passes or/and the temperature increases the possibility for food illness to arise increases, given that an *S. aureus* population of 10^5 cfu/g consists the threshold for alpha-staphylotoxin production in cheese [56]. In our study, the cheese was strongly protected from Staphylococcal intoxication even under abusing storage temperature of 10 °C. However, we should consider that the application of ALG/G/TO@NZ active films in fermented dairy products (such as cheeses) in doses enough to control pathogenic bacteria could also influence the growth and survival of lactic acid bacteria, which probably affect the post-acidification activity and improve the product's self-life [57].

3. Conclusions

In conclusion, we could say that NZ or TO@NZ nanostructures were incorporated perfectly with the ALG/G polymer matrix and provide a very promising active film for food packaging, which could extend the soft cheese shelf-life. Furthermore, the effort to develop a more environmentally friendly process seems to be successful because the new materials were based on natural raw materials with reduced use of chemicals. The overall success of this study was confirmed by the XRD and FTIR results while FTIR and DSC indicate a film rich in thymol oil physiosorbed in NZ nanocomposite. Consistent with the antimicrobial, antibacterial, and antioxidant measurements, the higher the thymol concentration, the better antimicrobial, antibacterial, and antioxidant results. Actually, there is a threshold of TO@NZ hybrid nanostructure concentration which is required for the film to exhibit inhibition and bactericidal activity against the tested bacteria and pathogens. This also improves the mechanical properties of the film because TO acts as a plasticizer. Nevertheless, in line with SEM and UV-vis results, the increase of TO@NZ concentration led to lower transparency and higher nanostructure aggregation in the film. Thus, there is an optimum for TO@NZ concentration in the polymer matrix and this is in the range between 5% wt and 10% wt. In this range, the D_{WV} and Pe_{O2} coefficients exhibit a local minimum which means the highest water and oxygen barrier.

4. Materials and Methods

4.1. Materials

Edible Natural Zeolite was purchased by a local pharmacy market. Sodium Alginate was purchased from Acros-Organics (Zeel West Zone 2, Janssen Pharmaceuticalaan 3a B2440 Geel, Belgium). Glycerol was purchased from Carlo-Erba (Denzlinger Str. 27, 79312 Emmendingen, Germany). Thyme oil was purchased from a local pharmacy market and produced by Chemco (Via Achille Grandi, 13–13/A, 42030 Vezzano sul Crostolo RE, Italy). The cottage cheese which used during the experimental process was purchased from a local Greek market with the brand name TYRAS, TYRAS S.A., Trikala, Thessalia, Greece, and immediately transferred to the laboratory refrigerator and stored under a temperature of 4 °C. According to the products specification, the expiration date of the product was 12 days after the production date. Moreover, the product expired in 4 days if the package film opened. Furthermore, the product could be stored for 75 days if it was freezed at temperature below −16°C and with closed packaging film. The PCA was found 7.52 cfu/g, while pH was determined at 4.9.

4.2. Thyme oil GC-MS Analysis

An amount of 1 μL thyme oil/pentane dilution with 2 mg/mL concentration was injected into the GC/MS (in splitless mode). The analysis of each sample was carried out three times. The used GC instrument was a Finnigan Trace GC Ultra 2000 equipped with a

Finnigan Trace DSQ MSD (Thermo Electron Corporation, Waltham, MA, USA) detector which was operated in EI mode. The conditions of separation process inside the capillary column HP-5MS (Agilent Technologies, USA) (30 m × 0.32 mm, 0.25 µm film thickness) were as follows: carrier gas Helium, flow rate 0.8 mL/min, initial temperature of column 60 °C, and the separation temperature ramp was achieved heating the column to 240 °C at a rate of 3 °C/min for 10 min. The detector voltage and temperature were kept at 70 eV and 250 °C respectively while the injector temperature was adjusted to 200 °C. ADAMS, Wiley275, NIST, and in-house created libraries were used for the compounds' identification. The method for this identification was by comparing the retention times and the mass spectra of volatiles [58,59].

4.3. Preparation of TO@NZ Hybrid Nanostructures

The preparation of TO@NZ bioactive hybrid nanostructures was carried out based on the method reported elsewhere for nanoclays modification [15] with some small changes. In a spherical flask, an amount of 20 mL TO were heated to boil. Before its use, the thyme oil was freed up from the Limonene and Cymene by distillation process at 190 °C. An amount of 6 mL was removed which is approximately 30% of the total amount. This result is in close agreement with the GC measurements data. The distillation process was the first stage of the overall process which was followed in this study. The distillation temperature was chosen a little higher than Limonene's and Cymene's boiling point temperatures (170–175 °C) and lower than thymol's boiling point temperature (230–235 °C). After the first stage, the remaining liquid colour was changed to dark brown and in line with LC-MS/MS measurements, it was rich of thymol oil. In the second step, an evaporation-adsorption process was carried out. The remained liquid from the first stage was boiled up to 250 °C. The produced thymol rich steam passed through a bed of 3 gr NZ which strongly adsorbed the essential oil. The remaining outlet steam was cooled down through a condenser and the final liquid was collected. At the same time, an amount of as received thyme oil was adsorbed by NZ material and reference hybrid nanostructured material TO_NZ was prepared. The %wt. adsorbed TO in both TO@NZ and TO_NZ hybrid nanostructures was calculated gravimetrically and was approx. 80% wt. for both hybrid nanostructures.

4.4. Preparation of ALG/G/NZ and ALG/G/TO@NZ Active Films

A quantity of 2 g ALG and of 1 g G was spread into 100 mL of distilled water and heated until a homogeneous hydrogel was obtained. An appropriate amount of NZ or TO@NZ powder was dispersed in 10 mL of distilled water using a glass beaker. Then, the homogeneous ALG/G hydrogel was added gradually in the NZ or TO@NZ suspension and vigorously stirred for 2 h. The obtained ALG/G/NZ and ALG/G/TO@NZ hydrogels were then spread in petri dishes of 11 cm diameter and dried at 25 °C. The obtained films were pilled of and stored further in a desiccator at 25 °C and 50 %RH. In Table 5, the code names and the amounts of ALG, G, NZ, and TO@NZ, which were used to obtain such films, are summarized.

Table 5. Code names and amounts of ALG, G, NZ and TO@NZ used for the preparation of films.

Code Name	ALG (g)	G (g)	NZ (g)	TO@NZ (g)
ALG/G	2	1	-	-
ALG/G/5NZ	2	1	0.15	-
ALG/G/10NZ	2	1	0.30	-
ALG/G/15NZ	2	1	0.45	-
ALG/G/5TO@NZ	2	1	-	0.15
ALG/G/10TO@NZ	2	1	-	0.30
ALG/G/15TO@NZ	2	1	-	0.45

4.5. DSC Measurements

A DSC214 Polyma Differential Scanning Calorimeter (NETZSCH manufacturer, Selb, Germany) was used to study the thermal behaviour of the obtained TO@NZ hybrid and of the pure NZ nanostructures. Quantity 1.2–3.3 mg of samples was heated from 50 °C to 300 °C, under nitrogen atmosphere, and an increasing temperature rate of 10 °C/min.

4.6. XRD Analysis

An Brüker D8 Advance X-ray diffractometer instrument (Brüker, Analytical Instruments, S.A., Athens, Greece) was used for XRD measurements on ALG/G, ALG/G/NZ, and ALG/G/TO@NZ active films, as well as on NZ and TO@NZ hybrid nanostructure powders. The instrument was equipped with a LINXEYE XE High-Resolution Energy-Dispersive detector, and the measurements were carried out in the range $2\theta = 0.5–30°$ and with an increment of 0.03.

4.7. FTIR Spectrometry

The chemical structure of both the NZ as received and the modified TO@NZ hybrid nanostructure, as well as of the obtained ALG/G, ALG/G/xNZ and ALG/G/xTO@NZ films, was analyzed with IR spectra measurements. An FT/IR-6000 JASCO Fourier transform spectrometer (JASCO, Interlab, S.A., Athens, Greece) was used and the measurements were carried out in the frequency range of 4000–400 cm^{-1}. The measured infrared (FT-IR) spectra resulted from 32 scans with 2 cm^{-1} resolution.

4.8. SEM Images

The surface morphology characterization of the obtained films was performed with SEM images. For SEM image measurements a JEOL JSM-6510 LV SEM Microscope (Ltd., Tokyo, Japan) was used equipped with an X-Act EDS-detector by Oxford Instruments, Abingdon, Oxfordshire, UK (an acceleration voltage of 20 kV was applied).

4.9. Tensile Properties

Tensile measurements of all prepared ALG/G, ALG/G/xNZ and ALG/G/xTO@NZ films were carried out using a SimantzüAX-G 5kNt instrument (Simandzu. Asteriadis, S.A., Athens, Greece) and as reported by the ASTM D638 method. Three to five samples of each film were used for statistical analysis of tensile properties. The samples, with a shape of dumb-bell and gauge dimensions of 10 × 3 × 0.22 mm, were tensioned at an across head speed of 2 mm/min. Stress, stain, and modulus of elasticity values were calculated using force (N) and deformation (mm) measurements, and the gauge dimensions.

4.10. UV-vis Transmittance of Films

Absorbance properties of ALG/G, ALG/G/xNZ, and ALG/G/xTO@NZ films were estimated via UV–vis measurements using a Shimatzu 1900 spectrophotometer. Measurements were carried out in the range of 200 to 800 nm.

4.11. Water Vapor Diffusivity

ASTM E96/E 96M-05 method was followed to estimate the Water-Vapor Transmission Rate (WVTR) of ALG/G/xNZ and ALG/G/xTO@NZ films The used apparatus was handmade, and the experiments were carried out at 38 °C and 95% RH according to literature [16,23,60,61]. Film samples was of 2.5 cm diameter and 0.09 mm average thickness and placed on the top of an one-open end cylindrical tube made of plexiglass. The cylinder, which contained dried silica gel inside, was sealed by a rubber O-ring. The test tube was placed in a glass desiccator under an environment of 95% relative humidity (RH) at 38 °C. Such conditions were obtained placing 200 mL of saturated magnesium nitrate solution in this desiccator. Each tube was weighed periodically for 24 h. The

WVTR ($g \cdot cm^{-2} \cdot s^{-1}$) values was calculated using such weighting measurements and according to Equations (1) and (2):

$$\overline{\left(\frac{\Delta G}{\Delta t}\right)} = \frac{1}{n}\sum_{i=1}^{n-1}\left(\frac{G_{i+1} - G_i}{t_{i+1} - t_i}\right) \tag{1}$$

$$WVTR = \overline{\left(\frac{\Delta G}{\Delta t}\right)} \cdot \frac{1}{A} \tag{2}$$

where: $\overline{\left(\frac{\Delta G}{\Delta t}\right)}$ ($g \cdot s^{-1}$) is the mean water vapor mass diffusion rate for the overall experiment duration, G_i (g) is the device weight which balanced at the elapsed time t_i (s) from the experiment start, $\frac{G_{i+1}-G_i}{t_{i+1}-t_i}$ ($g \cdot s^{-1}$) is the water vapour mass diffusion rate for each time interval, (n) is the number of weight measurements, and A (cm^2) is the permeation area of the film. Additionally, the tested films were weighed before and after the WVTR test to exclude any absorption phenomena of humidity in the film. $\Delta G/\Delta t$ (g/s) is the water transmission rate through the film which is calculated by the experimental points (G_i, t_i).

WVTR is equal to the specific mass flow rate for the diffusion process through a membrane which could be calculated using Fick's law [62]:

$$\frac{J}{A} = D \cdot \frac{\Delta C}{\Delta x} \tag{3}$$

where J (g/s) is the mass flow rate of a component through the membrane, A (cm^2) is the membrane cross-sectional area permeated by this component, ΔC ($g \cdot cm^{-3}$) is the concentration gradient of this component on the two sides of the membrane, and Δx (cm) is the membrane thickness. the humidity concentration in the outer side of the cylinder is 4.36509×10^{-5} $g \cdot cm^{-3}$ (95% RH at 38 °C) and in the opposite side of the film the silica gel absorbs the 100% of the permeated water vapor, which agrees with the ASTM E96/E 96M-05 method, then $\Delta C = 4.36509 \times 10^{-5}$ $g \cdot cm^{-3}$. Considering that WVTR = J/A we can calculate the diffusion coefficient D ($cm^2 \cdot s^{-1}$) for every film via the combination of Equations (2) and (3) which leads to Equation (4):

$$D_{WV} = WVTR \cdot \frac{\Delta x}{\Delta C} \tag{4}$$

where WVTR [$g \cdot cm^{-2} \cdot s^{-1}$)] is the water-vapour transmission rate, Δx (cm) is the film thickness, and ΔC ($g \cdot cm^{-3}$) is the humidity concentration gradient on the two opposite sides of the film.

4.12. Oxygen Permeability

An oxygen permeation analyzer (8001, Systech Illinois Instruments Co., Johnsburg, IL, USA) was used to measure the oxygen transmission rate (OTR) through the tested films. Following the ASTM D 3985 method the experimental conditions were regulated at 23 °C and 0% RH and the measured OTR values were expressed in cc O_2/m^2/day.

Gas permeability through polymers can be expressed according to the literature [63], as follows:

$$\frac{J}{A} = Pe_{gas} \cdot \frac{\Delta C}{\Delta x} \tag{5}$$

where J/A ($mol \cdot cm^{-2} \cdot s^{-1}$) is the specific molar rate of gas which permeates through the membrane, Pe_{gas} ($cm^2 \cdot s^{-1}$) is the permeability coefficient of the gas, ΔC ($mol \cdot cm^{-3}$ STP) is the concentration gradient between the two opposite sides of the membrane, and Δx (cm) is the membrane thickness.

Rearranging Equation (5) we can express the permeability coefficient as follows:

$$Pe_{gas} = \frac{J}{A \cdot \Delta C} \cdot \Delta x \quad (6)$$

By carrying out a dimensional analysis, the term J/(A*ΔC) (cm^3 STP·cm^{-2}·s^{-1})) is equal to the OTR measurements. Thus, in our case of oxygen gas,

$$Pe_{O_2} = OTR \cdot \Delta x \quad (7)$$

where Pe_{O2} (cm^2·s^{-1}) is the oxygen permeability coefficient, OTR (cm^3 STP·cm^{-2}·s^{-1}) is the experimentally measured by the instrument oxygen transmission rate, and Δx (cm) is the mean film thickness.

Thus, the oxygen permeability coefficient values (Pe_{O2}) of the tested samples were calculated by multiplying the OTR values by the average film thickness which was 0.035 mm. The OTR value for each kind of film was the mean value of measurements on three different pieces.

4.13. *Antioxidant Activity*

To evaluate the antioxidant activity of obtained ALG/G, ALG/G/xNZ, and ALG/G/xTO@NZ films, 300 mg of each film were cut into small pieces and placed inside dark glass bottles with 10 mL of a 40 ppm ethanolic solution of diphenyl-1-picrylhydrazyl (DPPH). A dark glass bottle with 10 mL of ethanolic DPPH solution without the addition of any film was used as the reference sample. The absorbance at 517 nm wavelength of the DPPH solution was measured at 0 h and after 24 h incubation using a Jasco V-530 UV-vis spectrophotometer. For each kind of film, three different samples were measured and the statistical mean was achieved as the final measurement.

The % antioxidant activity after 24 h incubation of films was calculated according to the following equation:

$$\% \text{ Antioxidant activity} = \frac{Abs_{ref.} - Abs_{sample}}{Abs_{ref.}} \times 100 \quad (8)$$

4.14. *Antimicrobial Activity Tests*

4.14.1. Antimicrobial Activity of TO@NZ Bioactive Hybrid Nanostructure

The antimicrobial activity of the TO@NZ hybrid nanostructure was estimated by the minimum inhibitory concentration (MIC) against several bacteria. The MICs of TO@NZ were studied by the broth dilution assay in Brain Heart Infusion broth (BHI, Oxoid, UK) and the selected bacteria were two lactic acid bacteria, which commonly are used as starter cultures in cheese manufacturing, namely *Lactococcus lactis* sp. *lactis* ACA-DC127 and *Streptococcus thermophilus* ST112, and 3 pathogenic strains, namely *Staphylococcus aureus* ATCC1538, *Listeria monocytogenes* NCTC10527, and *Enterococcus faecalis* EF1. All the strains were obtained from the ACA-DC Culture Collection of Food Science and Human Nutrition Department of the Agricultural University of Athens. The strains were preserved on cryobeads in cryovials of a cryopreservation system (TSC Ltd., Queensway Industrial Estate, 3–4 Arkwright Way, Scunthorpe DN16 1AL UK) at −80 °C, and overnight cultures were grown in BHI broth at 30 °C for *L. Lactis* ACA-DC127, and 37 °C for *S. thermophilus* ST112 and all pathogenic strains. For the broth dilution assay, five BHI broths with TO@NZ concentrations of 0.01, 0.025, 0.05, 0.075, and 0.1% (*w*/*v*) were made. The selected bacteria were inoculated into BHI broths, in a final concentration of approximately 10^6 cfu/mL in the presence of the above different concentrations of TO@NZ hybrid nanostructure, and incubation was followed for 48 h at the appropriate temperature for each strain. Growth was assessed by measuring the absorbance at 625 mM and agar plating after incubation for 48 h at 30 °C for the *L. lactis* sp. *lactis* AC A-DC127 strain and at 37° for the other bacteria to determine viable cfu/mL. The MIC values were read as the lowest concentrations of

TO@NZ at which bacterial growth was completely inhibited. In addition, to determine the minimal bactericidal concentration (MBC), the dilution representing the MIC and at least two of the more concentrated TO@NZ dilutions were plated and enumerated to determine viable cfu/mL. The MBC value was estimated as the lowest concentration that demonstrated a reduction (such as 99.9%) in cfu/mL of the strain tested.

4.14.2. Antimicrobial Activity of Active Films Application on Cheese against *S. aureus*

The antimicrobial activity of active films on toxigenic strain *S. aureus* ATCC1538 growth in cottage cheese was studied considering the possible diffusion of essential oil from the film matrix into the soft cheese. The cottage cheese divided into 5 treatments consisting of (1) control uncoated sample (C); (2) sample coated with ALG/G film without the addition of the hybrid nanostructure active hybrid nanostructure TO@NZ; (3) sample coated with ALG/G/5TO@NZ film; 4) sample coated with ALG/G/10 TO@NZ films; and (5) sample coated with ALG/G/15TO@NZ. Cottage cheese for each treatment was weighed into 50±0.5 g portions and placed in a stomacher bag. The cheese was inoculated with a fresh suspension of pathogenic *S. aureus* ATCC1538 strain in a final count of 10^4 cfu/g. The cheese was homogenized for 2 min by the stomacher (Seward 400; UK) and placed into sterile plastic petri-dish (9 cm diameter) as a thin layer. The film materials prepared as described above were applied to the upper surfaces of the cheese except for control (C); and the plates after being covered with their lid; were held at 10 °C for 10 days and sampled at days 0; 2; 5; 7; 10. The choice of time and temperature abuse of samples was selected as the most common conditions of unproperly preserved foodstuffs in the retail market; causing foodborne illness. At each sampling interval, duplicate plates from each treatment were aseptically opened and a 10 g-portion was weighed in a Stomacher-bag and homogenized in sterile peptone salt solution (Merck; Germany) in a stomacher (Seward 400; UK) for 1 min to make the initial dilution (10^{-1}). Appropriate serial dilutions (10^{-1}–10^{-5}) were spread plated on Baird Parker agar with egg yolk tellurite emulsion (BPA; Merck; Germany) and incubated at 37 °C for 48 h for *Staphylococcus aureus* ATCC1538 counts.

4.15. Statistical Analysis

All the results listed in Tables 1–4 are the mean values of measurements carried out to three samples for every kind of film. The statistical software SPSS ver. 20 was used for the interpretation of the experimental data. Mean values of all the measured properties were extracted based on the assumption of a confidence interval C.I. = 95% which is the most common value. Thus, the value of the statistical significance level was a = 0.05. Standard deviation values are also presented in such tables within parentheses beside the mean value. Finally, hypothesis tests were carried out to confirm that considering a different kind of film, every property has a statistically different mean value. Because of non-positive normality tests for all datasets, the non-parametric Kruskal–Wallis method was used. The results show that the mean values of all properties are statistically unequal.

Supplementary Materials: The following supporting information can be downloaded at: https://www.mdpi.com/article/10.3390/gels8090539/s1, Table S1. LC-MS/MS analysis of thyme oil as received used for the modification of NZ based hybrids; Table S2. LC-MS/MS analysis of remaining thyme oil after the first stage distillation process for the modification of NZ based hybrids.

Author Contributions: Synthesis experiment design, A.E.G. and C.E.S.; characterization measurements and interpretation, A.E.G., D.M., A.A. (Apostolos Avgeropoulos), N.E.Z., C.P., G.A., S.G., A.K. and C.E.S.; paper writing, A.E.G. and C.E.S.; overall evaluation of this work, A.E.G. and C.E.S.; experimental data analysis and interpretation, A.E.G., C.E.S., A.K. and C.P.; XRD, FTIR, OTR, tensile measurements, antioxidant activity, and WVTR experimental measurements, A.E.G., A.A. (Apostolos Avgeropoulos), N.E.Z., S.G. and C.E.S.; antimicrobial activity tests, K.Z., A.A. (Anastasios Aktypis) and G.-J.N. All authors have read and agreed to the published version of the manuscript.

Funding: This research was funded by the funding program "MEDICUS", Project F.K. 81541, of the University of Patras, Greece.

Data Availability Statement: The datasets generated for this study are available on request to the corresponding author.

Acknowledgments: The authors would like to thank the Department of Business Administration of Food and Agricultural Enterprises, University of Patras, Greece, for the given access to the XRD, OTR, and tensile equipment of the Food Technology Laboratory.

Conflicts of Interest: The authors declare no conflict of interest.

References

1. Otto, S.; Strenger, M.; Maier-Nöth, A.; Schmid, M. Food Packaging and sustainability—Consumer perception vs. correlated scientific facts: A review. *J. Clean. Prod.* **2021**, *298*, 126733. [CrossRef]
2. Sundqvist-Andberg, H.; Åkerman, M. Sustainability governance and contested plastic food packaging—An integrative review. *J. Clean. Prod.* **2021**, *306*, 127111. [CrossRef]
3. Asbahani, A.E.; Miladi, K.; Badri, W.; Sala, M.; Addi, E.H.A.; Casabianca, H.; Mousadik, A.E.; Hartmann, D.; Jilale, A.; Renaud, F.N.R.; et al. Essential oils: From extraction to encapsulation. *Int. J. Pharm.* **2015**, *483*, 220–243. [CrossRef] [PubMed]
4. Biopolymer Hybrid Materials: Development, Characterization, and Food Packaging Applications | Elsevier Enhanced Reader. Available online: https://reader.elsevier.com/reader/sd/pii/S2214289421000442?token=88623980430759D15A97DA076B67FCA8D2C2F53B55226D28DF548BEDE7B959F217E9E60E6EB57ADFDB2BC71882737DF3&originRegion=eu-west-1&originCreation=20220125174316 (accessed on 25 January 2022).
5. Li, D.; Wei, Z.; Xue, C. Alginate-based delivery systems for food bioactive ingredients: An overview of recent advances and future trends. *Compr. Rev. Food Sci. Food Saf.* **2021**, *20*, 5345–5369. [CrossRef] [PubMed]
6. Okolie, C.L.; Mason, B.; Mohan, A.; Pitts, N.; Udenigwe, C.C. Extraction technology impacts on the structure-function relationship between sodium alginate extracts and their in vitro prebiotic activity. *Food Biosci.* **2020**, *37*, 100672. [CrossRef]
7. Hecht, H.; Srebnik, S. Structural characterization of sodium alginate and calcium alginate. *Biomacromolecules* **2016**, *17*, 2160–2167. [CrossRef] [PubMed]
8. Gómez-Mascaraque, L.G.; Martínez-Sanz, M.; Hogan, S.A.; López-Rubio, A.; Brodkorb, A. Nano- and microstructural evolution of alginate beads in simulated gastrointestinal fluids. Impact of M/G ratio, molecular weight and pH. *Carbohydr. Polym.* **2019**, *223*, 115121. [CrossRef] [PubMed]
9. Sánchez-González, L.; Vargas, M.; González-Martínez, C.; Chiralt, A.; Cháfer, M. Use of essential oils in bioactive edible coatings: A review. *Food Eng. Rev.* **2011**, *3*, 1–16. [CrossRef]
10. Irkin, R.; Esmer, O.K. Novel Food packaging systems with natural antimicrobial agents. *J. Food Sci. Technol.* **2015**, *52*, 6095–6111. [CrossRef]
11. Jayakumar, A.; Heera, K.V.; Sumi, T.S.; Joseph, M.; Mathew, S.; Praveen, G.; Nair, I.C.; Radhakrishnan, E.K. Starch-PVA composite films with zinc-oxide nanoparticles and phytochemicals as intelligent pH sensing wraps for food packaging application. *Int. J. Biol. Macromol.* **2019**, *136*, 395–403. [CrossRef]
12. He, X.; Hwang, H.-M. Nanotechnology in food science: Functionality, applicability, and safety assessment. *J. Food Drug Anal.* **2016**, *24*, 671–681. [CrossRef]
13. Jagtiani, E. Advancements in nanotechnology for food science and industry. *Food Front.* **2021**, *3*, 56–82. [CrossRef]
14. Yu, H.; Park, J.-Y.; Kwon, C.W.; Hong, S.-C.; Park, K.-M.; Chang, P.-S. An overview of nanotechnology in food science: Preparative methods, practical applications, and safety. *J. Chem.* **2018**, *2018*, e5427978. [CrossRef]
15. Giannakas, A.; Tsagkalias, I.; Achilias, D.S.; Ladavos, A. A novel method for the preparation of inorganic and organo-modified montmorillonite essential oil hybrids. *Appl. Clay Sci.* **2017**, *146*, 362–370. [CrossRef]
16. Giannakas, A.; Stathopoulou, P.; Tsiamis, G.; Salmas, C. The effect of different preparation methods on the development of chitosan/thyme oil/montmorillonite nanocomposite active packaging films. *J. Food Process. Preserv.* **2019**, *44*, e14327. [CrossRef]
17. De Oliveira, L.H.; Trigueiro, P.; Souza, J.S.N.; de Carvalho, M.S.; Osajima, J.A.; da Silva-Filho, E.C.; Fonseca, M.G. Montmorillonite with essential oils as antimicrobial agents, packaging, repellents, and insecticides: An overview. *Colloids Surf. B Biointerfaces* **2022**, *209*, 112186. [CrossRef]
18. Biddeci, G.; Cavallaro, G.; Di Blasi, F.; Lazzara, G.; Massaro, M.; Milioto, S.; Parisi, F.; Riela, S.; Spinelli, G. Halloysite nanotubes loaded with peppermint essential oil as filler for functional biopolymer film. *Carbohydr. Polym.* **2016**, *152*, 548–557. [CrossRef] [PubMed]
19. Saucedo-Zuñiga, J.N.; Sánchez-Valdes, S.; Ramírez-Vargas, E.; Guillen, L.; Ramos-deValle, L.F.; Graciano-Verdugo, A.; Uribe-Calderón, J.A.; Valera-Zaragoza, M.; Lozano-Ramírez, T.; Rodríguez-González, J.A.; et al. Controlled release of essential oils using laminar nanoclay and porous halloysite/essential oil composites in a multilayer film reservoir. *Microporous Mesoporous Mater.* **2021**, *316*, 110882. [CrossRef]

20. Applications of Halloysite Nanotubes in Food Packaging for Improving Film Performance and Food Preservation—Science Direct. Available online: https://www.sciencedirect.com/science/article/pii/S0956713521000141?via%3Dihub (accessed on 26 January 2022).
21. Cheikh, D.; Majdoub, H.; Darder, M. An Overview of clay-polymer nanocomposites containing bioactive compounds for food packaging applications. *Appl. Clay Sci.* **2022**, *216*, 106335. [CrossRef]
22. Giannakas, A. Na-montmorillonite vs. organically modified montmorillonite as essential oil nanocarriers for melt-extruded low-density poly-ethylene nanocomposite active packaging films with a controllable and long-life antioxidant activity. *Nanomaterials* **2020**, *10*, 1027. [CrossRef]
23. Giannakas, A.E.; Salmas, C.E.; Karydis-Messinis, A.; Moschovas, D.; Kollia, E.; Tsigkou, V.; Proestos, C.; Avgeropoulos, A.; Zafeiropoulos, N.E. Nanoclay and polystyrene type efficiency on the development of polystyrene/montmorillonite/oregano oil antioxidant active packaging nanocomposite films. *Appl. Sci.* **2021**, *11*, 9364. [CrossRef]
24. Villa, C.C.; Valencia, G.A.; López Córdoba, A.; Ortega-Toro, R.; Ahmed, S.; Gutiérrez, T.J. Zeolites for Food applications: A review. *Food Biosci.* **2022**, *46*, 101577. [CrossRef]
25. Dogan, H.; Koral, M.; İnan, T.Y. Ag/Zn zeolite containing antibacterial coating for food-packaging substrates. *J. Plast. Film Sheeting* **2009**, *25*, 207–220. [CrossRef]
26. Eimontas, J.; Striūgas, N.; Abdelnaby, M.A.; Yousef, S. Catalytic pyrolysis kinetic behavior and TG-FTIR-GC–MS analysis of metallized food packaging plastics with different concentrations of ZSM-5 zeolite catalyst. *Polymers* **2021**, *13*, 702. [CrossRef]
27. Lee, J.; Lee, Y.-H.; Jones, K.; Sharek, E.; Pascall, M.A. Antimicrobial packaging of raw beef, pork and turkey using silver-zeolite incorporated into the material. *Int. J. Food Sci. Technol.* **2011**, *46*, 2382–2386. [CrossRef]
28. Reščec, A.; Katančić, Z.; Kratofil Krehula, L.; Ščetar, M.; Hrnjak-Murgić, Z.; Galić, K. Development of double-layered PE/PCL films for food packaging modified with zeolite and magnetite nanoparticles. *Adv. Polym. Technol.* **2018**, *37*, 837–842. [CrossRef]
29. Youssef, H.F.; El-Naggar, M.E.; Fouda, F.K.; Youssef, A.M. Antimicrobial packaging film based on biodegradable CMC/PVA-zeolite doped with noble metal cations. *Food Packag. Shelf Life* **2019**, *22*, 100378. [CrossRef]
30. Do Nascimento Sousa, S.D.; Santiago, R.G.; Soares Maia, D.A.; de Oliveira Silva, E.; Vieira, R.S.; Bastos-Neto, M. Ethylene adsorption on chitosan/zeolite composite films for packaging applications. *Food Packag. Shelf Life* **2020**, *26*, 100584. [CrossRef]
31. Makhal, S.; Kanawjia, S.K.; Giri, A. Effectiveness of thymol in extending keeping quality of cottage cheese. *J. Food Sci. Technol.* **2014**, *51*, 2022–2029. [CrossRef]
32. Kontogianni, V.G.; Kasapidou, E.; Mitlianga, P.; Mataragas, M.; Pappa, E.; Kondyli, E.; Bosnea, L. Production, characteristics and application of whey protein films activated with rosemary and sage extract in preserving soft cheese. *LWT* **2022**, *155*, 112996. [CrossRef]
33. PubChem. P-Cymene. Available online: https://pubchem.ncbi.nlm.nih.gov/compound/7463 (accessed on 8 May 2022).
34. PubChem. Limonene. Available online: https://pubchem.ncbi.nlm.nih.gov/compound/22311 (accessed on 8 May 2022).
35. PubChem. Thymol. Available online: https://pubchem.ncbi.nlm.nih.gov/compound/6989 (accessed on 8 May 2022).
36. Gao, C.; Pollet, E.; Avérous, L. Properties of glycerol-plasticized alginate films obtained by thermo-mechanical mixing. *Food Hydrocoll.* **2017**, *63*, 414–420. [CrossRef]
37. Paşcalău, V.; Popescu, V.; Popescu, G.L.; Dudescu, M.C.; Borodi, G.; Dinescu, A.; Perhaiţa, I.; Paul, M. The alginate/k-Carrageenan ratio's influence on the properties of the cross-linked composite films. *J. Alloys Compd.* **2012**, *536*, S418–S423. [CrossRef]
38. Kinninmonth, M.A.; Liauw, C.M.; Verran, J.; Taylor, R.; Edwards-Jones, V.; Shaw, D.; Webb, M. Investigation into the suitability of layered silicates as adsorption media for essential oils using FTIR and GC-MS. *Appl. Clay Sci.* **2013**, *83–84*, 415–425. [CrossRef]
39. Elaiopoulos, K.; Perraki, T.; Grigoropoulou, E. Monitoring the effect of hydrothermal treatments on the structure of a natural zeolite through a combined XRD, FTIR, XRF, SEM and N2-porosimetry analysis. *Microporous Mesoporous Mater.* **2010**, *134*, 29–43. [CrossRef]
40. Eimontas, J.; Yousef, S.; Striūgas, N.; Abdelnaby, M.A. Catalytic pyrolysis kinetic behaviour and TG-FTIR-GC–MS analysis of waste fishing nets over ZSM-5 zeolite catalyst for caprolactam recovery. *Renew. Energy* **2021**, *179*, 1385–1403. [CrossRef]
41. Olegario, E.M.; Mark Pelicano, C.; Cosiñero, H.S.; Sayson, L.V.; Chanlek, N.; Nakajima, H.; Santos, G.N. Facile synthesis and electrochemical characterization of novel metal oxide/philippine natural zeolite (MOPNZ) nanocomposites. *Mater. Lett.* **2021**, *294*, 129799. [CrossRef]
42. Extraction and Characterization of an Alginate from the Brown Seaweed Sargassum Turbinarioides Grunow | SpringerLink. Available online: https://link.springer.com/article/10.1007/s10811-009-9432-y (accessed on 31 January 2022).
43. Pereira, R.; Tojeira, A.; Vaz, D.C.; Mendes, A.; Bártolo, P. Preparation and characterization of films based on alginate and aloe vera. *Int. J. Polym. Anal. Charact.* **2011**, *16*, 449–464. [CrossRef]
44. Keshavarzi, N.; Mashayekhy Rad, F.; Mace, A.; Ansari, F.; Akhtar, F.; Nilsson, U.; Berglund, L.; Bergström, L. Nanocellulose–Zeolite composite films for odor elimination. *ACS Appl. Mater. Interfaces* **2015**, *7*, 14254–14262. [CrossRef]
45. Villegas, C.; Arrieta, M.P.; Rojas, A.; Torres, A.; Faba, S.; Toledo, M.J.; Gutierrez, M.A.; Zavalla, E.; Romero, J.; Galotto, M.J.; et al. PLA/organoclay bionanocomposites impregnated with thymol and cinnamaldehyde by supercritical impregnation for active and sustainable food packaging. *Compos. Part B Eng.* **2019**, *176*, 107336. [CrossRef]
46. Dou, L.; Li, B.; Zhang, K.; Chu, X.; Hou, H. Physical properties and antioxidant activity of gelatin-sodium alginate edible films with tea polyphenols. *Int. J. Biol. Macromol.* **2018**, *118*, 1377–1383. [CrossRef]

47. Structural, Physicochemical and Antioxidant Properties of Sodium Alginate Isolated from a Tunisian Brown Seaweed—ScienceDirect. Available online: https://www.sciencedirect.com/science/article/pii/S0141813014006928 (accessed on 3 April 2022).
48. Uning Rininingsih, E.M. Correlation Between Antioxidant Activity of Synthetic Zeolites Pillared Titanium Dioxide and Iron (III) Oxide with Adsorption DPPH. Masters' Thesis, Bogor Agricultural University, Bogor City, Indonesia, 2014.
49. Water | Free Full-Text | Evaluation of Different Clinoptilolite Zeolites as Adsorbent for Ammonium Removal from Highly Concentrated Synthetic Wastewater. Available online: https://www.mdpi.com/2073-4441/10/5/584 (accessed on 2 April 2022).
50. A Review on Applications and Uses of Thymus in the Food Industry—PMC. Available online: https://www.ncbi.nlm.nih.gov/pmc/articles/PMC7464319/ (accessed on 3 April 2022).
51. Comparative Inhibitory Effects of *Thymus vulgaris* L. Essential Oil against Staphylococcus aureus, Listeria Monocytogenes and Mesophilic Starter Co-Culture in Cheese-Mimicking Models—ScienceDirect. Available online: https://www.sciencedirect.com/science/article/pii/S0740002015001252?via%3Dihub (accessed on 3 April 2022).
52. Antimicrobial Efficacy of an Array of Essential Oils Against Lactic Acid Bacteria—Dunn—2016—Journal of Food Science—Wiley Online Library. Available online: https://ift.onlinelibrary.wiley.com/doi/10.1111/1750-3841.13181 (accessed on 3 April 2022).
53. Celikel, N.; Kavas, G. Antimicrobial properties of some essential oils against some pathogenic microorganisms. *Czech J. Food Sci.* **2008**, *26*, 174–181. [CrossRef]
54. Korukluoglu, M.; Gurbuz, O.; Sahan, Y.; Yigit, A.; Kacar, O.; Rouseff, R. Chemical characterization and antifungal activity of *Origanum onites* L. essential oils and extracts. *J. Food Saf.* **2009**, *29*, 144–161. [CrossRef]
55. Kavas, G.; Kavas, N.; Saygili, D. The effects of thyme and clove essential oil fortified edible films on the physical, chemical and microbiological characteristics of kashar cheese. *J. Food Qual.* **2015**, *38*, 405–412. [CrossRef]
56. Abril, A.G.; Villa, T.G.; Barros-Velázquez, J.; Cañas, B.; Sánchez-Pérez, A.; Calo-Mata, P.; Carrera, M. *Staphylococcus aureus* exotoxins and their detection in the dairy industry and mastitis. *Toxins* **2020**, *12*, 537. [CrossRef] [PubMed]
57. Leandro, O.; da Ramos, S.; Nuno, R.; Pereira, C.; Martins, J.T.; Malcata, F.X. Edible packaging for dairy products. In *Edible Food Packaging*; CRC Press: Boca Raton, FL, USA, 2016; ISBN 978-1-315-37317-1.
58. Pasias, I.N.; Ntakoulas, D.D.; Raptopoulou, K.; Gardeli, C.; Proestos, C. Chemical composition of essential oils of aromatic and medicinal herbs cultivated in Greece—Benefits and drawbacks. *Foods* **2021**, *10*, 2354. [CrossRef] [PubMed]
59. Mitropoulou, G.; Sidira, M.; Skitsa, M.; Tsochantaridis, I.; Pappa, A.; Dimtsoudis, C.; Proestos, C.; Kourkoutas, Y. Assessment of the antimicrobial, antioxidant, and antiproliferative potential of *Sideritis raeseri* Subps. *Raeseri* essential oil. *Foods* **2020**, *9*, 860. [CrossRef] [PubMed]
60. Giannakas, A.; Vlacha, M.; Salmas, C.; Leontiou, A.; Katapodis, P.; Stamatis, H.; Barkoula, N.-M.; Ladavos, A. Preparation, characterization, mechanical, barrier and antimicrobial properties of chitosan/PVOH/clay nanocomposites. *Carbohydr. Polym.* **2016**, *140*, 408–415. [CrossRef]
61. Salmas, C.E.; Giannakas, A.E.; Baikousi, M.; Kollia, E.; Tsigkou, V.; Proestos, C. Effect of copper and titanium-exchanged montmorillonite nanostructures on the packaging performance of chitosan/poly-vinyl-alcohol-based active packaging nanocomposite films. *Foods* **2021**, *10*, 3038. [CrossRef]
62. Bastarrachea, L.; Dhawan, S.; Sablani, S.S. Engineering properties of polymeric-based antimicrobial films for food packaging: A review. *Food Eng. Rev.* **2011**, *3*, 79–93. [CrossRef]
63. Yasuda, H. Units of gas permeability constants. *J. Appl. Polym. Sci.* **1975**, *19*, 2529–2536. [CrossRef]

gels

MDPI

Article

The Use of Biopolymers as a Natural Matrix for Incorporation of Essential Oils of Medicinal Plants

Roxana Gheorghita Puscaselu [1,2,*], Andrei Lobiuc [2], Ioan Ovidiu Sirbu [1,3] and Mihai Covasa [2,4,*]

1 Department of Biochemistry, Victor Babeş University of Medicine and Pharmacy, 300041 Timisoara, Romania
2 Department of Medicine and Biomedical Sciences, College of Medicine and Biological Science, University of Suceava, 720229 Suceava, Romania
3 Center for Complex Network Science, Victor Babes University of Medicine and Pharmacy, 300041 Timisoara, Romania
4 Department of Basic Medical Sciences, College of Osteopathic Medicine, Western University of Health Sciences, Pomona, CA 91766, USA
* Correspondence: roxana.puscaselu@usm.ro (R.G.P.); mcovasa@westernu.edu (M.C.)

Abstract: The benefits of using biopolymers for the development of films and coatings are well known. The enrichment of these material properties through various natural additions has led to their applicability in various fields. Essential oils, which are well-known for their beneficial properties, are widely used as encapsulating agents in films based on biopolymers. In this study, we developed biopolymer-based films and tested their properties following the addition of 7.5% and 15% (w/v) essential oils of lemon, orange, grapefruit, cinnamon, clove, chamomile, ginger, eucalyptus or mint. The samples were tested immediately after development and after one year of storage in order to examine possible long-term property changes. All films showed reductions in mass, thickness and microstructure, as well as mechanical properties. The most considerable variations in physical properties were observed in the 7.5% lemon oil sample and the 15% grapefruit oil sample, with the largest reductions in mass (23.13%), thickness (from 109.67 μm to 81.67 μm) and density (from 0.75 g/cm^3 to 0.43 g/cm^3). However, the microstructure of the sample was considerably improved. Although the addition of lemon essential oil prevented the reduction in mass during the storage period, it favored the degradation of the microstructure and the loss of elasticity (from 16.7% to 1.51% for the sample with 7.5% lemon EO and from 18.28% to 1.91% for the sample with 15% lemon EO). Although the addition of essential oils of mint and ginger resulted in films with a more homogeneous microstructure, the increase in concentration favored the appearance of pores and modifications of color parameters. With the exception of films with added orange, cinnamon and clove EOs, the antioxidant capacity of the films decreased during storage. The most obvious variations were identified in the samples with lemon, mint and clove EOs. The most unstable samples were those with added ginger (95.01%), lemon (92%) and mint (90.22%).

Keywords: biopolymers; citrus; cinnamon; clove; eucalyptus; mint; ginger; chamomile

Citation: Gheorghita Puscaselu, R.; Lobiuc, A.; Sirbu, I.O.; Covasa, M. The Use of Biopolymers as a Natural Matrix for Incorporation of Essential Oils of Medicinal Plants. *Gels* **2022**, *8*, 756. https://doi.org/10.3390/gels8110756

Academic Editors: Aris E. Giannakas, Constantinos Salmas and Charalampos Proestos

Received: 19 October 2022
Accepted: 19 November 2022
Published: 21 November 2022

Publisher's Note: MDPI stays neutral with regard to jurisdictional claims in published maps and institutional affiliations.

1. Introduction

For the past decade, the use of biopolymers materials instead of conventional and highly polluting materials, has increased substantially. Biopolymers can be obtained from various sources; the most used are those based on polysaccharides (sodium alginate, agar, chitosan, carrageenan, starch and cellulose), lipids (waxes and fatty acids) and proteins (gelatin, collagen and soy protein isolate) [1]. They are extensively used, owing to the properties of biobased materials, such as mechanical performance, physicochemical characteristics comparable to those of conventional materials [2], antibacterial and antioxidant properties, high biocompatibility, compostability, non-toxicity, non-immunogenicity, non-carcinogenicity, and non-inflammatory and non-allergenic character [3,4]. Multiple manufacturing methods have increased their widespread use. They can be used for the development of films or coatings, capsules, gels or hydrogels. Owing to their matrix, which has a compact structure and high retention capacity, they can be enriched by the addition of various biologically active substances. Furthermore, they are environmentally friendly, completely edible and produce zero waste after use [5].

Essential oils (EOs) have been used since ancient times, owing to their health benefits and sensory and antioxidant properties. Thus, essential oils, such as those extracted from medicinal plants, contain active compounds, such as limonene, pinene, myrcene (orange, lemon, grapefruit and chamomile) [6,7], eucalyptol, eugenol, gingerol, and cedrene (ginger, cinnamon, clove and mint) compound [8], with beneficial effects on the human body. Thus, orange, lemon and ginger EOs have been used for their relaxant effect or pain relief [9,10], whereas grapefruit, cinnamon and mint EOs have been used for their anti-inflammatory, antimycotoxigenic, antitumoral and antigenotoxic effects [11–13]. In medicine, lemon EO has been used to relieve the symptoms of gastrointestinal diseases and depression [14]; ginger EO has been used for sore throat, cough, cold, dyspepsia, gastritis and gastric ulcerations [15]; clove EO has been used for urinary tract infections, digestive disorders, athlete's foot disease, in dentistry and as a kidney tonic [16,17].

Several studies have demonstrated the positive effects of the use of essential oils, for example, in preventing infection with the SARS-CoV-2 virus. Senthil Kumar et al. [18] showed that lemon essential oil, through its compounds, such as citronellol, geraniol, and neryl acetate, was capable of preventing replication of coronaviruses by blocking the entry of the virus into the host cells through downregulation of ACE2 receptor expression in epithelial cells.

Owing to their benefits and consumers predisposition toward products that are as natural as possible, research in the field has been oriented towards the development of innovative materials obtained from renewable resources. Thus, materials based on biopolymers with incorporated essential oils have been developed. Initially, they were used in the food industry to improve the sensorial quality or to increase the shelf life of various food products, such as those based on meat or fish [19–22], fruits or vegetables [23–25], sweets [26,27], beverages [28] or dairy products [29,30]. Later, they were added to increase the nutritional value and bioavailability of products, especially those with beneficial effects on health. Thus, numerous essential oils, such as those from citrus fruits (lemon, orange and grapefruit) or medicinal EOs (ginger, chamomile, mint, eucalyptus, cinnamon or cloves) have been incorporated into biopolymeric materials. However, the development, characterization and testing of these films under various experimental conditions using several well-known plant-based oils with applicability in biomedical, cosmetic or food industries are still needed. Therefore, in this study, we tested biofilms incorporated with two concentrations (7.5% and 15%, *v*/*w*) of essential oils of lemon, orange, grapefruit, cinnamon, clove, chamomile, ginger, eucalyptus or mint at the time of development and after one year of storage under normal temperature and humidity conditions.

2. Results and Discussion

In the present study, we examined the preservation capacity of biopolymeric films with various incorporated plant-derived essential oils. Numerous studies have examined the benefits of using films and coatings based on biopolymers and essential oils; however, the long-term effect on their properties is not entirely known. Our results show that the storage of films considerably influences their properties. Although the films were kept under temperature- and humidity-controlled conditions, their mass varied after one year (Figure 1).

As depicted in Figure 1, the mass of all samples was reduced during the test period. Although samples were kept in silicone paper packaging, simulating a real storage environment, significant moisture loss had occurred after one year. The highest mass reduction was observed in sample 4 with 15% grapefruit EO (23.13%), and the lowest mass reduction was observed in sample 1 with 15% lemon EO (3.54%). The mass loss was associated with a significant reduction in film thickness (Figure 2).

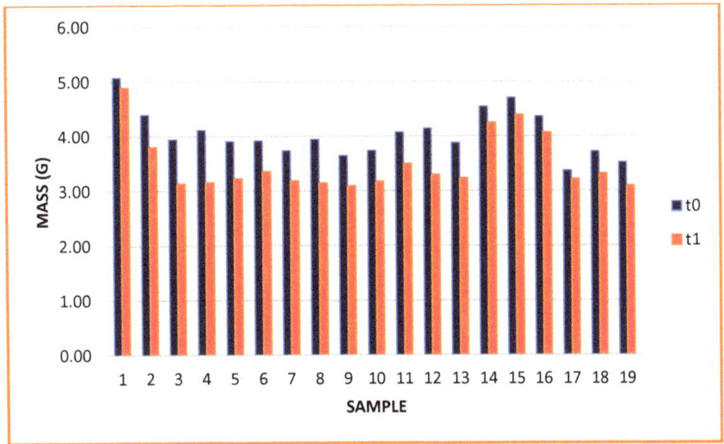

Figure 1. Mass of biopolymeric films.

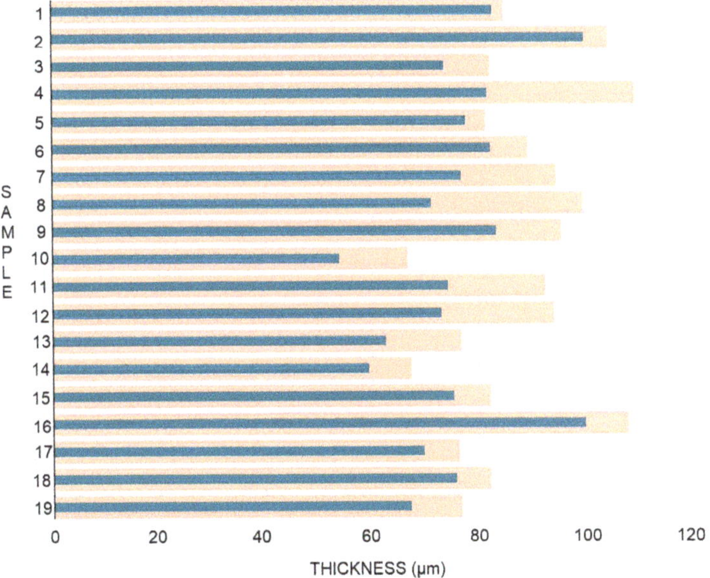

Figure 2. Sample thickness after one year of storage. T0, light brown bars; T1, blue bars.

The thickness of the films did not undergo major changes after one year compared to the initial period. The largest variation was observed in the films with the addition of cinnamon (from 94.67 to 76.67 μm and 99.67 to 71.00 μm, respectively) and cloves (from 95.67 μm to 83.33 μm (9) and from 66.67 μm to 53.67 μm for sample 10) oils. The thickness of the control sample was reduced by 9.7 μm. The determination and the results are of interest to developers who may want to produce such materials on a large scale.

The mass reduction in the films can be attributed to the elimination of water from the film matrix, which was also indicated by the reduced values of the water activity index. These values decreased by at least 50% for all samples (Table 1). A decrease in the water activity index can be beneficial in terms of preventing the development of microorganisms (which require at least $a_w > 0.7$), but it can also alter the film microstructure, making it more brittle and fragile.

Table 1. Optical parameters and water activity index at the time of development (t0) and after one year of storage (t1).

Sample	Density, g * cm^{-1}		Opacity, A * mm^{-1}		Water Activity Index, aw	
	t0	t1	t0	t1	t0	t1
1	0.71 ± 0.05 ^{a}	0.66 ± 0.12 ^{a}	3.31 ± 0.15 ^{h}	4.01 ± 0.28 ^{e}	0.786 ± 0.05 ^{a}	0.334 ± 0.15 ^{c}
2	0.76 ± 0.06 ^{a}	0.65 ± 0.17 ^{a}	3.82 ± 0.01 ^{e}	3.61 ± 0.25 ^{j}	0.743 ± 0.05 ^{b}	0.286 ± 0.59 ^{d}
3	0.55 ± 0.10 ^{e,f}	0.39 ± 0.16 ^{d,e,f}	7.77 ± 0.01 ^{a}	3.83 ± 0.42 ^{g,h}	0.714 ± 0.01 ^{d}	0.304 ± 0.15 ^{c,d}
4	0.74 ± 0.02 ^{a}	0.42 ± 0.11 ^{c,d}	4.85 ± 0.09 ^{c}	3.86 ± 0.18 ^{f,g}	0.737 ± 0.05 ^{c}	0.298 ± 0.15 ^{c,d}
5	0.54 ± 0.12 ^{e,f}	0.43 ± 0.12 ^{c,d}	2.07 ± 0.12 ^{j}	3.62 ± 0.02 ^{i,j}	0.738 ± 0.07 ^{c}	0.302 ± 0.10 ^{c,d}
6	0.56 ± 0.17 ^{d,e,f}	0.46 ± 0.54 ^{c}	3.66 ± 0.17 ^{f,g}	3.97 ± 0.14 ^{e,f}	0.700 ± 0.03 ^{e}	0.295 ± 0.03 ^{c,d}
7	0.58 ± 0.10 ^{c,d,e}	0.42 ± 0.12 ^{c,d}	1.47 ± 0.02 ^{l}	4.21 ± 0.06 ^{d}	0.702 ± 0.05 ^{e}	0.332 ± 0.10 ^{c}
8	0.64 ± 0.19 ^{b,c}	0.38 ± 0.81 ^{d,e,f}	1.84 ± 0.01 ^{k}	4.32 ± 0.02 ^{d}	0.685 ± 0.07 ^{g}	0.304 ± 0.26 ^{c,d}
9	0.57 ± 0.16 ^{d,e,f}	0.43 ± 0.21 ^{c,d}	1.73 ± 0.04 ^{k}	3.73 ± 0.02 ^{h,i}	0.574 ± 0.02 ^{k}	0.292 ± 0.15 ^{c,d}
10	0.42 ± 0.13 ^{g}	0.25 ± 0.38 ^{g}	1.85 ± 0.01 ^{k}	5.71 ± 0.10 ^{b}	0.571 ± 0.06 ^{k}	0.272 ± 0.05 ^{d}
11	0.62 ± 0.10 ^{b,c,d}	0.43 ± 0.15 ^{c,d}	7.03 ± 0.01 ^{b}	2.03 ± 0.02 ^{l,m}	0.587 ± 0.75 ^{j}	0.517 ± 0.02 ^{b}
12	0.65 ± 0.15 ^{b}	0.40 ± 0.17 ^{d,e}	3.57 ± 0.03 ^{g}	1.97 ± 0.08 ^{m,n}	0.751 ± 0.01 ^{b}	0.524 ± 0.02 ^{a,b}
13	0.51 ± 0.17 ^{f}	0.34 ± 0.15 ^{f}	3.74 ± 0.03 ^{e,f}	6.47 ± 0.02 ^{a}	0.706 ± 0.75 ^{e}	0.564 ± 0.04 ^{a}
14	0.53 ± 0.15 ^{f}	0.42 ± 0.10 ^{c,d}	1.83 ± 0.03 ^{k}	5.27 ± 0.03 ^{c}	0.684 ± 0.69 ^{g}	0.562 ± 0.02 ^{a}
15	0.64 ± 0.30 ^{b,c}	0.54 ± 0.21 ^{b}	1.55 ± 0.07 ^{l}	1.83 ± 0.03 ^{o}	0.656 ± 0.01 ^{h}	0.561 ± 0.02 ^{a}
16	0.76 ± 0.42 ^{a}	0.66 ± 0.26 ^{a}	0.95 ±0.01 ^{m}	1.14 ± 0.02 ^{p}	0.692 ± 0.05 ^{f}	0.546 ± 0.05 ^{a,b}
17	0.43 ± 0.02 ^{g}	0.35 ± 0.02 ^{e,f}	1.53 ± 0.31 ^{l}	2.12 ± 0.05 ^{l}	0.655 ± 0.75 ^{h}	0.534 ± 0.03 ^{a,b}
18	0.52 ± 0.15 ^{e,f}	0.42 ± 0.13 ^{c,d}	2.54 ± 0.03 ^{i}	1.92 ± 0.10 ^{n,o}	0.645 ± 0.05 ^{i}	0.527 ± 0.03 ^{a,b}
19	0.44 ± 0.35 ^{g}	0.34 ± 0.21 ^{f}	4.41 ± 0.11 ^{d}	2.51 ± 0.10 ^{k}	0.646 ± 0.01 ^{i}	0.281 ± 0.01 ^{d}

* The values represent the mean ± SD. Means that do not share the same superscript (a–p) are significantly different ($p < 0.05$).

In addition to dehydration, a reduction in sample density also occurred, totaling approximately one unit for samples 5, 6 and 14–18 with incorporated orange, chamomile, ginger and eucalyptus essential oils. Additionally, the density of the control sample (19) decreased by one unit (from 0.44 to 0.34 g/cm^3). The greatest reduction in density was observed in sample 4 (56.75%) composed of 15% grapefruit essential oil. Except for grapefruit (3, 4), mint (11, 12) and the control (19), all films exhibited increased opacity after one year of storage. These results were confirmed by a corresponding decrease in the transmittance parameter evaluated before and after storage (Figures 3 and 4).

As depicted in Figure 3, the transmittance values increased within 300–700 nm range. At 800 nm (visible light), films with 7.5% lemon (1), 15% cinnamon (8), 7.5% mint (11), 7.5 and 15% ginger (15,16), 15% eucalyptus (18) EOs and the control (19) exhibited increased transmittance, whereas in the remainder of the films, the transmission was decreased. This determination is important because it evaluates the ability of the material to resist UV radiation can degrade the incorporated product and even the foil, especially when it contains essential oil. The smallest variations were observed in the control sample (19) both before and after storage. After storage, the films were more resistant to the action of UV radiation, an effect strengthened by the increased opacity values after storage (Table 1).

The sample color was also influenced by storage. Except for samples with added mint oil (11, 12), in which the luminosity values increased during the storage period (from 90.61 and 91.9 to 92.25 and 92.18, respectively), for all other samples, the values were reduced but did not change significantly. The color deviation was more evident for samples 1–10 (ΔE = 6.24–7.55) (Table 2). For samples 11–17, ΔE did not exceed 3.63. The smallest color deviation was observed in sample 12 with 15% mint oil (ΔE = 0.244), and the largest color deviation was observed in sample 18 with 15% eucalyptus oil (ΔE = 11.35).

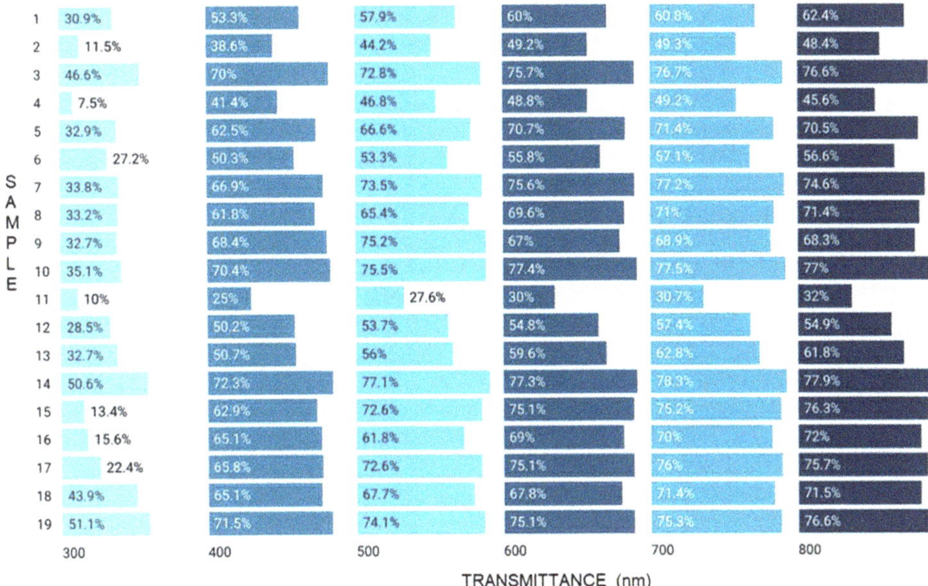

Figure 3. Transmittance values of samples (t0).

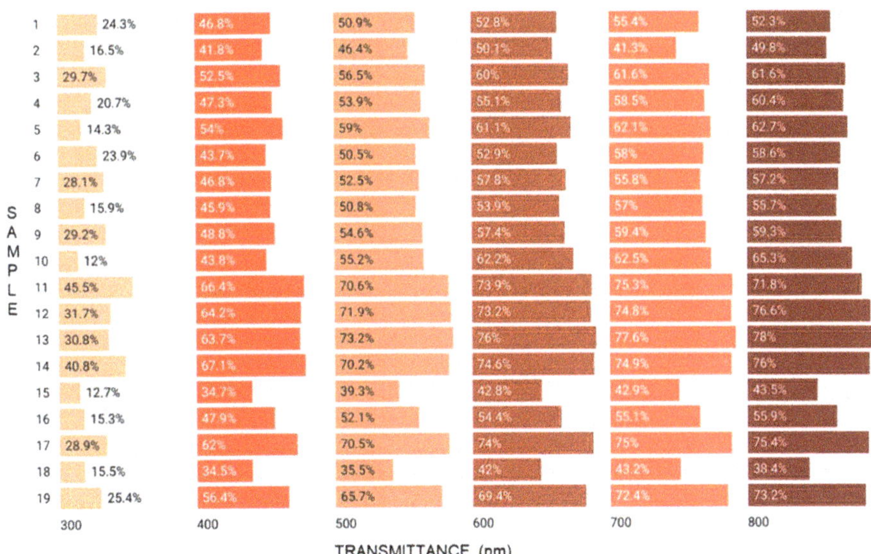

Figure 4. The transmittance values of samples after one year of storage (t1).

With some exceptions, the surface morphology of films with added oils was influenced by prolonged storage (Figure S1, Supplementary Files). The microstructure of films with 7.5 and 15% cinnamon (7,8), 7.5 and 15% clove (9,10) and 15% chamomile essential oils was negatively modified during storage. Changes also occurred in the structure of the control sample. The film with 15% grapefruit EO became rougher, indicating changes in the matrix, which was more homogeneous and compact prior to storage. The same pattern was observed for samples with 15% concentration of orange (6) and cinnamon (8) EOs. The film with 15% added clove oil (10) exhibited pores in its microstructure.

Table 2. Evaluation of color parameters before (t0) and after storage (t1).

Sample	L*		a*		b*		ΔE
	t0	t1	t0	t1	t0	t1	
1	92.73 ± 0.13 ^{a,b}	88.79 ± 0.18 ^{b,c}	−5.51 ± 0.04 ^{a}	−0.49 ± 0.02 ^{a}	11.42 ± 0.17 ^{h}	10.76 ± 0.31 ^{f}	6.42 ± 0.01 ^{f}
2	92.23 ± 0.5 ^{a,b,c}	88.99 ± 0.16 ^{b,c}	−5.82 ± 0.08 ^{f,g}	−0.42 ± 0.24 ^{a}	13.43 ± 0.67 ^{a,b,c}	10.61 ± 0.30 ^{f}	6.53 ± 0.04 ^{f}
3	92.12 ± 0.16 ^{a,b,c}	88.30 ± 0.40 ^{b,c,d}	−5.90 ± 0.02 ^{g,h}	−0.54 ± 0.02 ^{a}	12.93 ± 0.33 ^{c,d}	11.33 ± 0.48 ^{d,e,f}	6.92 ± 0.01 ^{d}
4	92.42 ± 0.28 ^{a,b,c}	88.87 ± 0.96 ^{b,c}	−5.97 ± 0.08 ^{h}	−0.61 ± 0.05 ^{a}	14.08 ± 0.5 ^{a}	10.72 ± 0.53 ^{f}	7.55 ± 0.02 ^{b}
5	92.45 ± 0.28 ^{a,b,c}	88.88 ± 0.73 ^{b,c}	−5.78 ± 0.04 ^{e,f}	−0.49 ± 0.04 ^{a}	12.5 ± 0.19 ^{d,e,f}	8.85 ± 0.20 ^{g}	7.25 ± 0.05 ^{c}
6	92.92 ± 0.31 ^{a}	87.17 ± 0.63 ^{b}	−5.73 ± 0.02 ^{d,e,f}	−0.48 ± 0.15 ^{a}	12.82 ± 0.08 ^{c,d,e}	9.85 ± 0.73 ^{f,g}	6.97 ± 0.02 ^{d}
7	92.47± 0.32 ^{a,b,c}	89.17 ± 0.30 ^{b}	−5.68 ± 0.04 ^{c,d,e}	−0.48 ± 0.01 ^{a}	11.36 ± 0.30 ^{h}	10.18 ± 0.08 ^{d,e,f}	6.51 ± 0.03 ^{f}
8	92.57 ± 0.16 ^{a,b,c}	88.58 ± 0.92 ^{b,c}	−5.66 ± 0.01 ^{b,c,d}	−0.49 ± 0.02 ^{a}	11.84 ± 0.30 ^{f,g,h}	10.95 ± 0.42 ^{f}	6.52 ± 0.03 ^{f}
9	92.41 ± 0.39 ^{a,b,c}	88.50 ± 0.94 ^{b,c,d}	−5.60 ± 0.01 ^{b,c,d}	−0.47 ± 0.03 ^{a}	12.50 ± 0.27 ^{d,e,f}	11.14 ± 1.00 ^{e,f}	6.74 ± 0.02 ^{e}
10	92.46 ± 0.21 ^{a,b,c}	88.31 ± 0.68 ^{b,c,d}	−5.64 ± 0.03 ^{b,c,d}	−0.52 ± 0.01 ^{a}	11.34 ± 0.31 ^{h}	11.22 ± 0.40 ^{d,e,f}	6.24 ± 0.06 ^{g}
11	90.61 ± 0.64 ^{d}	92.25 ± 0.80 ^{a}	−5.52 ± 0.03 ^{a}	−5.78 ± 0.07 ^{c}	13.30 ± 0.04 ^{b,c}	13.70 ± 0.53 ^{b,c}	1.25 ± 0.01 ^{k}
12	91.90 ± 0.48 ^{b,c}	92.18 ± 0.15 ^{a}	−5.60 ± 0.03 ^{a,b,c}	−5.79 ± 0.02 ^{c}	12.50 ± 0.13 ^{d,e,f}	12.84 ± 0.52 ^{c,d}	0.24 ± 0.15 ^{m}
13	92.65 ± 0.24 ^{a,b}	91.00 ± 0.64 ^{a}	−5.59 ± 0.02 ^{a,b,c}	−5.76 ± 0.01 ^{c}	12.51 ± 0.05 ^{d,e,f}	15.01 ± 0.40 ^{c}	3.63 ± 0.02 ^{h}
14	92.34 ± 0.11 ^{a,b,c}	92.08 ± 0.28 ^{a}	−5.67 ± 0.03 ^{b,c,d}	−5.85 ± 0.03 ^{c}	12.18 ± 0.01 ^{e,f,g}	13.61 ± 0.37 ^{b,c}	0.85 ± 0.01 ^{l}
15	92.54 ± 0.14 ^{a,b,c}	92.27 ± 0.24 ^{a}	−5.68 ± 0.06 ^{c,d,e}	−5.85 ± 0.03 ^{c}	12.50 ± 0.38 ^{d,e,f}	13.17 ± 0.40 ^{c}	0.83 ± 0.02 ^{l}
16	91.80 ± 0.40 ^{c}	91.83 ± 0.15 ^{a}	−5.81 ± 0.01 ^{f,g}	−5.76 ± 0.08 ^{c}	13.82 ± 0.02 ^{a,b}	15.06 ± 0.27 ^{b}	1.61 ± 0.02 ^{j}
17	92.41 ± 0.19 ^{a,b,c}	92.09 ± 0.44 ^{a}	−5.68 ± 0.03 ^{c,d,e}	−5.91 ± 0.10 ^{c}	11.53 ± 0.15 ^{g,h}	12.84 ± 0.67 ^{c,d,e}	2.25 ± 0.04 ^{i}
18	92.42 ± 0.33 ^{a,b,c}	86.97 ± 1.09 ^{d}	−5.58 ± 0.04 ^{a,b}	−5.18 ± 0.23 ^{b}	11.75 ± 0.18 ^{g,h}	21.97 ± 0.8 ^{a}	11.35 ± 0.21 ^{a}
19	92.51 ± 0.12 ^{a,b,c}	87.47 ± 0.22 ^{c,d}	−5.64 ± 0.02 ^{b,c,d}	−0.62 ± 0.02 ^{a}	11.32 ± 0.09 ^{h}	10.32 ± 0.56 ^{f,g}	7.14 ± 0.02 ^{c}

L*, lightness; a*, green-to-red parameter; b*, blue-to-yellow parameter. The values represent mean ± SD. Means that do not share the same superscript (a–m) are significantly different ($p < 0.05$).

Except for samples with added mint and ginger oil (11–14), all other films with essential oil presented a less homogeneous microstructure than the control sample. Thus, in order to obtain materials with superior physical properties, the matrix could be improved by increasing the amount of emulsifier in the film-forming solution (i.e., Tween 80).

The mechanical characteristics of the films were not influenced by the storage period. All samples subjected to testing showed a increased tensile strength after storage (Figure 5). The increase in breaking strength may be due to the amount of essential oil lost during storage in association with the overall decrease in the antioxidant activity of most films with added oil. The breaking strength of biopolymeric films has been shown to be affected by the addition of natural substances. For example, the mechanical performance was decreased with increasing content of natural substances added to films [31,32]. This effect was attributed to the plasticizing effect of EOs weakening the intermolecular interaction between polymer chains and increasing the ductility of the film [32,33].

Sample 14 (15% ginger) could not be tested to evaluate its mechanical properties as a result of the drying method used. Thus, a 15% concentration of ginger oil may require a increased amount of plasticizer or emulsifier in the composition.

As expected, the elongation at break of the films was negatively influenced during the storage period. The loss of water from the films unbalanced the matrix, so despite strengthened tear resistance, they showed considerably reduced elasticity (Figure 6).

The largest variation in elongation was observed in the lemon oil samples (16.7 (t0) to 1.51% (t1) and 18.28 (t0) to 1.91% (t1) for samples 1 and 2, respectively), regardless of the concentration. Large variations were also observed in the grapefruit, ginger and eucalyptus samples, with variations ranging between 37% and 50%. High variation was also observed in the control sample without the addition of essential oil. Therefore, the dehydration of the material did not depend on the nature of oil added but on the biopolymeric composition or environmental factors. However, the addition of orange, cinnamon, clove, mint and chamomile essential oils maintained the elasticity of the samples, with minimal change relative to the initial period.

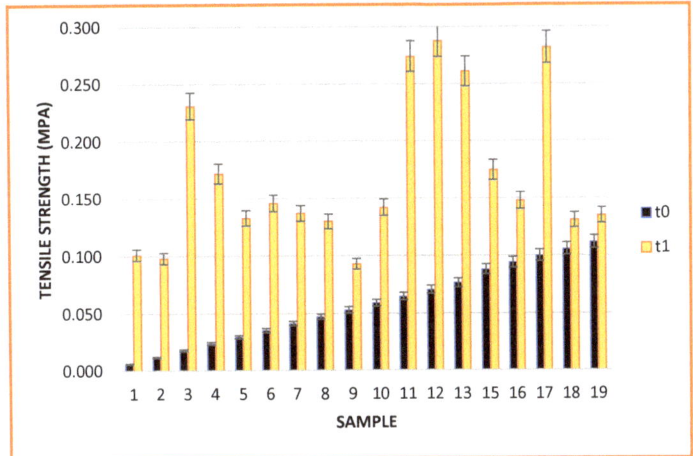

Figure 5. Sample tensile strength.

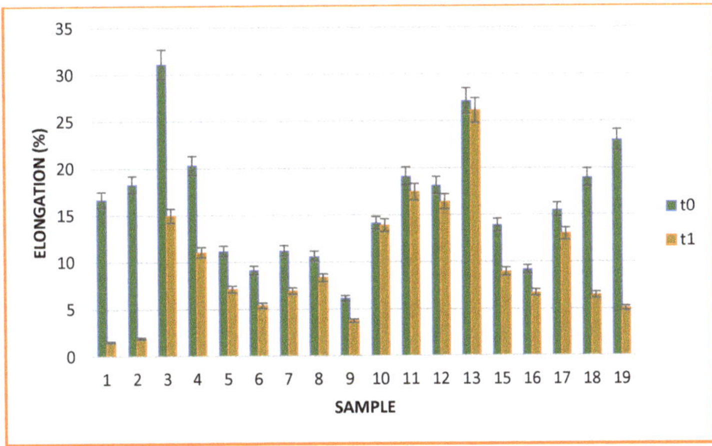

Figure 6. Sample elongation.

The antioxidant activity of most samples was significantly diminished after one year of storage (Figure 7). For example, the antioxidant activity of sample 2 with 15% lemon oil decreased from 26% to 1.9%, which was the most significant decrease among all samples tested. A significant reduction in the antioxidant capacity was also observed in sample 8 with 15% cinnamon oil (from 23.36% to 2.28%). A modest decrease was also observed in sample 5 (7.5% orange oil) when the antioxidant activity dropped from 4.72% at $t0$ to 3.5% at $t1$. However, for a few samples, the antioxidant activity of the films increased with the addition of 15% orange (6), 7.5% cinnamon (7) and 7.5% clove (9) essential oils. As such, the incorporation of orange oil increased antioxidant activity from 11.09% to 17.73%, cinnamon oil increased antioxidant activity from 9.36% to 15.98% and clove oil increased said activity from 8.36% to 17.27%.

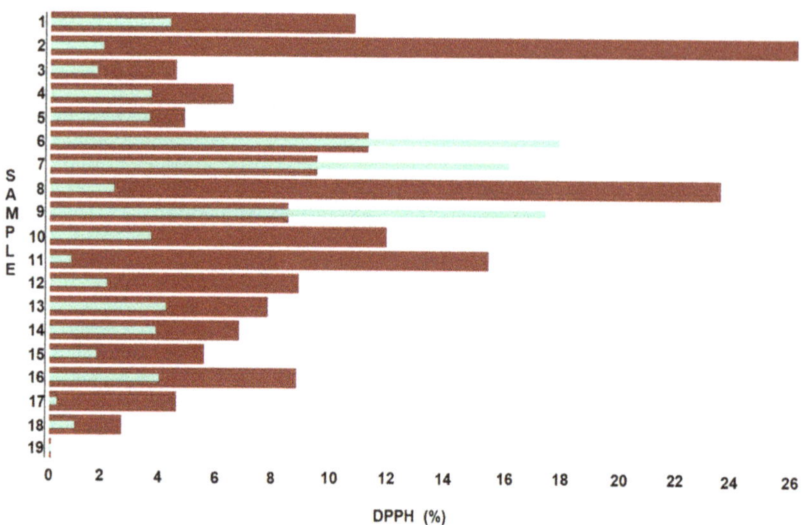

Figure 7. Graphical representation of DPPH radical scavenging activity of biopolymeric films before (red) and after storage (turquoise).

The antioxidant effects of various essential oils on films are not entirely known, given the scarcity of studies examining their effects on biopolymeric matrices. However, the differential antioxidant effects of various natural oils have been well documented. For example, the antioxidant properties of clove and cinnamon essential oils are comparable to those of BHT (butylhydroxytoluene), a chemical substance known for its protective effect against oxidation. These results were attributed to the high eugenol and β-caryophyllene content of these oils [34], with clove essential oil being one of the most powerful natural antioxidants, with effects even superior to those of BHT or butylated hydroxyanisole [35]. The high antioxidant activity and radical scavenging effect of eugenol is the result of its phenolic hydroxyl groups, which remain stable during the film development process [36]. Similarly, linalool, another chemical compound present in orange and cinnamon oils (Table S1), has been recognized for its antioxidant and antibacterial activity [37,38]. When examining the antioxidant activity of 25 essential oils used for medicinal purposes, Wei and Shibamoto [39] showed that clove (52%) and cinnamon (23%) oils had antioxidant activity superior to that of chamomile (20%), anise (20%), rosemary (10%) and orange (9%) EOs. Moreover, according to their study, mint essential oil, along with sandalwood and bergamot, had exhibited pro-oxidant activity. Therefore, our results of increased antioxidant activity caused by clove, cinnamon and orange essential oils are somewhat in line with the general antioxidant activity effects of these oils, albeit in a different model and likely due to the high eugenol and linalool contents (Table S1, Supplementary Files), as well as the high antioxidant capacity of EOs. Our data cannot explain the concentration differences in the antioxidant effects of these oils, which may be attributed, in part, to their varied embedding properties into the a biopolymeric matrix; however, this would require further investigation.

As shown in Table 3, microorganisms proliferated (*TC*) from t0 to t1 (samples 1–7, 9–14 and 19). Except for samples 15–18, which did not show a microbiological load before or after storage, the other films showed colony-forming units on their surfaces. Only samples with 7.5% essential oil of lemon, clove and mint showed contamination with coliforms bacteria, whereas samples 4 (15% grapefruit oil) and 13 (7.5% chamomile oil) were contaminated with *Staphylococcus aureus*. No *Escherichia coli*, enterococcus, *Listeria monocytogenes*, yeasts or molds developed on the surfaces of the biopolymeric materials.

Table 3. Microbiological assessments before (t0) and after storage (t1).

Sample	TC		EC		ETC		CF		YM		X-SA		LM	
	t0	t1	t0	t1	t0	t1	t0	t1	t0	t1	t0	t1	t0	t1
1	1	15	-	-	-	-	1	13	-	-	-	-	-	-
2	-	7	-	-	-	-	-	-	-	-	-	-	-	-
3	15	22	-	-	-	-	-	-	-	-	-	-	-	-
4	21	42	-	-	-	-	-	-	-	-	1	6	-	-
5	-	6	-	-	-	-	-	-	-	-	-	-	-	-
6	7	31	-	-	-	-	-	-	-	-	-	-	-	-
7	2	4	-	-	-	-	-	-	-	-	-	-	-	-
8	-	-	-	-	-	-	-	-	-	-	-	-	-	-
9	13	48	-	-	-	-	2	7	-	-	-	-	-	-
10	1	13	-	-	-	-	-	-	-	-	-	-	-	-
11	-	9	-	-	-	-	2	5	-	-	-	-	-	-
12	8	11	-	-	-	-	-	-	-	-	-	-	-	-
13	19	64	-	-	-	-	-	-	-	-	1	4	-	-
14	6	34	-	-	-	-	-	-	-	-	-	-	-	-
15	-	-	-	-	-	-	-	-	-	-	-	-	-	-
16	-	-	-	-	-	-	-	-	-	-	-	-	-	-
17	-	-	-	-	-	-	-	-	-	-	-	-	-	-
18	-	-	-	-	-	-	-	-	-	-	-	-	-	-
19	28	114	-	-	-	-	-	-	-	-	-	-	-	-

TC, total count; EC, Escherichia coli; ETC, Enterococcus; CF, coliform; YM, yeasts and molds; X-SA, Staphylococcus aureus; LM, Listeria monocytogenes. All results are expressed in CFU/g.

3. Conclusions

Films based on biopolymers have proven their applicability in various fields and can replace conventional materials based on non-renewable resources. The ease of incorporating various active substances into the matrix of films has contributed to their success. Essential oils present numerous benefits, both for the products packaged in such materials and in terms of the health of those who consume the edible films and coatings. Our results highlighted the use of essential oil favoring stability of the materials, even after one year of storage. The films with incorporated essential oils were preserved their properties after one year storage compared to the initial period. These positive effects were not present in the control samples. Therefore, the addition of plant-extracted oils to biopolymer products is associated with significant benefits, making them suitable for use in several industries. However, additional research is required to determine their mechanism of action, their prolonged effects and the synergistic effects that may occur when used in combination with other essential oils or biologically active substances, as well as their adverse effects.

4. Materials and Methods

The aim of this study was to develop, test and characterize new biopolymeric based films with incorporated EOs (Table 4). All substances used to produce 19 film-forming solutions (agar, sodium alginate, glycerol, emulsifier and Tween 80) were obtained from Sigma Aldrich (Germany). The commercially available EOs from medicinal plants, such as lemon, grapefruit, orange, cinnamon, clove, mint, eucalypt and chamomile, were obtained from Carl Roth (Germany). Ginger oil was purchased from Merk (Germany). According to the manufacturer data sheets, the essential oils were obtained by steam distillation; their chemical composition is summarized in Table S1.

Biopolymeric materials were obtained based on a previously developed and tested modified methodology [40]. Briefly, a ratio of 2:1:1 agar: alginate: glycerol was used. The film-forming solution was mixed for 20 min at 90 ± 2 °C and 450 rpm, cooled to 40 °C and 7.5 and 15% w/v essential oils were added. The solutions were obtained through the cast method after maintenance for 38–42 h at room temperature (26 ± 3 °C) and rH = 52 ± 3% until complete drying was achieved.

Table 4. Composition of new biopolymer-based films.

Sample	7.5% EO (w/v)	15% EO (w/v)	Sample	7.5% EO (w/v)	15% EO (w/v)
1	lemon	x	11	mint	x
2	x	lemon	12	x	mint
3	grapefruit	x	13	chamomile	x
4	x	grapefruit	14	x	chamomile
5	orange	x	15	ginger	x
6	x	orange	16	x	ginger
7	cinnamon	x	17	eucalyptus	x
8	x	cinnamon	18	x	eucalyptus
9	clove	x	19	Control, no essential oil added	
10	X	clove			

The films were kept for one year in silicone paper packaging, protected from humidity and sunlight. They were tested immediately after development and after one year for physical and optical properties, as well as antioxidant and antimicrobial capacity. In order to observe possible variations in their mass, samples were weighed using an analytical balance. Film thickness (t, μm) was measured after five readings in different areas of the material surface using a Yato micrometer (Shanghai, China). The density (D, g/cm^3) of the films was calculated by relating their mass (w) to the thickness (t) and surface (a) [41]:

$$\boldsymbol{Density,}\left(\frac{\mathbf{g}}{\mathbf{cm^3}}\right) = \frac{w}{a*t} \quad (1)$$

Transmittance (T, %) and absorbance (A) were read spectrophotometrically (Epoch, BioTek Instruments, Winooski, VT, USA) in triplicate using 1 cm × 3 cm film samples. The transmittance was read in the wavelength range of 300 to 800 nm, with absorbance at 600 nm. An empty cuvette was used as a standard. The opacity of the material (O) was calculated according to the following formula:

$$\boldsymbol{Opacity,}\left(\frac{\mathbf{A}}{\mathbf{mm}}\right) = A/t \quad (2)$$

where A = absorbance, and t = thickness (mm).

The water activity index (a_w) was determined with AquaLab 4TE equipment (Meter Group, München, Germany) at 25 ± 0.7 °C. The results indicate the average of 5 readings of the tested materials. The evaluation of this parameter is of interest when film dehydration occurs. Additionally, a low value of the water activity index favors the prevention of the development and proliferation of microorganisms. A water activity index score above 0.7 is required to survive the environmental conditions.

The sample color was evaluated using a CIELab system with a Chroma Meter CR400 colorimeter (Konika Minolta, Tokyo, Japan). The results represent the arithmetic mean of ten readings taken over the entire surface of the material. In order to test the color difference between the samples tested before and after storage, the color deviation (ΔE) was calculated according to the following formula [42]:

$$\Delta E = \sqrt{(\Delta L*\Delta L) + (\Delta a*\Delta a) + (\Delta b*\Delta b)} \quad (3)$$

The microstructure of images was visualized with a Celena microscope, and images and microtopographs were analyzed using Mountains Premium 9 (Digital Surf, Lavoisier, France).

The tensile strength (*TS*) and elongation (*E*) were tested with an ESM Mark 10 texturometer according to STAS ASTM D882 (Standard Test Method for Tensile Properties of Thin Plastic Sheeting) [43] and calculated according to Formulae (4) and (5) [44]. As such, a 5 KN cell and special grips for thin films and foils were attached, and 1cm × 10 cm film samples were tested. The travel speed was set at 10 mm/min, and the working temperature was 27.2 ± 0.2 °C.

$$\boldsymbol{TS},\ (\mathbf{MPa}) = \frac{\boldsymbol{F}}{\boldsymbol{a}} \quad (4)$$

where *F* is the maximum load (kN), and *a* is the surface (mm^2). A travel speed of 10 mm/min was chosen based on standard requirements for testing films and foils of 5 to 10 mm/min, as well as on the published evidence [45–50].

The elongation at break (*E*) is the ratio between the final (Δl) and initial length (l) after test specimen breakage.

$$\boldsymbol{E},\ (\%) = \frac{\boldsymbol{\Delta l}}{\boldsymbol{l}} * \mathbf{100} \quad (5)$$

Antioxidants were assessed using a DPPH (2,2-diphenyl-1-picrylhydrazyl) radical scavenging assay according to the method described by Aloui et al. [51], with some modifications. Briefly, a film sample was cut into 20 mm × 20 mm, and 2 mL of DPPH was added, mixed for 1 min at 500 rpm and incubated at 35 °C for 30 min. After incubation, the absorbance was read at 517 nm using an Epoq spectrophotometer (BioTek Instruments, Winooski, VT, USA). The experiment was carried out in triplicate, and the radical scavenging activity was calculated according to Formula (6), where *Ac* is the absorbance of the DPPH solution without film, and *As* is absorbance of the sample:

$$\boldsymbol{Radical\ scaveging\ activity},\ (\%) = \frac{\boldsymbol{Ac} - \boldsymbol{As}}{\boldsymbol{Ac}} * \mathbf{100} \quad (6)$$

The antimicrobial activity of the films was tested using specific plates (NISSUI Pharmaceutical, Tokyo, Japan) with dehydrated culture media. Thus, total count (*TC*), coliforms (*CF*), *Escherichia coli* (*EC*), *Staphylococcus aureus* (*X-SA*), *Listeria monocytogenes* (*LM*) and yeasts and molds (*YM*) were evaluated. The proposed method it is useful, as it faster than traditional methods and eliminates the risks that can intervene in the manipulation of strains of pathological microorganisms. It can also be used in laboratories in which the use of strains of pathogenic microorganisms is not allowed, as is our case. For the assay, 1 g of film was mixed with 9 mL saline solution at 500 rpm. Then, 1 mL of the solution was used to hydrate the culture medium. Later, plates with culture media were thermostated for 36–72 h at 37 °C. The results are expressed in CFU/g.

Statistical analyses were performed with one-way analysis of variance (ANOVA) and Tukey's test. Significance was set at $p < 0.05$. Data analysis was performed using MiniTAB statistical software (MiniTAB Ltd., Coventry, UK). All determinations were made after sample development (*t*0) and after one year of storage (*t*1).

Supplementary Materials: The following supporting information can be downloaded at: https://www.mdpi.com/article/10.3390/gels8110756/s1, Figure S1: Microstructure images of tested films, obtained at initial time (*t*0) and after one year storage (*t*1); Table S1. Essential oils' chemical composition, according to the manufacturer data sheet

Author Contributions: Conceptualization, R.G.P.; methodology, R.G.P., A.L., I.O.S. and M.C.; software, R.G.P. and M.C.; validation, I.O.S. and M.C.; formal analysis, R.G.P. and A.L.; investigation, R.G.P. and A.L.; resources, R.G.P., A.L., I.O.S. and M.C.; writing—original draft preparation, R.G.P. and A.L.; writing—review and editing, I.O.S. and M.C.; supervision, I.O.S. and M.C.; project administration, R.G.P. and M.C.; funding acquisition, R.G.P. and A.L. All authors have read and agreed to the published version of the manuscript.

Funding: This work was supported by a grant of the Romanian Ministry of Research, Innovation and Digitization, CNCS/CCCDI–UEFISCDI, project number PN-III-P1-1.1-PD-2019-0793, within PNCDI III.

Institutional Review Board Statement: Not applicable.

Informed Consent Statement: Not applicable.

Data Availability Statement: Not applicable.

Conflicts of Interest: The authors declare no conflict of interest.

References

1. Patti, A.; Acierno, D. Towards the Sustainability of the Plastic Industry through Biopolymers: Properties and Potential Applications to the Textiles World. *Polymers* **2022**, *14*, 692. [CrossRef]
2. Agarwal, A.; Shaida, B.; Rastogi, M.; Singh, N.B. Food Packaging Materials with Special Reference to Biopolymers-Properties and Applications. *Chem. Afr.* **2022**, *1*, 1–28. [CrossRef]
3. Jose, A.A.; Hazeena, S.H.; Lakshmi, N.M.; B, A.K.; Madhavan, A.; Sirohi, R.; Tarafdar, A.; Sindhu, R.; Awasthi, M.K.; Pandey, A.; et al. Bacterial biopolymers: From production to applications in biomedicine. *Sustain. Chem. Pharm.* **2022**, *25*, 100582. [CrossRef]
4. Rao, S.S.; Rekha, P.D. Biopolymers in Cosmetics, Pharmaceutical, and Biomedical Applications. *Biopolymer*, 2022; 223–244. [CrossRef]
5. Dogra, V.; Verma, D.; Fortunati, E. Biopolymers and nanomaterials in food packaging and applications. *Nanotechnol.-Based Sustain. Altern. Manag. Plant Dis.* **2022**, 355–374. [CrossRef]
6. Essential Oil Safety–New Edition by Robert Tisserand & Rodney Young. Available online: https://roberttisserand.com/essential-oil-safety-2nd-edition/ (accessed on 31 August 2022).
7. Singh, B.; Singh, J.P.; Kaur, A.; Yadav, M.P. Insights into the chemical composition and bioactivities of citrus peel essential oils. *Food Res. Int.* **2021**, *143*, 110231. [CrossRef]
8. Goñi, P.; López, P.; Sánchez, C.; Gómez-Lus, R.; Becerril, R.; Nerín, C. Antimicrobial activity in the vapour phase of a combination of cinnamon and clove essential oils. *Food Chem.* **2009**, *116*, 982–989. [CrossRef]
9. Cheung, K.S.; Hung, I.F.N.; Chan, P.P.Y.; Lung, K.C.; Tso, E.; Liu, R.; Ng, Y.Y.; Chu, M.Y.; Chung, T.W.; Tam, A.R.; et al. Gastrointestinal Manifestations of SARS-CoV-2 Infection and Virus Load in Fecal Samples from a Hong Kong Cohort: Systematic Review and Meta-analysis. *Gastroenterology* **2020**, *159*, 81–95. [CrossRef]
10. Dosoky, N.S.; Setzer, W.N. Biological Activities and Safety of Citrus sEssential Oils. *Int. J. Mol. Sci.* **2018**, *19*, 1966. [CrossRef]
11. Hosseini, M.; Jamshidi, A.; Raeisi, M.; Azizzadeh, M. Effect of sodium alginate coating containing clove (*Syzygium Aromaticum*) and lemon verbena (*Aloysia Citriodora*) essential oils and different packaging treatments on shelf-life extension of refrigerated chicken breast. *J. Food Process. Preserv.* **2021**, *45*, e14946. [CrossRef]
12. Dagli, N.; Dagli, R.; Mahmoud, R.S.; Baroudi, K. Essential oils, their therapeutic properties, and implication in dentistry: A review. *J. Int. Soc. Prev. Community Dent.* **2015**, *5*, 335–340. [CrossRef]
13. Denkova-Kostova, R.; Teneva, D.; Tomova, T.; Goranov, B.; Denkova, Z.; Shopska, V.; Slavchev, A.; Hristova-Ivanova, Y. Chemical composition, antioxidant and antimicrobial activity of essential oils from tangerine (*Citrus reticulata L.*), grapefruit (*Citrus paradisi L.*), lemon (*Citrus lemon L.*) and cinnamon (*Cinnamomum zeylanicum Blume*). *Z. Naturforsch. C* **2020**, *76*, 175–185. [CrossRef]
14. Yazgan, H.; Ozogul, Y.; Kuley, E. Antimicrobial influence of nanoemulsified lemon essential oil and pure lemon essential oil on food-borne pathogens and fish spoilage bacteria. *Int. J. Food Microbiol.* **2019**, *306*, 108266. [CrossRef]
15. Ahmed, L.I.; Ibrahim, N.; Abdel-Salam, A.B.; Fahim, K.M. Potential application of ginger, clove and thyme essential oils to improve soft cheese microbial safety and sensory characteristics. *Food Biosci.* **2021**, *42*, 101177. [CrossRef]
16. Rosarior, V.L.; Lim, P.S.; Wong, W.K.; Yue, C.S.; Yam, H.C.; Tan, S.A. Antioxidant-rich Clove Extract, A Strong Antimicrobial Agent against Urinary Tract Infections-causing Bacteria in vitro. *Trop. Life Sci. Res.* **2021**, *32*, 45–63. [CrossRef]
17. Kumar Dey, B.; Sanyal Mukherjee, S. Potential of clove and its nutritional benefits in physiological perspective: A review. *Int. J. Physiol. Nutr. Phys. Educ.* **2021**, *6*, 103–106.
18. Senthil Kumar, K.J.; Gokila Vani, M.; Wang, C.-S.; Chen, C.-C.; Chen, Y.-C.; Lu, L.-P.; Huang, C.-H.; Lai, C.-S.; Wang, S.-Y. Geranium and Lemon Essential Oils and Their Active Compounds Downregulate Angiotensin-Converting Enzyme 2 (ACE2), a SARS-CoV-2 Spike Receptor-Binding Domain, in Epithelial Cells. *Plants* **2020**, *9*, 770. [CrossRef]
19. Razmjoo, F.; Sadeghi, E.; Alizadeh-Sani, M.; Noroozi, R.; Azizi-Lalabadi, M. Fabrication and application of functional active packaging material based on carbohydrate biopolymers formulated with Lemon verbena/Ferulago angulata extracts for the preservation of raw chicken meat. *J. Food Process. Preserv.* **2022**, *46*, e16830. [CrossRef]
20. Marzlan, A.A.; Muhialdin, B.J.; Abedin, N.H.Z.; Manshoor, N.; Ranjith, F.H.; Anzian, A.; Hussin, A.S.M. Incorporating torch ginger (*Etlingera elatior Jack*) inflorescence essential oil onto starch-based edible film towards sustainable active packaging for chicken meat. *Ind. Crops Prod.* **2022**, *184*, 115058. [CrossRef]
21. Tsironi, M.; Kosma, I.S.; Badeka, A.V. The Effect of Whey Protein Films with Ginger and Rosemary Essential Oils on Microbiological Quality and Physicochemical Properties of Minced Lamb Meat. *Sustainability* **2022**, *14*, 3434. [CrossRef]
22. Abeltia, A.L.; Teka, T.A.; Forsido, S.F.; Tamiru, M.; Bultosa, G.; Alkhtib, A.; Burton, E. Bio-based smart materials for fish product packaging: A review. *Int. J. Food Prop.* **2022**, *25*, 857–871. [CrossRef]

23. Zehra, A.; Wani, S.M.; Bhat, T.A.; Jan, N.; Hussain, S.Z.; Naik, H.R. Preparation of a biodegradable chitosan packaging film based on zinc oxide, calcium chloride, nano clay and poly ethylene glycol incorporated with thyme oil for shelf-life prolongation of sweet cherry. *Int. J. Biol. Macromol.* **2022**, *217*, 572–582. [CrossRef]
24. de Souza, A.G.; da SBarbosa, R.F.; Quispe, Y.M.; dos SRosa, D. Essential oil microencapsulation with biodegradable polymer for food packaging application. *J. Polym. Environ.* **2022**, *30*, 3307–3315. [CrossRef]
25. Pandey, V.K.; Islam, R.U.; Shams, R.; Dar, A.H. A comprehensive review on the application of essential oils as bioactive compounds in Nano-emulsion based edible coatings of fruits and vegetables. *Appl. Food Res.* **2022**, *2*, 100042. [CrossRef]
26. Bauer, A.-S.; Leppik, K.; Galić, K.; Anestopoulos, I.; Panayiotidis, M.I.; Agriopoulou, S.; Milousi, M.; Uysal-Unalan, I.; Varzakas, T.; Krauter, V. Cereal and Confectionary Packaging: Background, Application and Shelf-Life Extension. *Foods* **2022**, *11*, 697. [CrossRef]
27. Banerjee, S.; Prakash, P.C.; Kadeppagari, R.K. Zein-Based Nanoproducts in Nutrition and Food Sectors. In *Handbook of Consumer Nanoproducts*; Springer: Singapore, 2021; pp. 1–16.
28. Garcia-Oliveira, P.; Pereira, A.G.; Carpena, M.; Carreira-Casais, A.; Fraga-Corral, M.; Prieto, M.A.; Simal-Gandara, J. Application of Releasing Packaging in Beverages. In *Releasing Systems in Active Food Packaging*; Springer: Cham, Switzerland, 2022; pp. 373–401.
29. Smaoui, S.; Ben Hlima, H.; Tavares, L.; Ennouri, K.; Ben Braiek, O.; Mellouli, L.; Abdelkafi, S.; Khaneghah, A.M. Application of essential oils in meat packaging: A systemic review of recent literature. *Food Control* **2022**, *132*, 108566. [CrossRef]
30. Volpe, S.; Valentino, M.; Muhammad, R.; Torrieri, E. Application of Releasing Systems in Active Packaging for Dairy Products. In *Releasing Systems in Active Food Packaging*; Springer: Cham, Switzerland, 2022; pp. 353–372.
31. Shakouri, M.; Salami, M.; Lim, L.T.; Ekrami, M.; Mohammadian, M.; Askari, G.; Emam-Djomeh, Z.; McClements, D.J. Development of active and intelligent colorimetric biopolymer indicator: Anthocyanin-loaded gelatin-basil seed gum films. *J. Food Meas. Charact.* **2022**. [CrossRef]
32. Chen, H.; Hu, X.; Chen, E.; Wu, S.; McClements, D.J.; Liu, S.; Li, B.; Li, Y. Preparation, characterization, and properties of chitosan films with cinnamaldehyde nanoemulsions. *Food Hydrocoll.* **2016**, *61*, 662–671. [CrossRef]
33. Shen, Y.; Ni, Z.J.; Thakur, K.; Zhang, J.G.; Hu, F.; Wei, Z.J. Preparation and characterization of clove essential oil loaded nanoemulsion and pickering emulsion activated pullulan-gelatin based edible film. *Int. J. Biol. Macromol.* **2021**, *181*, 528–539. [CrossRef]
34. Hadidi, M.; Pouramin, S.; Adinepour, F.; Haghani, S.; Jafari, S.M. Chitosan nanoparticles loaded with clove essential oil: Characterization, antioxidant and antibacterial activities. *Carbohydr. Polym.* **2020**, *236*, 116075. [CrossRef]
35. Navikaite-Snipaitinene, V.; Ivanauskas, L.; Jakstas, V.; Ruegg, N.; Rutkaite, R.; Wolfram, E.; Yildirim, S. Development of antioxidant food packaging materials containing eugenol for extending display life of fresh beef. *Meat Sci.* **2018**, *145*, 9–15. [CrossRef]
36. Li, M.; Yu, H.; Xie, Y.; Guo, Y.; Cheng, Y.; Qian, H.; Yao, M. Fabrication of eugenol loaded gelatin nanofibers by electrospinning technique as active packaging material. *LWT* **2021**, *139*, 110800. [CrossRef]
37. Silva, C.G.; Yudice, E.D.C.; Campini, P.A.L.; Rosa, D.S. The performance evaluation of Eugenol and Linalool microencapsulated by PLA on their activities against pathogenic bacteria. *Mater. Today Chem.* **2021**, *21*, 100493. [CrossRef]
38. Parasuraman, V.; Sharmin, A.M.; Vijaia Anand, M.A.; Sivakumar, A.S.; Surendhiran, D.; Sharesh, G.; Kim, S. Fabrication and bacterial inhibitory activity of essential oil linalool loaded biocapsules against *Escherichia coli*. *J. Drug Deliv. Sci. Technol.* **2022**, *74*, 103495. [CrossRef]
39. Wei, A.; Shiabamoto, T. Antioxidant/Lipoxygenase Inhibitory Activities and Chemical Compositions of Selected Essential Oils. *J. Agric. Food Chem.* **2010**, *58*, 7218–7225. [CrossRef]
40. Puscaselu, R.G.; Besliu, I.; Gutt, G. Edible Biopolymers-Based Materials for Food Applications—The Eco Alternative to Conventional Synthetic Packaging. *Polymers* **2021**, *13*, 3779. [CrossRef]
41. Teleky, B.E.; Mitrea, L.; Plamada, D.; Nemes, S.A.; Calinoiu, L.F.; Pascuta, M.S.; Varvara, R.A.; Szabo, K.; Vajda, P.; Szekely, C.; et al. Development of Pectin and Poly (vinyl alcohol)-Based Active Packaging Enriched with Itaconic Acid and Apple Pomace-Derived Antioxidants. *Antioxidants* **2022**, *11*, 1729.
42. Mellinas, C.; Ramos, M.; Grau-Atineza, A.; Jorda, A.; Burgos, N.; Jimenez, A.; Serrano, E.; Garrigos, M. Biodegradable Poly(ε-Caprolactone) Active Films Loaded with MSU-X Mesoporous Silica for the Release of α-Tocopherol. *Polymers* **2020**, *12*, 137. [CrossRef]
43. Available online: https://petro-pack.com/wp-content/uploads/ASTM-D882_2010_513100089384.pdf (accessed on 10 October 2022).
44. da Silva, A.O.; Cortez-Vega, W.R.; Prentice, C.; Fonseca, G.G. Development and characterization of biopolymer films based on bocaiuva (Acromonia aculeata) flour. *Int. J. Biol. Macromol.* **2020**, *155*, 1157–1168. [CrossRef]
45. Rincón, E.; Bautista, J.M.; Espinosa, E.; Serrano, L. Biopolymer-based sachets enriched with acorn shell extracts produced by ultrasound-assisted extraction for active packaging. *J. APoly. Sci.* **2022**, *139*, e53102.
46. Shekh, M.I.; Zhu, G.; Xiong, W.; Wu, W.; Stadler, F.J.; Patel, D.; Zhu, C. Dynamically bonded, tough, and conductive MXene@oxidized sodium alginate: Chitosan based multi-networked elastomeric hydrogels for physical motion detection. *Int. J. Biol. Macromol.* **2022**, *in press*. [CrossRef]
47. Yakimets, I.; Paes, S.S.; Wellner, N.; Smith, A.C.; Wilson, R.H.; Mitchell, J.R. Effect of Water Content on the Structural Reorganization and Elastic Properties of Biopolymer Films: A Comparative Study. *Biomacromolecules* **2007**, *8*, 1710–1722. [CrossRef]

48. Venegas, R.; Torres, A.; Rueda, A.M.; Morales, M.A.; Arias, M.J.; Porras, A. Development and Characterization of Plantain (*Musa paradisiaca*) Flour-Based Biopolymer Films Reinforced with Plantain Fibers. *Polymers* **2022**, *14*, 748. [CrossRef]
49. Sukhawipat, N.; Saengdee, L.; Pasetto, P. *Caesalpinia sappan L.* wood fiber: Bio-reinforcement for polybutylene succinate-based biocomposite film. *Cellulose* **2022**, *29*, 3375–3387. [CrossRef]
50. Lima, D.B.; Almeida, R.D.; Pasquali, M.; Borges, S.; Fook, M.L.; Lisbo, H.M. Physical characterization and modeling of chitosan/peg blends for injectable scaffolds. *Carbohydr. Polym.* **2018**, *189*, 238–249. [CrossRef]
51. Aloui, H.; Deshmukh, A.R.; Khomlaem, C.; Kim, B.S. Novel composite films based on sodium alginate and gallnut extract with enhanced antioxidant, antimicrobial, barrier and mechanical properties. *Food Hydrocoll.* **2021**, *113*, 106508. [CrossRef]

gels

MDPI

Article

The Future Packaging of the Food Industry: The Development and Characterization of Innovative Biobased Materials with Essential Oils Added

Roxana Gheorghita Puscaselu [1], Andrei Lobiuc [1,*] and Gheorghe Gutt [2]

[1] Faculty of Medicine and Biological Sciences, Stefan Cel Mare University of Suceava, 720229 Suceava, Romania
[2] Faculty of Food Engineering, Stefan Cel Mare University of Suceava, 720229 Suceava, Romania
* Correspondence: andrei.lobiuc@usm.ro

Abstract: The need to replace conventional, usually single-use, packaging materials, so important for the future of resources and of the environment, has propelled research towards the development of packaging-based on biopolymers, fully biodegradable and even edible. The current study furthers the research on development of such films and tests the modification of the properties of the previously developed biopolymeric material, by adding 10, respectively 20% *w/v* essential oils of lemon, grapefruit, orange, cinnamon, clove, mint, ginger, eucalypt, and chamomile. Films with a thickness between 53 and 102 μm were obtained, with a roughness ranging between 147 and 366 nm. Most films had a water activity index significantly below what is required for microorganism growth, as low as 0.27, while all essential oils induced microbial growth reduction or 100% inhibition. Tested for the evaluation of physical, optical, microbiological or solubility properties, all the films with the addition of essential oil in the composition showed improved properties compared to the control sample.

Keywords: biopolymers; antimicrobial; environment; film; coating

Citation: Puscaselu, R.G.; Lobiuc, A.; Gutt, G. The Future Packaging of the Food Industry: The Development and Characterization of Innovative Biobased Materials with Essential Oils Added. *Gels* **2022**, *8*, 505. https://doi.org/10.3390/gels8080505

Academic Editors: Aris E. Giannakas, Constantinos E. Salmas and Charalampos Proestos

Received: 24 July 2022
Accepted: 12 August 2022
Published: 14 August 2022

1. Introduction

The use of biopolymers in the development of packaging materials is increasingly common due to the properties they possess: biodegradability, compostability, non-toxicity, reuseability, ease of handling, the possibility of incorporating various active substances, but also relatively low costs [1,2]. Their use has been transposed to other fields, such as the biomedical one. Due to high biocompatibility, biopolymers are today successfully used in tissue engineering or wound dressing development. Numerous hydrocolloids have been used as food packaging film, with high performances in terms of the ability to withstand environments with high humidity, preserving or even improving sensory characteristics, extending shelf-life, physico-chemical, optical or mechanical properties [3–5]. Their edible character was another advantage in obtaining various types of films and coatings, used successfully for packaging food or coating fresh fruits and vegetables [6–8]. At the moment, the most used biopolymers are those based on polysaccharides (agar, sodium alginate, carrageenan, chitosan, etc.) [9], followed by those based on proteins (collagen, casein, soy or wheat protein, etc.) and lipid ones (waxes, fatty acids, acylglycerols) [10]. Since a biopolymer usually does not exhibit high physico-chemical or mechanical performances at the same time, the possibility of combining them represents the best alternative in obtaining a material with characteristics similar to plastic [11].

Research in the field has demonstrated the fact that biopolymers possess biological properties, as they are biocompatible, non-toxic, and non-immunogenic. They do not induce allergic reactions after ingestion, psychological properties, as they contribute to the stability of the products they contain and increase the shelf life of them. They also allow the transport of various solutes, gases, water vapors, and organic molecules.

Polysaccharides such as sodium alginate or agar have been intensively studied due to their synergistic character and the ability to obtain films with very good properties: resistant, flexible, with high elasticity, glossy, with a homogeneous microstructure, without pores or fissures in the composition [12].

Alginate is a marine biopolymer that can form soluble films when obtained by casting or drying, or insoluble films when crosslinked with calcium salt [13]. The biggest advantages in using alginate are the low cost of obtaining them and the multiple possibilities of use in the food and biomedical/pharmaceutical fields [14,15], due to characteristics such as good gelling and film forming properties, pH-sensitivity, mucoadhesiveness, or crosslinking capacity [16,17].

Agar is a polysaccharide extracted from red algae. In 1972, it received the status as *Generally Recognized as Safe* by GRAS and, as well as alginate, can be used in quantum statis. Almost 80% of globally produced agar is used for human consumption. Used as a film-forming component, it forms brittle material, with poor mechanical properties, moisture barrier and thermal stabilities [18]. Therefore, for use for this purpose, it is combined with other biopolymers or plasticizers [19].

Active and smart packaging are increasingly used all over the world, especially in Japan, the USA, and Australia. In Europe, their use was regulated in 2004 by Regulation EC, 1935/2004. In 2009, the EU established "Commission Regulation (EC) No 450/2009, on active and intelligent materials and articles intended to come into contact with food". Thus, according to the European Union, active packaging systems are designed "to deliberately incorporate components that would release or absorb substances into or from the packaged food or the environment surrounding the food" [20]. If active films based on biopolymers are used in many fields, the coatings find their applicability to fruits and vegetables. Their use is important if we take into account that most losses occur during storage, handling or transport, especially due to the fact that, even after harvesting, fruits and vegetables continue their respiratory process; applying such a coating will inhibit the ripening processes [21].

Biopolymer films and coatings can be used as active delivery systems when they contain substances with an antimicrobial or antioxidant effect [22,23]. Thus, films containing gold, silver, or copper ions were effective against the development and sporulation of microorganisms, inhibiting the growth of *Staphylococcus aureus*, *Escherichia coli*, *Candida albicans*, and *Aspergillus niger* [24]. Furthermore, the physicochemical properties were improved [25]. Recently, however, the addition of essential oils in the biopolymeric matrix has attracted the attention of researchers, who intensified their studies on the development of materials with superior characteristics. Thus, essential oils embedded in films showed inhibitory activity against numerous foodborne pathogens (Table 1).

Table 1. Examples of bio-based films incorporating essential oils (EOs) with antimicrobial properties.

Biopolymer	EO	Food Product	Beneficial Effect	References
Starch	Cinnamon	Active packaging	Thermal stability, but porous microstructure	[26]
Chitosan	Lemon	Citrus	Extended shelf -life, improved food storage quality. antimicrobial effect	[27]
Collagen/ chitosan	Lemon	Pork meat	Prolonged the shelf life for 21 days, inhibited lipid oxidation, and prevent microbial proliferation	[28]
Pectin	Clove	Fish (bream)	Inhibited the growth of Gram-negative bacteria; the level of lactic acid bacteria remained constant	[29]
Whey protein isolate	Clove	Cheese	Positive effects of the physical-chemical properties of cheese. *E. coli*, *S. aureus*, and *L. monocytogenes* decreased during 60 days of testing	[30]

Table 1. *Cont.*

Biopolymer	EO	Food Product	Beneficial Effect	References
Alginate/k-carrageenan	Clove	Food packaging	Antimicrobial and antioxidant effect; the addition of EO reduced the mechanical properties of film, but improved the flexibility	[31]
Alginate/clay	Clove, coriander, cinnamon, cumin, caraway, marjoram	Food packaging	Inhibited the growth of Gram-negative bacteria	[32]
Alginate/CMC	Clove	Fish	Without loss of color, odor, texture during storage (16 days)	[33]
Chitosan/pectin/starch	Mint	Food packaging	Antioxidant and antimicrobial effect; improved barrier properties, tensile strength, and thermal stability	[34]
Sodium alginate	Carvacrol	White mushrooms	Improved mechanical properties, water resistance, light barrier property, and antiviral properties	[35]
Chitosan	Eucalyptus	Sliced sausages	Good antimicrobial activity	[36]
Gelatin	Chamomile and peppermint	Edible packaging	The antioxidant activity and bioactivity were improved	[37]
Cellulose	Ginger	Barbecue chicken	Improved spoilage control; the addition of essential oils prolonged the meat shelf life by more than 6 days.	[38]
Chitosan	Ginger	Fresh poultry meat	Reduced coliforms proliferation; the addition of EO showed minimum changes than the product uncoated	[39]
Chitosan	Ginger	Active packaging	Strong antimicrobial activity	[40]
Chitosan/protein	Ginger	Fish	Stored at 4 °C, the shelf life has been extended	[41]

Due to the biologically active compounds, essential oils are increasingly recognized and used in the packaging materials industry. However, usually, the properties of the films change after the addition of essential oils. For example, clove oil, when embedded in films based on alginate and carrageenan, showed strong antimicrobial activity against *E. coli*, although the tensile strength of the films was lower. The inhibition zone increased directly proportional with clove oil concentration [31]. *Mostaghimi* et al. identified that the addition of clove oil to the film based on sodium alginate was more effective against *Bacillus cereus* than *E. coli* and that the properties of the material were not negatively influenced by the incorporation of essential oil [42]. Even so, if the safety of the ingested food is taken into account, it is much more important to pursue the microbial inhibition, since the properties of the films can be improved by various other additions. The antimicrobial activity of essential oils is high. They are effective against the development and proliferation of several pathogenic microorganisms (Table 2).

Table 2. Essential oils displaying antimicrobial properties against relevant microorganism categories and species.

EO	M	TC	*LM*	CF	*SA*	*EC*	References
Cinnamon	√	√	√	√	√	√	[43–46]
Lemon	√	√		√	√	√	[27,46–48]
Grapefruit	√	√		√	√		[48–50]
Orange	√	√		√	√	√	[48,50]
Clove		√	√	√	√	√	[29–32]
Peppermint	√	√	√			√	[34,51]
Eucalyptus		√			√	√	[36]
Chamomile		√			√	√	[37]
Ginger		√	√	√	√	√	[39,52]

M—molds, TC—total count, *LM*—*Listeria monocytogenes*, CF—coliforms, *SA*—*Staphylococcus aureus*, *EC*—*Escherichia coli*.

The current study followed the development of films based on sodium alginate, agar, and water, according to a composition that represents, at the time of writing this study, a patent proposal sent to the Romanian State Office for Inventions and Trademarks. Since specialized studies have highlighted the active nature of films with the addition of essential oil, we used lemon, grapefruit, orange, cinnamon, clove, mint, ginger, eucalypt, and chamomile oils and we tested their effect on the films' properties. The obtained results strengthen the specialized studies and recommend their use in the development of materials with superior characteristics.

2. Results and Discussion

The films obtained by the casting method were observed immediately after drying and subsequently tested. All the samples that contained essential oil in the composition presented a specific taste and smell, the intensity of which varied with the volume of oil added to the film-forming solution. All the films were glossy, pleasant to the touch, soft, with regular edges and without undissolved particles. The absence of insoluble particles and fissures or pores can also be seen from the microscopic images and microtopographies obtained (Figure 1).

With the exception of the film with the addition of ginger essential oil, which broke during drying and showed extremely low flexibility, and the film with 10% mint EO, which was also less flexible, all other films were very flexible and allowed multiple bendings. The sample with 10% essential oil of mint showed high adhesiveness and can be used as a self-adhesive film. The sample with 20% essential oil of eucalyptus showed a tendency to tighten upon drying, which proves the need to reduce the addition of essential oil or increase the amount of glycerol or Tween 80. According to the images in Figure 1, the most homogeneous structure can be highlighted in the case of the film with the addition of 10% chamomile EO (**B17**). The result is also confirmed by its reduced roughness, unlike other samples-146.90 nm (Table 3).

According to the data in the table above, the mechanical properties of the films changed with the increase in the volume of essential oil. Thus, the films with 20% lemon, grapefruit, cinnamon, ginger, and eucalyptus EO presented lower values of tensile strength, unlike those with 10% addition. Increasing the volume of orange, clove, and mint essential oil favored the development of stronger films. The most resistant films were those with the addition of 10, respectively 20% mint EO (**B11**—0.274 MPa and **B12**—0.288 MPa), the one with the addition of 10% essential oil of chamomile (**B13**—0.261 MPa), and the one with 10% eucalyptus EO (**B17**—0.282 MPa). The addition of orange, cinnamon, cloves and 20% eucalyptus EO had a negative effect on the tensile strength, the values obtained being below that of the control sample (**B19**—0.135 MPa).

Similar to the resistance to breaking, the elongation of the samples with the addition of lemon oil showed reduced values. Thus, they were approximately 30% lower than the control sample (1.51, respectively 1.91% compared to 5.04%), while the film with 10% clove EO also showed lower values than the control sample. Even if some films showed lower elongation values with the increase in the volume of added essential oil, the values are still higher than the control sample (see films with lemon (**B1**, **B2**), cinnamon (**B7**, **B8**) and clove (**B9**, **B10**) EO).

The film with 10% essential chamomile EO (**B13**) presented the best mechanical properties, both tensile strength (0.261 MPa) and elasticity (26.11%). Unfortunately, the sample with 20% essential chamomile oil in the composition could not be sized according to the standards used. For the development of a film with even more improved properties, it is necessary to use a higher amount of plasticizer in the composition.

The highest roughness value (366.20 nm) was identified in **B3**—the sample with 10% grapefruit EO.

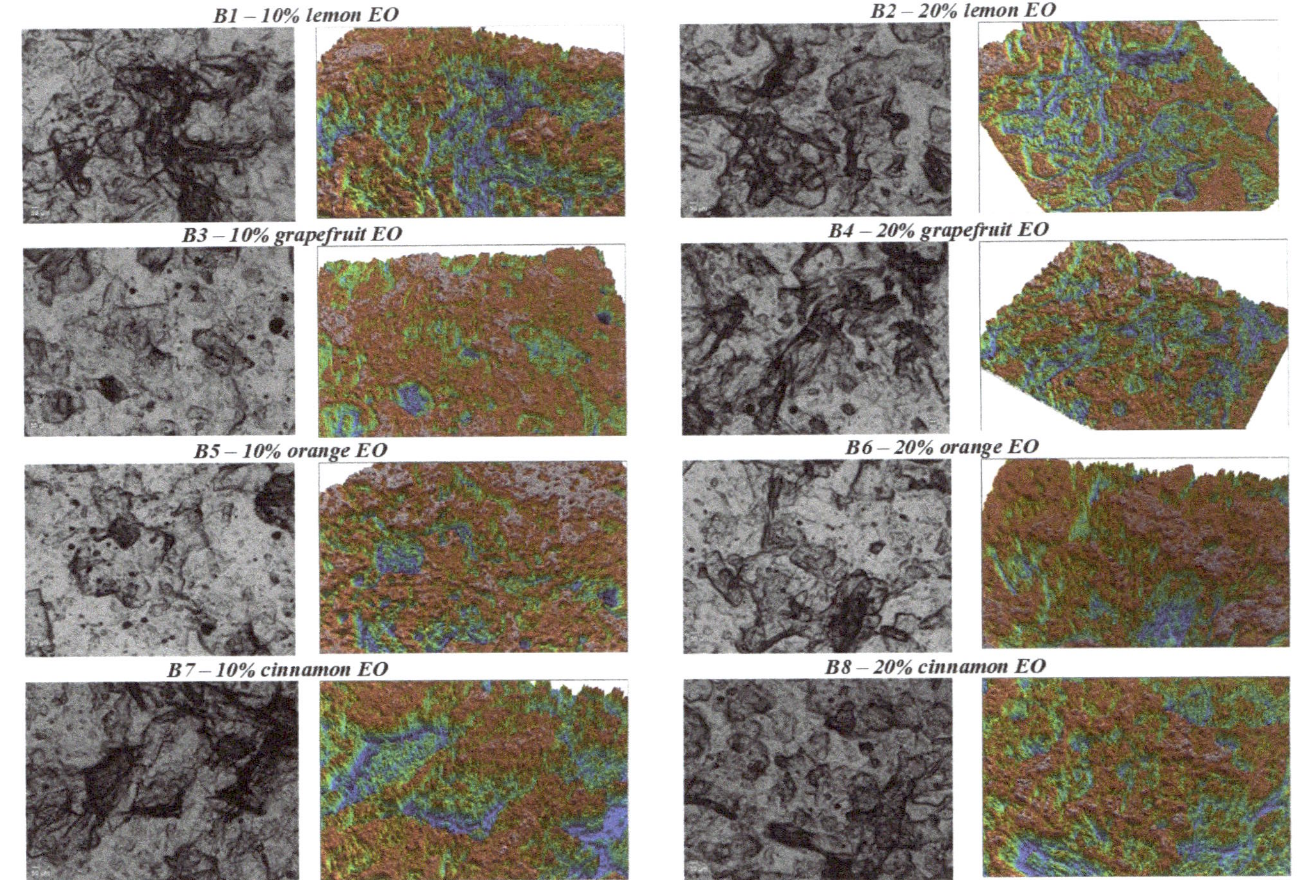

Figure 1. *Cont.*

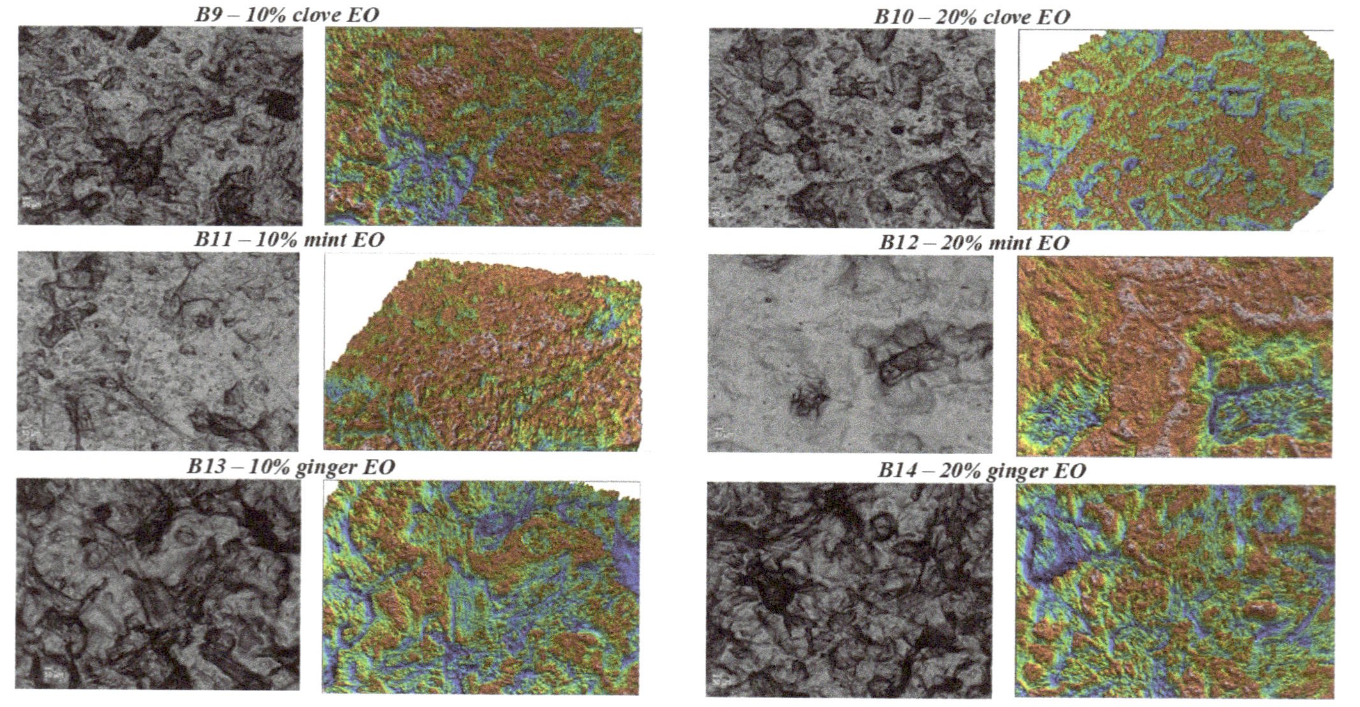

Figure 1. *Cont.*

Figure 1. Microstructural and 3D microtopographical aspects of films developed with essential oils included.

Table 3. Structural characteristics of the developed films with essential oils incorporated.

Sample	Thickness, µm	Retraction Ratio, %	Roughness, nm	Tensile Strength, MPa	Elongation, %	Moisture Content, %	Water Activity Index
B1	99.80 ^{b,c} ± 2.40	4.95 ^{i,j} ± 2.28	253.10 ^{b,c,d,e,f} ± 2.99	0.101 ^{l} ± 0.10	1.51 ^{r} ± 0.01	11.95 ^{e} ± 0.03	0.33 ^{c} ± 0.01
B2	100.60 ^{b} ± 1.85	4.19 ^{j} ± 1.76	333.20 ^{a,b} ± 3.55	0.098 ^{l} ± 0.05	1.91 ^{q} ± 0.05	11.24 ^{f} ± 0.01	0.29 ^{d} ± 0.04
B3	88.80 ^{f,g} ± 1.47	15.43 ^{e,f} ± 1.40	366.20 ^{a} ± 3.03	0.231 ^{e} ± 0.05	14.97 ^{d} ± 0.20	13.24 ^{c} ± 0.22	0.30 ^{c,d} ± 0.01
B4	90.80 ^{e,f} ± 1.17	13.52 ^{f,g} ± 1.11	243.70 ^{b,c,d,e,f,g} ± 4.16	0.172 ^{f} ± 0.01	11.04 ^{g} ± 0.33	11.27 ^{f} ± 0.61	0.30 ^{c,d} ± 0.01
B5	81.00 ^{h,i} ± 2.68	22.86 ^{c,d} ± 2.55	255.00 ^{b,c,d,e,f} ± 1.03	0.133 ^{l} ± 0.01	7.10 ^{j} ± 0.15	13.73 ^{b} ±0.13	0.30 ^{c,d} ± 0.08
B6	93.60 ^{d,e} ± 1.85	10.86 ^{g,h} ± 1.76	304.87 ^{a,b,c,d} ± 0.85	0.146 ^{g} ± 0.01	5.37 ^{n} ± 0.11	13.53 ^{b} ±0.29	0.30 ^{c,d} ± 0.03
B7	98.80 ^{b,c} ± 1.94	5.90 ^{i,j} ± 1.84	307.47 ^{a,b,c,d} ± 1.96	0.137 ^{i} ± 0.01	6.92 ^{k} ± 0.10	13.13 ^{c} ± 0.14	0.33 ^{c} ± 0.01
B8	101.00 ^{b} ± 0.89	3.81 ^{j} ± 0.85	308.97 ^{a,b,c,d} ± 0.47	0.130 ^{k} ± 0.05	8.33 ^{i} ± 0.21	12.74 ^{d} ±0.68	0.30 ^{c,d} ± 0.02
B9	91.60 ^{e,f} ± 1.50	12.76 ^{f,g} ± 1.42	293.67 ^{a,b,c,d} ± 0.59	0.093 ^{m} ± 0.01	3.74 ^{p} ±0.07	13.24 ^{c} ± 0.76	0.29 ^{c,d} ± 0.01
B10	98.80 ^{a,b} ± 1.67	4.76 ^{i,j} ± 1.59	314.33 ^{a,b,c} ± 0.51	0.142 ^{h} ± 0.01	13.87 ^{e} ± 0.07	13.04 ^{c} ± 0.75	0.27 ^{d} ± 0.01
B11	59.40 ^{j} ± 1.36	43.43 ^{b} ± 1.29	216.17 ^{d,e,f,g} ± 0.68	0.274 ^{c} ± 0.05	17.45 ^{b} ± 0.18	11.26 ^{f} ± 0.25	0.52 ^{b} ± 0.02
B12	76.80 ^{i} ± 1.60	26.86 ^{c} ± 1.52	169.23 ^{f,g} ± 0.57	0.288 ^{a} ± 0.15	16.42 ^{c} ± 0.02	10.34 ^{g} ±0.20	0.52 ^{a,b} ± 0.02
B13	101.20 ^{b} ± 1.33	3.62 ^{j} ± 1.26	248.37 ^{b,c,d,e,f} ± 4.46	0.261 ^{d} ± 0.10	26.11 ^{a} ± 0.03	11.73 ^{e} ± 0.19	0.56 ^{a} ± 0.03
B14	102.00 ^{a,b} ±1.10	2.86 ^{j} ± 1.04	288.27 ^{a,b,c,d,e} ± 4.02	*	*	8.73 ^{h} ± 0.20	0.56 ^{a} ± 0.02
B15	92.00 ^{d,e,f} ± 0.89	12.38 ^{f,g,h} ± 0.85	230.60 ^{c,d,e,f,g} ± 3.04	0.175 ^{g} ± 0.10	8.93 ^{h} ± 0.02	8.51 ^{h} ± 0.43	0.56 ^{a} ± 0.02
B16	98.80 ^{b,c} ± 1.33	5.90 ^{i,j} ± 1.26	192.03 ^{e,f,g} ± 3.61	0.148 ^{g} ± 0.05	6.71 ^{l} ± 0.06	19.34 ^{a} ±0.11	0.55 ^{a,b} ± 0.01
B17	53.00 ^{k} ± 1.67	49.52 ^{a} ± 1.59	146.90 ^{g} ± 0.57	0.282 ^{b} ± 0.02	12.98 ^{f} ± 0.01	2.43 ^{j} ± 0.19	0.53 ^{a,b} ± 0.02
B18	96.20 ^{c,d} ± 1.17	8.38 ^{h,i} ± 1.11	222.60 ^{c,d,e,f,g} ± 3.63	0.131 ^{k} ± 0.10	6.44 ^{m} ± 0.01	4.54 ^{i} ± 0.36	0.53 ^{a,b} ± 0.01
B19	85.20 ^{g,h} ± 1.94	18.86 ^{d,e} ± 1.84	176.93 ^{f,g} ± 1.17	0.135 ^{i,j} ± 0.10	5.04 ^{o} ± 0.02	11.44 ^{f} ± 0.18	0.28 ^{d} ± 0.02

* The testing could not be carried out because the sample could not be sized according to STAS ASTM D882. Means that do not share a letter are significantly different. a–j, Significance level $\alpha = 0.05$.

The thickness of all the films increased with the increase in the volume of added essential oil (Table 3). The biggest difference can be observed in the case of films with the addition of chamomile EO, when the thickness varied by 43.20 µm. Thus, sample **B17**, with 10% addition of EO, had a thickness of 53 µm, and **B18**, with 20% EO, had 96.20 µm. The smallest difference can be observed in the case of samples with ginger essential oil (**B13**, **B14**), when the increase in the volume of essential oil led to an increase in thickness by only 0.8 µm. The same thickness value can be observed in case of **B1** and **B2** samples, with lemon EO added. The samples with 10% orange EO (**B5**), 10 and 20% mint EO (**B11**, **B12**) and 10% chamomile EO (**B17**) had lower thicknesses than the control sample.

The values of the retraction ratio are directly correlated with those of the thickness. This means that, when it will be replicated on an industrial scale, the processor can establish the final composition of the materials knowing the retraction ratio and thickness values. The lowest retraction ratio is observed for ginger EO films (**B13**, **B14**). For all developed films, the reattraction ratio values decrease with the increase in the volume of essential oil added to the composition. This can be attributed to their hydrophobic character, so that the matrix becomes more compact and prevents the evaporation of water molecules from the composition and, implicitly, the withdrawal of the film during drying.

Except for samples **B15–B18**, all the films presented lower values of moisture content with the increase in the volume of EO. Compared to the control sample, most films had minor (approximately 2%) moisture content variation, except for sample B16 (20% eucalyptus), where MC was higher with 8 percentual points and samples B17 and B18 (10 and, respectively, 20% chamomile), where MC was lower with as much as 9 percentual points.

The water activity index showed similar values for the samples with the addition of lemon, grapefruit, orange, cinnamon, or clove essential oil (0.30 ± 0.03). Those with additions of mint, ginger, eucalyptus, or chamomile showed similar values (0.50 ± 0.6), but higher than the others. The control sample showed reduced values of the water activity index (0.28), similar to those of the films with the addition of clove EO (**B9**, **B10** with 0.29, respectively 0.27).

According to the images in Figure 1, the addition of essential oil had negative effects on the microstructure of the films. Thus, all samples with 10% EO volume in the composition presented more homogeneous structures than those with 20%. This may indicate the need to increase the amount of plasticizer in the composition or increase the mixing speed during the homogenization of the film-forming solution.

The transmittance of the samples varied depending on the essential oil added to the film-forming solution (Table 4). Thus, the addition of lemon, grapefruit, orange, clove, cinnamon, and ginger oil favored the develompent of more transparent films than the control sample (transmittance 68.9%). A quantity of 10% essential oil of chamomile added had the effect of obtaining a film with the highest transmittance (77%), but the addition of a further 10% essential oil favoured the development of a film with the lowest value of transmittance, 18.33%. Mint essential oil, regardless of the added concentration, had the effect of increasing the transmittance by 2.8, respectively 4.6% compared to the control sample.

Table 4. Optical properties of films with essential oils incorporated.

Sample	Transmittance, %	Opacity, A × mm^{-1}	Color		
			L*	a*	b*
B1	49.40 ^{i} ± 0.10	3.31 ^{e} ± 0.04	88.79 ^{b,c} ± 0.18	−0.49 ^{a} ± 0.02	10.76 ^{e} ± 0.31
B2	40.66 ^{k} ± 0.75	3.59 ^{b} ± 0.12	88.99 ^{b,c} ± 0.16	−0.42 ^{a} ± 0.24	10.61 ^{e} ± 0.29
B3	51.00 ^{g} ± 0.20	3.21 ^{f} ± 0.14	88.30 ^{b,c,d} ± 0.39	−0.54 ^{a} ± 0.02	11.33 ^{d,e} ± 0.48
B4	52.60 ^{e} ± 0.20	3.49 ^{c} ± 0.22	88.87 ^{b,c} ± 0.96	−0.61 ^{a} ± 0.05	10.72 ^{e} ± 0.53
B5	53.10 ^{e} ± 0.10	3.47 ^{c} ± 0.15	88.88 ^{b,c} ± 0.73	−0.51 ^{a} ± 0.86	9.35 ^{e} ± 0.65
B6	51.73 ^{f} ± 0.15	3.50 ^{c} ± 0.10	89.17 ^{b} ± 0.63	−0.48 ^{a} ± 0.15	9.85 ^{e} ± 0.73
B7	50.16 ^{h} ± 0.11	3.26 ^{e,f} ± 0.20	89.17 ^{b} ± 0.30	−0.48 ^{a} ± 0.16	10.03 ^{e} ± 0.28
B8	50.83 ^{g,h} ± 0.11	3.02 ^{h} ± 0.09	88.58 ^{b,c} ± 0.92	−0.49 ^{a} ± 0.02	11.21 ^{d,e} ± 0.83
B9	50.43 ^{g,h} ± 0.15	3.39 ^{d} ± 0.08	88.50 ^{b,c,d} ± 0.93	−0.47 ^{a} ± 0.03	11.16 ^{d,e} ± 1.02
B10	52.73 ^{e} ± 0.11	3.03 ^{g,h} ± 0.16	88.31 ^{b,c,d} ± 0.68	−0.52 ^{a} ± 0.01	10.73 ^{e} ± 0.65
B11	71.63 ^{c} ± 0.11	2.53 ^{j} ± 0.08	92.25 ^{a} ± 0.79	−5.78 ^{c} ± 0.07	13.70 ^{b,c} ± 0.53
B12	73.56 ^{b} ± 0.12	1.88 ^{l} ± 0.21	92.18 ^{a} ± 0.15	−5.79 ^{c} ± 0.02	12.84 ^{c,d} ± 0.52
B13	70.01 ^{a} ± 0.03	3.99 ^{a} ± 0.12	91.00 ^{a} ± 0.64	−5.76 ^{c} ± 0.01	15.26 ^{b} ± 0.94
B14	18.33 ^{l} ± 0.21	3.08 ^{g} ± 0.12	92.07 ^{a} ± 0.28	−5.85 ^{c} ± 0.03	13.61 ^{b,c} ± 0.37
B15	41.30 ^{k} ± 0.10	1.52 ^{m} ± 0.17	92.19 ^{a} ± 0.28	−5.83 ^{c} ± 0.03	13.11 ^{c,d} ± 0.42
B16	46.60 ^{j} ± 0.20	1.15 ^{n} ± 0.12	91.83 ^{a} ± 0.15	−5.76 ^{c} ± 0.08	14.31 ^{b,c} ± 1.17
B17	73.5 ^{b} ± 0.20	2.87 ^{i} ± 0.23	92.09 ^{a} ± 0.44	−5.91 ^{c} ± 0.10	12.84 ^{c,d} ± 0.67
B18	40.73 ^{k} ± 0.11	1.5 ^{m} ± 0.13	86.97 ^{d} ± 1.09	−5.18 ^{b} ± 0.22	21.97 ^{a} ± 1.09
B19	68.83 ^{d} ± 0.11	1.97 ^{k} ± 0.11	87.63 ^{c,d} ± 0.41	−0.62 ^{a} ± 0.02	10.32 ^{e} ± 0.56

* Means that do not share a letter are significantly different. a–n, Significance level $\alpha = 0.05$.

The color was influenced by the type of essential oil used, although it did not vary much from the control sample. All films showed high luminosity.

The swelling index values were inversely proportional to the volume of EO added (Figure 2). This is perfectly normal if we take into account the strong hydrophobic nature of essential oils. The obtained results confirm the fact that, when it is desired to use these materials for the packaging of products with high humidity, the solubility can be improved by increasing the volume of EO added to the composition. According to Figure 2, all the samples showed a swelling tendency in the first 7 min after immersion in water (room temperature), followed by a much lower increase and even a plateau phase up to 20 min after immersion. It is very likely that the matrix will saturate and prevent additional liquid absorption. This is beneficial because the tested samples did not disintegrate, so there was no complete solubility of them.

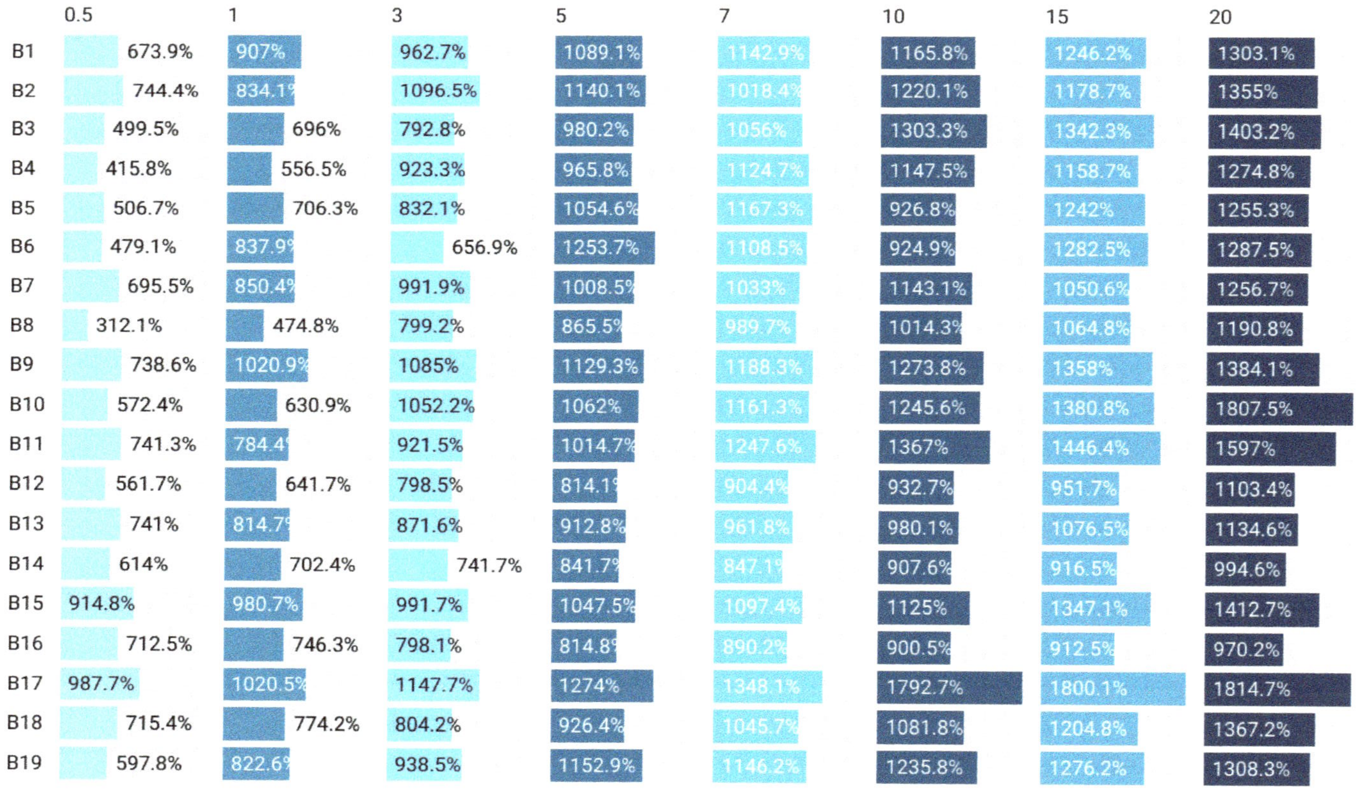

	0.5	1	3	5	7	10	15	20
B1	673.9%	907%	962.7%	1089.1%	1142.9%	1165.8%	1246.2%	1303.1%
B2	744.4%	834.1%	1096.5%	1140.1%	1018.4%	1220.1%	1178.7%	1355%
B3	499.5%	696%	792.8%	980.2%	1056%	1303.3%	1342.3%	1403.2%
B4	415.8%	556.5%	923.3%	965.8%	1124.7%	1147.5%	1158.7%	1274.8%
B5	506.7%	706.3%	832.1%	1054.6%	1167.3%	926.8%	1242%	1255.3%
B6	479.1%	837.9%	656.9%	1253.7%	1108.5%	924.9%	1282.5%	1287.5%
B7	695.5%	850.4%	991.9%	1008.5%	1033%	1143.1%	1050.6%	1256.7%
B8	312.1%	474.8%	799.2%	865.5%	989.7%	1014.3%	1064.8%	1190.8%
B9	738.6%	1020.9%	1085%	1129.3%	1188.3%	1273.8%	1358%	1384.1%
B10	572.4%	630.9%	1052.2%	1062%	1161.3%	1245.6%	1380.8%	1807.5%
B11	741.3%	784.4%	921.5%	1014.7%	1247.6%	1367%	1446.4%	1597%
B12	561.7%	641.7%	798.5%	814.1%	904.4%	932.7%	951.7%	1103.4%
B13	741%	814.7%	871.6%	912.8%	961.8%	980.1%	1076.5%	1134.6%
B14	614%	702.4%	741.7%	841.7%	847.1%	907.6%	916.5%	994.6%
B15	914.8%	980.7%	991.7%	1047.5%	1097.4%	1125%	1347.1%	1412.7%
B16	712.5%	746.3%	798.1%	814.8%	890.2%	900.5%	912.5%	970.2%
B17	987.7%	1020.5%	1147.7%	1274%	1348.1%	1792.7%	1800.1%	1814.7%
B18	715.4%	774.2%	804.2%	926.4%	1045.7%	1081.8%	1204.8%	1367.2%
B19	597.8%	822.6%	938.5%	1152.9%	1146.2%	1235.8%	1276.2%	1308.3%

Figure 2. Swelling ratio index of films developed with EO included. B1–B19 sample designation, 0.5–20 represents the time, in minutes, of maintaining the sample in the liquid.

The results of microbiological determinations showed a reduction in the total number of forming colony units in all samples, unlike the control sample (Table 5). Only the films with the addition of eucalyptus and chamomile EO did not show any contamination. It is very important that, before use, essential oils to be tested against microbiological contamination and to be purchased only from reliable producers. Even if, in general, they have an antimicrobial effect, they can become contaminated after being obtained.

Table 5. Microbiological assessment (ufc/g) of films developed with essential oils incorporated for relevant microorganism categories and species.

Sample	TC	EC	ETC	CF	YM	X-SA	LM
B1	1	-	-	1	-	-	-
B2	-	-	-	-	-	-	-
B3	15	-	-	-	-	-	-
B4	21	-	-	-	-	1	-
B5	-	-	-	-	-	-	-
B6	7	-	-	-	-	-	-
B7	2	-	-	-	-	-	-
B8	-	-	-	-	-	-	-
B9	13	-	-	2	-	-	-
B10	1	-	-	-	-	-	-
B11	-	-	-	2	-	-	-
B12	8	-	-	-	-	-	-
B13	19	-	-	-	-	1	-
B14	6	-	-	-	-	-	-
B15	-	-	-	-	-	-	-
B16	-	-	-	-	-	-	-
B17	-	-	-	-	-	-	-
B18	-	-	-	-	-	-	-
B19	28	-	-	-	-	-	-

TC—total count, EC—*Escherichia coli*, ETC—*Enterococcus*, CF—*coliforms*, YM—yeasts and molds, X-SA—*Staphylococcus aureus*, LM—*Listeria monocytogenes*.

The obtained results reinforce the possibility that these types of films can be successfully used as packaging materials in the food industry (Figure 3). Depending on the composition, the characteristics of the material may differ, so that it can be used for a wide range of products, with different characteristics.

Figure 3. Examples of potential applications of developed films: protein powder packed in biopolymeric foils with orange (**B6**) and cinnamon (**B8**) EO added. Each package contains a measure of protein powder, according to the manufacturer. The packages are glued by hot thermowelding (180 °C, 20 s). The addition of orange and cinnamon EO may improve the sensorial properties of such powders and may sustain beneficial effect on health.

3. Conclusions

Agar-alginate films developed with essential oils incorporated have shown adequate structural properties for use in food packaging applications, especially at 10% EO concentration. All films have totally inhibited or significantly reduced microbial growth, compared to control films, without EO included. Films with 10 and 20% eucalyptus oil showed best performance when considering both structural (thickness, water activity index etc.) and antimicrobial properties. Other well-performing composites were 10% mint and 10% and 20% orange EO films, while the rest of the films had good structural and optical properties and partial reductions of microbiological counts.

The obtained results bring to the fore the advantages of using essential oils in the development of biopolymer films. Even if, for similar polymeric composites, reductions of physico-chemical performances of the films were reported after the addition of such compounds, the results obtained in this case are much better than those of the control sample. Films based on biopolymers have shown their benefits in use. These can be extended and greatly improved by adding essential oils. Easy to obtain, use, and handle, these materials can successfully replace the conventional ones, based on petroleum, difficult to sort and almost impossible to recycle, so harmful for the environment. Future research involves other practical applications of these materials, with time testing of the properties of the films, but also of the packaged products, offering promising avenues for packaging food products.

4. Materials and Methods

Agar, alginate, glycerol, and Tween 80 were purchased from Sigma Aldrich (Romanian branch, Bucharest). All the essential oils used—lemon, grapefruit, orange, cinnamon, clove, mint, ginger, eucalypt, and chamomile—were purchased from Carl Roth (Germany). For the development of biopolymer films, the previously developed and tested composition, with some changes, was used [12]. Thus, agar: alginate: glycerol in a ratio of 2: 1: 1 was used for the film-forming solution. After obtaining the film-forming solution followed by stirring for 20 min (90 ± 2 °C and 450 rpm), the solution was cooled to a temperature of 40 °C and 10, respectively 20% v/v essential oil was added. Then the solutions were poured on silicone support and maintained for 38–42 h, until complete drying, at room temperature (24 ± 3 °C) and rH = 47 ± 3% (Figure 4). A total of 18 samples with the addition of essential oil and one control were obtained.

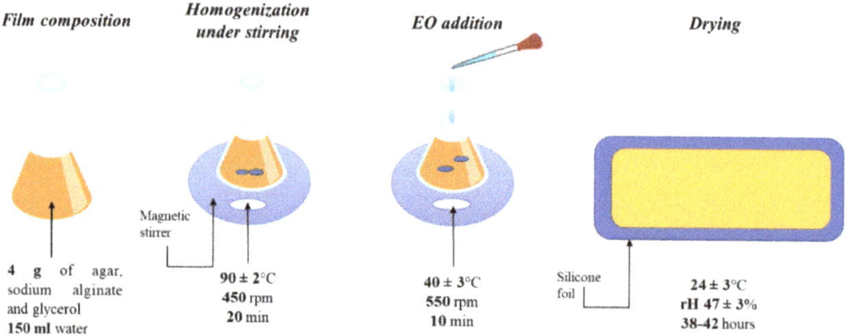

Figure 4. Diagram of schematic preparation process.

After development, the samples were observed to identify the regularity of the edges, the presence of pores and fissures, the uniformity of the film or the degree of homogenization of the substances used. The photos of the developed films are available in supplementary file (Figure S1—Photos of the films).

In order to evaluate the physical and optical properties, the thickness, the retraction ratio, the transmittance, the opacity, and the color were tested. The thickness was measured with a Yato micrometer (Shanghai), with an accuracy of 0.001 mm and a test range of

0–25 mm. For the evaluation of the film thickness, five readings were made, in different areas of the surface, and the value noted in Table 3 represents the average. For the retraction ratio (RR) values, the thickness of the film-forming solution (*T1*, 1050 μm), poured on the silicone support immediately after obtaining and the final thickness of the film (*T2*, μm), were taken into account. Thus, the result was calculated by the formula:

$$RR, (\%) = \frac{T1 - T2}{T1} * 100 \tag{1}$$

Transmittance (*T*, %) and absorbance were read spectrophotometrically, using Epoch equipment (BioTek Instruments, Winooski, VT, USA). For the determination, 1 cm × 3 cm film samples were used. The determinations were performed in triplicate. The transmittance was read at a wavelength of 660 nm and the absorbance at 600 nm. The read absorbance was used to determine the opacity, after the following formula was applied:

$$Opacity, \ (A/mm) = \frac{A}{T} \tag{2}$$

where *A*—absorbance and *T*—thickness, mm.

The color was evaluated using the Konica Minolta CR 400 colorimeter. The value taken into account represents the average of at least five readings, made in different areas of the film surface. The parameters considered were L*, a*, b*, and the blank values were L* = 94.12, a* = −5.52, and b* = 9.27.

The method used in previous research [53] was used to determine the humidity. Thus, 3 cm × 3 cm film samples were weighed before and after maintaining for 24 h at a temperature of 110 ± 2 °C. The determination was made in triplicate and the result was calculated with Formula (3):

$$MC, \ (\%) = \frac{W0 - W1}{W0} * 100 \tag{3}$$

where *W0* represent the sample mass before drying (g) and *W1* the dried mass (g).

In order to evaluate the solubility, the swelling ratio index was taken into account. This evaluation is important when the material is an edible one and it is intended to be consumed with the product it contains. For determination, 3 × 3 cm film samples were weighed (*M0*), immersed in water at room temperature and maintained for 0.5, 1, 3, 5, 7, 10, 15, 20 min. After this time, the excess water was removed using filter paper, and the samples were re-weighed (*M1*). The determinations were made as single experiment.

The water activity index (a_w) was determined with AquaLab 4TE equipment (Meter Group, München, Germany) and the results were taken into account after at least 5 readings performed in different areas of the film surface. The determinations were made at 25 ± 0.7 °C.

STAS ASTM D882 ((Standard Test Method for Tensile Properties of Thin Plastic Sheeting) [54] was used to determine the mechanical properties. Thus, for the evaluation of tear resistance and elongation, 1 cm × 10 cm film samples were tested using the ESM Mark 10 universal texturometer loaded with a 5 kN cell. Special grips for thin films and foils were attached. The travel speed was set at 10 mm/min and temperature was 25.5 ± 0.2 °C.

Tensile strength (TS) was calculated according to the Formula (4):

$$TS, \ (MPa) = \frac{F}{S} \tag{4}$$

where *F* is the maximum load (kN) and *S* represent the surface (mm^2).

The elongation at break represents the ratio between final length and initial length after breakage of the test specimen. It was calculated according to the following formula:

$$E, \ (\%) = \frac{\Delta l}{l} * 100 \tag{5}$$

where Δl represent the distance between final length and initial length, l, in mm.

The mechanical tests were performed in triplicate and the results are noted in Table 3.

Microbiological analysis of the films involved the use of Compact Dry (NISSUI Pharma) type plates, with dehydrated culture medium. Thus, total count (TC), *Escherichia coli* (EC), enterococcus (ETC), coliforms (CF), *Listeria monocytogenes* (LM), *Staphylococcus aureus* (XSA), and yeasts and molds (YM) were tested.

For determination, 1 g of film was solubilized in 9 mL of saline. From the solution thus obtained, 1 mL was poured on the culture medium. After rehydration, the plates were thermostated at 37 degrees for 36 h for TC, EC, ETC, CF, and LM, respectively 72 h for yeasts and molds.

The microstructure of images and 3D topographies were analyzed using Mountains 9 software (Digital Surf, Lavoisier, France). The roughness was calculated using the operators of the same software, taking into account the highest point, the lowest point and other areas on the surface of the films.

Significant differences of data were evaluated by carrying out one-way analysis of variance (ANOVA) and Tukey's test at $p < 0.05$. Data analysis was performed using MiniTAB statistics software (MiniTAB Ltd., Coventry, UK).

5. Patents

The composition used as a matrix for the incorporation of essential oils was proposed for patenting at the State Office for Inventions and Trademarks in Romania, registered with number A000737/06.12.2021.

Supplementary Materials: The following supporting information can be downloaded at: https://www.mdpi.com/article/10.3390/gels8080505/s1, Figure S1: Photos of the films;

Author Contributions: Conceptualization, R.G.P.; methodology, R.G.P.; software, A.L.; validation, R.G.P., A.L. and G.G.; writing—original draft preparation, R.G.P.; writing—review and editing, A.L.; supervision, G.G.; project administration, R.G.P. and G.G. All authors have read and agreed to the published version of the manuscript.

Funding: This work was supported by a grant of the Romanian Ministry of Research, Innovation and Digitization, CNCS/CCCDI–UEFISCDI, project number PN-III-P1-1.1-PD-2019-0793, within PNCDI III.

Institutional Review Board Statement: Not applicable.

Informed Consent Statement: Not applicable.

Data Availability Statement: Not applicable.

Conflicts of Interest: The authors declare no conflict of interest.

References

1. Lim, L.I.; Tan, H.L.; Pui, L.P. Development and characterization of alginate-based edible film incorporated with hawthorn berry (*Crataegus pinnatifida*) extract. *J. Food Meas. Charact.* **2021**, *15*, 2540–2548. [CrossRef]
2. Azmin, S.N.H.M.; Hayat, N.A.; Mat Nor, M.S. Development and characterization of food packaging bioplastic film from cocoa pod husk cellulose incorporated with sugarcane bagasse fibre. *J. Bioresour. Bioprod.* **2020**, *5*, 248–255. [CrossRef]
3. Maan, A.A.; Reiad Ahmed, Z.F.; Iqbal Khan, M.K.; Riaz, A.; Nazir, A. Aloe vera gel, an excellent base material for edible films and coatings. *Trends Food Sci. Technol.* **2021**, *116*, 329–341. [CrossRef]
4. Basavegowda, N.; Baek, K.H.; Roy, S.; Rhim, J.W. Advances in Functional Biopolymer-Based Nanocomposites for Active Food Packaging Applications. *Polymers* **2021**, *13*, 4198. [CrossRef] [PubMed]
5. Tan, C.; Han, F.; Zhang, S.; Li, P.; Shang, N. Novel Bio-Based Materials and Applications in Antimicrobial Food Packaging: Recent Advances and Future Trends. *Int. J. Mol. Sci.* **2021**, *22*, 9663. [CrossRef] [PubMed]
6. Ribeiro, A.M.; Estevinho, B.N.; Rocha, F. Preparation and Incorporation of Functional Ingredients in Edible Films and Coatings. *Food Bioprocess Technol.* **2021**, *14*, 209–231. [CrossRef]
7. Díaz-Montes, E.; Castro-Muñoz, R. Edible Films and Coatings as Food-Quality Preservers: An Overview. *Foods* **2021**, *10*, 249. [CrossRef]
8. Rajesh Kumar Dora, T.; Ghosh, S.; Damodar, R. Synthesis and evaluation of physical properties of Agar biopolymer film coating; an alternative for food packaging industry. *Mater. Res. Express* **2020**, *7*, 095307. [CrossRef]

9. Kumar, L.; Ramakanth, D.; Akhila, K.; Gaikwad, K.K. Edible films and coatings for food packaging applications: A review. *Environ. Chem. Lett.* **2021**, *20*, 875–900. [CrossRef]
10. Atta, O.M.; Manan, S.; Shahzad, A.; Ul-Islam, M.; Ullah, M.W.; Yang, G. Biobased materials for active food packaging: A review. *Food Hydrocoll.* **2022**, *125*, 107419. [CrossRef]
11. Sani, M.A.; Azizi-Lalabadi, M.; Tavassoli, M.; Mohammadi, K.; McClements, D.J. Recent Advances in the Development of Smart and Active Biodegradable Packaging Materials. *Nanomaterials* **2021**, *11*, 1331. [CrossRef] [PubMed]
12. Gheorghita Puscaselu, R.; Besliu, I.; Gutt, G. Edible Biopolymers-Based Materials for Food Applications—The Eco Alternative to Conventional Synthetic Packaging. *Polymers* **2021**, *13*, 3779. [CrossRef]
13. Mahmud, N.; Islam, J.; Tahergorabi, R. Marine Biopolymers: Applications in Food Packaging. *Processes* **2021**, *9*, 2245. [CrossRef]
14. Shen, S.; Chen, X.; Shen, Z.; Chen, H. Marine Polysaccharides for Wound Dressings Application: An Overview. *Pharmaceutics* **2021**, *13*, 1666. [CrossRef] [PubMed]
15. Dupuis, V.; Cerbu, C.; Witkovski, L.; Potarniche, A.V.; Timar, M.C.; Zhyska, M.; Sabliov, C. Nanodelivery of essential oils as efficient tools against antimicrobial resistance: A review of the type and physical-chemical properties of the delivery systems and applications. *Drug Deliv.* **2022**, *29*, 1007–1024. [CrossRef] [PubMed]
16. Teixeira-Costa, B.E.; Andrade, C.T. Natural Polymers Used in Edible Food Packaging; History, Function and Application Trends as a Sustainable Alternative to Synthetic Plastic. *Polysaccharides* **2022**, *3*, 32–58. [CrossRef]
17. Vieira, T.M.; Moldão-Martins, M.; Alves, V.D. Design of Chitosan and Alginate Emulsion-Based Formulations for the Production of Monolayer Crosslinked Edible Films and Coatings. *Foods* **2021**, *10*, 1654. [CrossRef]
18. Anis, A.; Pal, K.; Al-Zahrani, S.M. Essential Oil-Containing Polysaccharide-Based Edible Films and Coatings for Food Security Applications. *Polymers* **2021**, *13*, 575. [CrossRef]
19. Jayakody, M.M.; Vanniarachchy, M.P.G.; Wijesekara, I. Seaweed derived alginate, agar, and carrageenan based edible coatings and films for the food industry: A review. *J. Food Meas. Charact.* **2022**, *16*, 1195–1227. [CrossRef]
20. EUR-Lex—32009R0450—EN-EUR-Lex. Available online: https://eur-lex.europa.eu/legal-content/EN/TXT/?uri=celex%3A32009R0450 (accessed on 23 July 2022).
21. Trajkovska Petkoska, A.; Daniloski, D.; D'Cunha, N.M.; Naumovski, N.; Broach, A.T. Edible packaging: Sustainable solutions and novel trends in food packaging. *Food Res. Int.* **2021**, *140*, 109981. [CrossRef]
22. Perera, K.Y.; Sharma, S.; Pradhan, D.; Jaiswal, A.K.; Jaiswal, S. Seaweed Polysaccharide in Food Contact Materials (Active Packaging, Intelligent Packaging, Edible Films, and Coatings). *Foods* **2021**, *10*, 2088. [CrossRef]
23. Zhao, L.; Duan, G.; Zhang, G.; Yang, H.; He, S.; Jiang, S. Electrospun Functional Materials toward Food Packaging Applications: A Review. *Nanomaterials* **2020**, *10*, 150. [CrossRef]
24. Santhosh, R.; Nath, D.; Sarkar, P. Novel food packaging materials including plant-based byproducts: A review. *Trends Food Sci. Technol.* **2021**, *118*, 471–489. [CrossRef]
25. Dash, K.K.; Kumar, A.; Kumari, S.; Malik, M.A. Silver Nanoparticle Incorporated Flaxseed Protein-Alginate Composite Films: Effect on Physicochemical, Mechanical, and Thermal Properties. *J. Polym. Environ.* **2021**, *29*, 3649–3659. [CrossRef]
26. Syafiq, R.M.O.; Sapuan, S.M.; Zuhri, M.R.M. Effect of cinnamon essential oil on morphological, flammability and thermal properties of nanocellulose fibre-reinforced starch biopolymer composites. *Nanotechnol. Rev.* **2020**, *9*, 1147–1159. [CrossRef]
27. Zhang, W.; Lin, M.; Feng, X.; Yao, Z.; Wang, T.; Xu, C. Effect of lemon essential oil-enriched coating on the postharvest storage quality of citrus fruits. *Food Sci. Technol.* **2022**, *42*, e125421. [CrossRef]
28. Jiang, Y.; Lan, W.; Sarmeen, D.; Ahmed, S.; Qin, W.; Zhang, Q.; Chen, H.; Dai, J.; He, L.; Liu, Y. Preparation and characterization of grass carp collagen-chitosan-lemon essential oil composite films for application as food packaging. *Int. J. Biol. Macromol.* **2020**, *60*, 340–351. [CrossRef]
29. Nisar, T.; Yang, X.; Alim, A.; Iqbal, M.; Wang, Z.C.; Guo, Y. Physicochemical responses and microbiological changes of bream (*Megalobrama ambycephala*) to pectin based coatings enriched with clove essential oil during refrigeration. *Int. J. Biol. Macromol.* **2019**, *124*, 1156–1166. [CrossRef] [PubMed]
30. Kavas, G.; Kavas, N.; Saygili, D. The Effects of Thyme and Clove Essential Oil Fortified Edible Films on the Physical, Chemical and Microbiological Characteristics of Kashar Cheese. *J. Food Qual.* **2015**, *38*, 405–412. [CrossRef]
31. Prasetyaningrum, A.; Utomo, D.P.; Raemas, A.F.; Kusworo, T.D.; Jos, B.; Djaeni, M. Alginate/κ-Carrageenan-Based Edible Films Incorporated with Clove Essential Oil: Physico-Chemical Characterization and Antioxidant-Antimicrobial Activity. *Polymers* **2021**, *13*, 354. [CrossRef]
32. Alboofetileh, M.; Rezaei, M.; Hosseini, H.; Abdollahi, M. Antimicrobial activity of alginate/clay nanocomposite films enriched with essential oils against three common foodborne pathogens. *Food Control* **2014**, *36*, 1–7. [CrossRef]
33. Jalali, N.; Ariiai, P.; Fattahi, E. Effect of alginate/carboxyl methyl cellulose composite coating incorporated with clove essential oil on the quality of silver carp fillet and Escherichia coli O157:H7 inhibition during refrigerated storage. *J. Food Sci. Technol.* **2016**, *53*, 757–765. [CrossRef] [PubMed]
34. Akhter, R.; Masoodi, F.A.; Wani, T.A.; Rather, S.A. Functional characterization of biopolymer based composite film: Incorporation of natural essential oils and antimicrobial agents. *Int. J. Biol. Macromol.* **2019**, *137*, 1245–1255. [CrossRef]
35. Cheng, M.; Wang, J.; Zhang, R.; Kong, R.; Lu, W.; Wang, X. Characterization and application of the microencapsulated carvacrol/sodium alginate films as food packaging materials. *Int. J. Biol. Macromol.* **2019**, *141*, 259–267. [CrossRef]

36. Azadbakht, E.; Maghsoudlou, Y.; Khomiri, M.; Kashiri, M. Development and structural characterization of chitosan films containing *Eucalyptus globulus* essential oil: Potential as an antimicrobial carrier for packaging of sliced sausage. *Food Packag. Shelf Life* **2018**, *7*, 65–72. [CrossRef]
37. Tang, Y.; Zhou, Y.; Lan, X.; Huang, D.; Luo, T.; Ji, J.; Mafang, Z.; Maio, X.; Wang, H.; Wang, W. Electrospun Gelatin Nanofibers Encapsulated with Peppermint and Chamomile Essential Oils as Potential Edible Packaging. *J. Agric. Food Chem.* **2019**, *67*, 2227–2234. [CrossRef] [PubMed]
38. Khaledian, Y.; Pajohi-Alamoti, M.; Bazargani-Gilani, B. Development of cellulose nanofibers coating incorporated with ginger essential oil and citric acid to extend the shelf life of ready-to-cook barbecue chicken. *J. Food Process. Preserv.* **2019**, *43*, e14114. [CrossRef]
39. Souza, V.G.L.; Pires, J.R.A.; Vieira, E.T.; Coelhoso, I.M.; Duarte, M.P.; Fernando, A.L. Shelf life assessment of fresh poultry meat packaged in novel bionanocomposite of chitosan/montmorillonite incorporated with ginger essential oil. *Coatings* **2018**, *8*, 177. [CrossRef]
40. Al-Hilifi, S.A.; Al-Ali, R.M.; Petkoska, A.T. Ginger Essential Oil as an Active Addition to Composite Chitosan Films: Development and Characterization. *Gels* **2022**, *8*, 327. [CrossRef]
41. Cai, L.; Wang, Y.; Cao, A. The physiochemical and preservation properties of fish sarcoplasmic protein/chitosan composite films containing ginger essential oil emulsions. *J. Food Process Eng.* **2020**, *43*, e13495. [CrossRef]
42. Mostaghimi, M.; Majdinasab, M.; Hosseini, S.M. Characterization of Alginate Hydrogel Beads Loaded with Thyme and Clove Essential Oils Nanoemulsions. *J. Polym. Environ.* **2022**, *30*, 1647–1661. [CrossRef]
43. Valizadeh, A.; Shirzad, M.; Esmaeili, F.; Amani, A. Increased antibacterial activity of Cinnamon Oil Microemulsionin Comparison with Cinnamon Oil Bulk and Nanoemulsion. *Nanomed. Res. J.* **2018**, *3*, 37–43.
44. Lim, Z.Q.; Tong, S.Y.; Wang, K.; Lim, P.N.; Thian, E.S. Cinnamon oil incorporated polymeric films for active food packaging. *Mater. Lett.* **2022**, *313*, 131744. [CrossRef]
45. Abdollahzadeh, E.; Mahmoodzadeh Hosseini, H.; Imani Fooladi, A.A. Antibacterial activity of agar-based films containing nisin, cinnamon EO, and ZnO nanoparticles. *J. Food Saf.* **2018**, *38*, e12440. [CrossRef]
46. Kunová, S.; Sendra, E.; Haščík, P.; Vukovic, N.L.; Vukic, M.; Kačániová, M. Influence of Essential Oils on the Microbiological Quality of Fish Meat during Storage. *Animals* **2021**, *11*, 3145. [CrossRef] [PubMed]
47. Peng, Y.; Li, Y. Combined effects of two kinds of essential oils on physical, mechanical and structural properties of chitosan films. *Food Hydrocoll.* **2014**, *36*, 287–293. [CrossRef]
48. Viuda-Martos, M.; Ruiz-Navajas, Y.; Fernández-López, J.; Pérez-Álvarez, J. Antifungal activity of lemon (*Citrus lemon* L.), mandarin (*Citrus reticulata* L.), grapefruit (*Citrus paradisi* L.) and orange (*Citrus sinensis* L.) essential oils. *Food Control* **2008**, *19*, 1130–1138. [CrossRef]
49. Özogul, Y.; Özogul, F.; Kulawik, P. The antimicrobial effect of grapefruit peel essential oil and its nanoemulsion on fish spoilage bacteria and food-borne pathogens. *LWT* **2021**, *136*, 110362. [CrossRef]
50. Khalid, K.A.; Darwesh, O.M.; Ahmed, A.M. Peel Essential Oils of Citrus Types and Their Antimicrobial Activities in Response to Various Growth Locations. *J. Essent. Oil-Bear. Plants* **2021**, *24*, 480–499. [CrossRef]
51. de Oliveira Filho, J.G.; Albiero, B.; Cipriano, L.; de Oliviera, C.C.; Bezerra, N.; Oldoni, F.; Egea, M.; de Azeredo, H.M.C.; Ferreira, M.D. Arrowroot starch-based films incorporated with a carnauba wax nanoemulsion, cellulose nanocrystals, and essential oils: A new functional material for food packaging applications. *Cellulose* **2021**, *28*, 6499–6511. [CrossRef]
52. Bag, A.; aChattopadhyay, R.R. Evaluation of Synergistic Antibacterial and Antioxidant Efficacy of Essential Oils of Spices and Herbs in Combination. *PLoS ONE* **2015**, *10*, 131321. [CrossRef] [PubMed]
53. Gheorghita Puscaselu, R.; Amariei, S.; Norocel, L.; Gutt, G. New edible packaging material with function in shelf life extension: Applications for the meat and cheese industries. *Foods* **2020**, *9*, 562. [CrossRef] [PubMed]
54. "Tensile Testing of Thin Plastic Sheeting (Film) ASTM D882". Available online: https://www.intertek.com/polymers/tensile-testing/astm-d882/ (accessed on 8 August 2022).

 MDPI

Article

Effect of Composite Chitosan/Sodium Alginate Gel Coatings on the Quality of Fresh-Cut Purple-Flesh Sweet Potato

Chit-Swe Chit [1,†], Ibukunoluwa Fola Olawuyi [1,†], Jong Jin Park [1,2] and Won Young Lee [1,3,*]

1 School of Food Science and Biotechnology, Kyungpook National University, Daegu 41566, Republic of Korea
2 Coastal Agricultural Research Institute, Kyungpook National University, Daegu 41566, Republic of Korea
3 Research Institute of Tailored Food Technology, Kyungpook National University, Daegu 41566, Republic of Korea
* Correspondence: wonyoung@knu.ac.kr; Tel.: +82-53-950-7761
† These authors are considered the first author.

Abstract: In this study, single-layer coating using chitosan (Ch) and sodium alginate (SA) solutions and their gel coating (ChCSA) formed by layer-by-layer (LbL) electrostatic deposition using calcium chloride (C) as a cross linking agent were prepared to improve storage qualities and shelf-life of fresh-cut purple-flesh sweet potatoes (PFSP). The preservative effects of single-layer coating in comparison with LbL on the quality parameters of fresh-cut PFSP, including color change, weight loss, firmness, microbial analysis, CO_2 production, pH, solid content, total anthocyanin content (TAC), and total phenolic content (TPC) were evaluated during 16 days of storage at 5 °C. Uncoated samples were applicable as a control. The result established the effectiveness of coating in reducing microbial proliferation (~2 times), color changes (~3 times), and weight loss (~4 times) with negligible firmness losses after the storage period. In addition, TAC and TPC were better retained in the coated samples than in the uncoated samples. In contrast, quality deterioration was observed in the uncoated fresh cuts, which progressed with storage time. Relatively, gel-coating ChCSA showed superior effects in preserving the quality of fresh-cut PFSP and could be suggested as a commercial method for preserving fresh-cut purple-flesh sweet potato and other similar roots.

Keywords: purple-flesh sweet-potato; chitosan; sodium alginate; gel coating; preservation

Citation: Chit, C.-S.; Olawuyi, I.F.; Park, J.J.; Lee, W.Y. Effect of Composite Chitosan/Sodium Alginate Gel Coatings on the Quality of Fresh-Cut Purple-Flesh Sweet Potato. *Gels* **2022**, *8*, 747. https://doi.org/10.3390/gels8110747

Academic Editors: Aris E. Giannakas, Constantinos Salmas and Charalampos Proestos

Received: 31 October 2022
Accepted: 16 November 2022
Published: 17 November 2022

1. Introduction

Purple flesh sweet potato (*Ipomoea batatas*) is a very nutritious root vegetable native to the tropical regions of America. They are an abundant source of carbohydrates, dietary fiber, vitamins including A, B1, B2, C, and E, and minerals including Ca, Mg, K, and Zn [1]. In addition, purple flesh sweet potatoes (PFSP) contain a large amount of anthocyanins, an antioxidant whose long-term dietary intake can prevent cancer, cardiovascular diseases, viral infections, Alzheimer's disease, and diabetes [2]. The growing consciousness among consumers about what they eat, especially the health benefits, has led to an increase in the consumption of fruits and vegetables. Combined with busy lifestyle patterns, the demand for fresh-cut produce has increased significantly in recent years [3]. Ready-to-use fresh-cut produce is convenient, eliminates consumer waste, and saves time. However, the minimal processing of fresh-cut produce results in tissue softening and discoloration. It increases microbiological deterioration due to the exposed tissues, which makes them vulnerable to metabolism, microbial invasion, and mechanical damage [4]. These factors impact a product's storage and shelf life [5,6]. Therefore, a suitable packaging technique effective to reduce these factors influence and preserve the quality of fresh-cut produce during marketing and storage is required [4].

Antimicrobial coatings/films (inedible or edible) and modified atmosphere packaging (MAP) have been applied to fresh produce to maintain their qualities and extend their shelf life [7–10]. In particular, edible coatings have been investigated for their potential to enhance the quality and shelf life of food items [3,11]. Edible coatings can preserve fresh-cut food from mechanical and microbial damage, delay biochemical changes, and enhance their surface appearance [12]. Moreover, edible coatings can meet additional requirements, such as having antimicrobial activity and acting as good moisture and oxygen barriers. These requirements are beneficial for whole or fresh-cut fruits and vegetables that are often prone to microbial harm and highly susceptible to water loss, which causes size shrinkage and texture degradation [5]. Thus, coatings intended for fruit and vegetable preservation are expected to have good gas permeability for typical CO_2/O_2 exchange, low water vapor permeability to minimize moisture leakage, and antibacterial properties to inhibit microbial proliferation. It is, however, challenging for a single coating material to satisfy all these requirements [13]

The composite layer-by-layer (LbL) coating technique, which is based on electrostatic deposition technology, was developed to incorporate numerous preservatives derived from various polymer components [5,14]. This approach is based on the alternate deposition of oppositely charged polyelectrolytes in the presence of a cross linking agent, resulting in a novel gel coating with improved properties and functionalities [11]. Due to the effectiveness of the LbL coating technique, its commercial implementation has been suggested for preserving minimally processed fruits. Cationic biopolymers such as chitosan and poly-L-lysine, and anionic biopolymers such as pectin and alginate are commonly used for LbL coating of foods [15]. Alginate is a hydrophilic biopolymer with excellent film-forming properties due to its unusual colloidal properties, including thickening, suspension formation, gel formation, and emulsion stabilization [16]. In addition, sodium alginate coating was beneficial in preserving the post-harvest quality of tomatoes [17] and peaches [18]. However, alginate has no antimicrobial properties, and their poor mechanical properties and water vapor resistance has limited their industrial applications [19]. In contrast, chitosan, a cationic polysaccharide with a high molecular weight and soluble in organic acids, is applicable as a preservative coating material for fruits due to its anti fungal mechanisms [20–23].

Some studies have examined the effect of alginate and chitosan on fresh-cut melon, mangoes, blueberries [24], guavas, and nectarines [25]. The combination of alginate and chitosan displayed various preservative effects depending on the fresh-cut fruit. However, the application of the sequential coating of chitosan and alginate on fresh-cut purple sweet potatoes has not been studied. This study aimed to investigate the effect of chitosan coating (Ch), sodium alginate gel coating (SA + C), and their composite gel coating (ChCSA) on the quality and shelf life of fresh-cut purple sweet potatoes during refrigerated storage at 5 °C for 16 days.

2. Results and Discussion

2.1. *Effect of Coatings on the Color Change during Storage*

Color is one of the significant visual characteristics of fresh-cut food items. Excessive discoloration often impacts consumer acceptance, and indicates poor performance packaging techniques used to preserve products [26]. The color change (ΔE) value of the samples was used to evaluate discoloration in samples during storage (Figure 1). Change in color was observed in all samples, which was more pronounced in CON (uncoated fresh-cuts). During the first 12 days of storage, no significant difference was observed in coated samples (Ch, SA + C, and ChCSA). However, at the end of storage, notable differences were observed between all samples. The ΔE values for CON, Ch, SA + C, and ChCSA coated fresh-cuts were 22.90, 16.86, 13.05, and 8.97, respectively, indicating that ChCSA gel coating was more efficient in retaining the color of fresh-cut purple sweet potatoes than their single coatings. Biochemical reactions responsible for the degradation of color pigments in sweet potatoes require oxygen and light [1]. The inner and outer film layers of chitosan and alginate, respectively, form a protective barrier on the surface of the coated fresh cuts, which impacts the selective permeability of gas and light [27,28]. Moreover, Ch and SA coatings have been reported to improve the storage quality of various fruits by inhibiting color changes such as browning in papaya, apple, and melon [29–31].

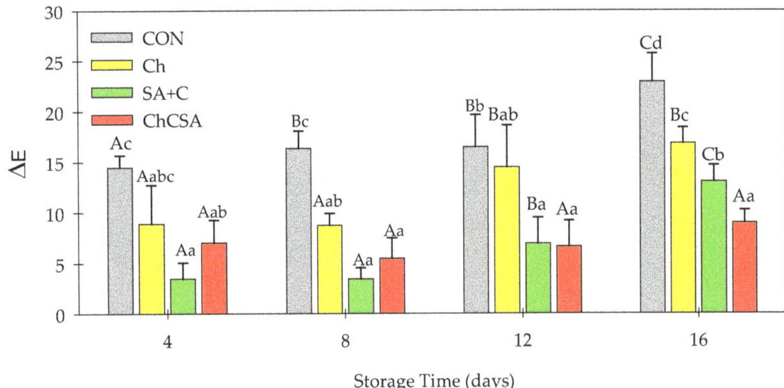

Figure 1. The effect of single-layer and gel coatings on the changes in total color difference value (ΔE) of fresh-cut purple sweet potatoes. Vertical bars represent means and standard deviation. Bars with different alphabets within the same storage day (lower case) or same treatment group at different storage days (upper case) are significantly different (Tukey's HSD Test, $p \leq 0.05$).

2.2. Effect of Coatings on Weight Loss during Storage

Fresh-cut products are susceptible to weight loss by transpiration [32]. In addition, excessive weight loss reduces valuation and consumers' perception of purchasing a product [33]. Thus, evaluating weight loss during storage is crucial. Weight loss was gradually increased in all samples according to the storage time (Figure 2). Higher weight loss was recorded in the control samples throughout the storage period, whereas the coated samples had minor weight losses. Significantly, SA + C and ChCSA gel coatings slowed down the weight loss during storage, having the lowest weight loss value (~1.4%) after 16 days of storage. The formation of gel films on the surface of fresh-cut samples improved moisture retention and prevented excess transpiration. Similar to this study, weight loss reduction in coated fresh-cut nectarines [34] and blueberries [24] have been reported.

2.3. Effect of Composite Edible Coatings on Firmness

The firmness of roots and vegetables is also an indicative quality parameter significant for consumer acceptance. The firmness of the control sample decreased throughout the storage period from 341.96 to 254.30 N, whereas CH- and SA + C-coated samples retained their hardness until day 12 (Figure 3). After 16 days of storage, a slight decrease in firmness was observed in Ch (from 384.42 to 314.19 N) and SA + C (411.21 to 306.02 N). However, no noticeable decrease was observed for the ChCSA-coated samples, indicating the beneficial and synergetic effect of multilayer gel coating over their single-layer film coatings. Previous studies reported that layer-by-layer coating enhanced the cell-wall structure and slowed down the cell degradation of fresh-cut products [35,36]. In addition, the combined antimicrobial and adhesion effects of Ch and SA inhibited the production and activities of microbial hydrolytic enzymes associated with cell wall components hydrolysis [11]. Moreover, the use of calcium chloride as a cross linking agent in ChCSA could have further enhanced firmness of coated samples [24].

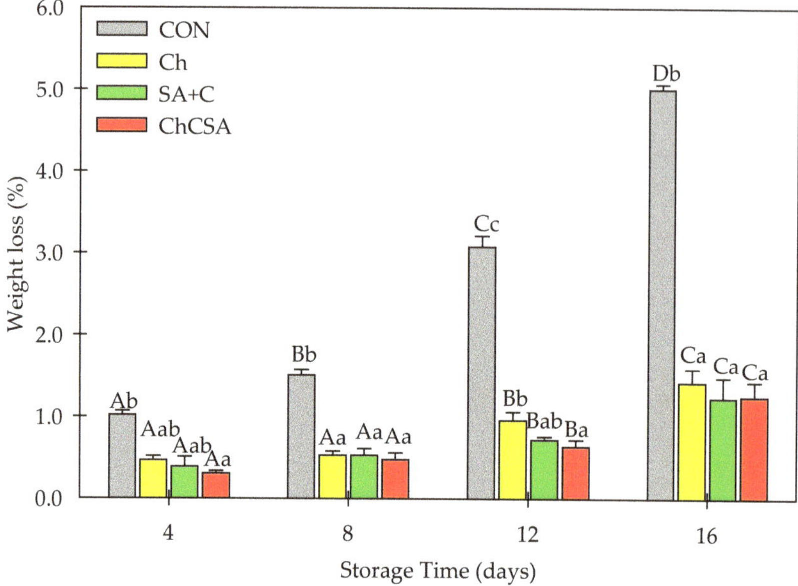

Figure 2. The effect of single-layer and gel coatings on the percentage of weight loss of fresh-cut purple sweet potatoes. Vertical bars represent means and standard deviation. Bars with different alphabets within the same storage day (lower case) or same treatment group at different storage days (upper case) are significantly different (Tukey's HSD Test, $p \leq 0.05$).

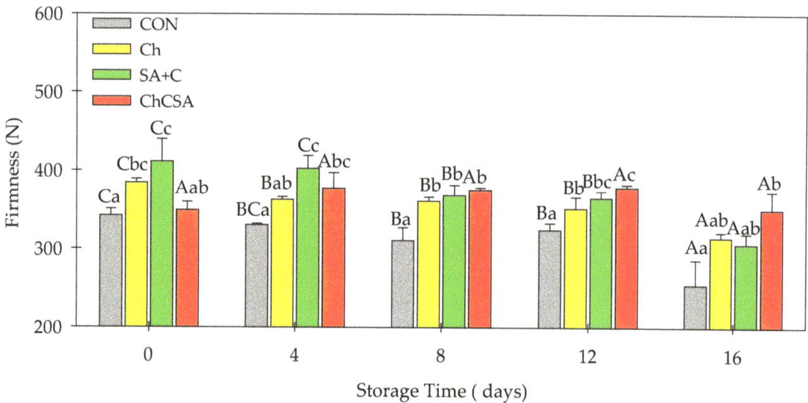

Figure 3. The effect of single-layer and gel coatings on the flesh firmness of fresh-cut purple sweet potatoes. Vertical bars represent means and standard deviation. Bars with different alphabets within the same storage day (lower case) or same treatment group at different storage days (upper case) are significantly different (Tukey's HSD Test, $p \leq 0.05$).

2.4. Effect of Composite Edible Coatings on Microbial Growth

Microbial contamination is the major reason for the deterioration of fresh-cut products. The presence and growth of microorganisms during product storage and distribution affects food quality and safety [5]. However, some edible coatings have shown barrier properties, inhibiting their proliferation in coated foods [35]. Notably, the application of coatings reduced the initial population of aerobic bacteria (Figure 4) and total fungi (Figure 5). However, an increase in bacteria (3.48 log CFU/mL in CON) and fungi (up to

~4.57 log CFU/mL in CON) were observed in samples at the end of storage. However, all coated samples showed lower microbial concentration. For instance, after 16 days of storage, ChCSA-coated samples had aerobic bacteria and total fungi counts of 2.44 log CFU/mL, and 2.37 log CFU/mL, respectively. The antimicrobial properties of ChCSA coatings could be attributed to the intrinsic bacteriostatic and fungistatic characteristics of chitosan, combined with the oxygen barrier properties of Ch and SA coatings which limited oxygen requirement for microbial proliferation [3,5]

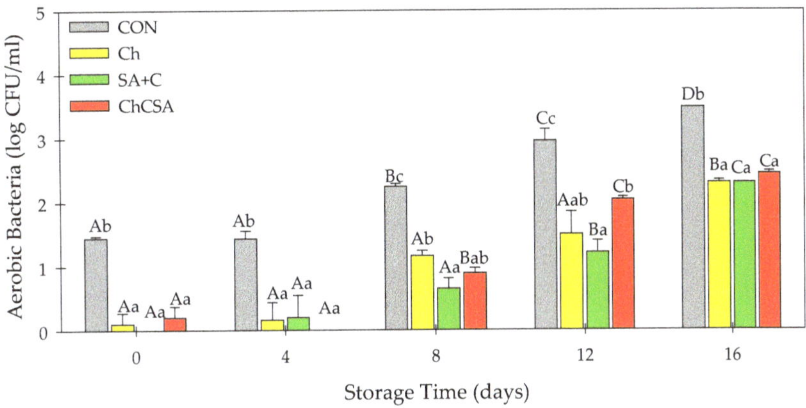

Figure 4. The effect of single-layer and gel coatings on the aerobic bacteria on fresh-cut purple sweet potatoes. Vertical bars represent means and standard deviation. Bars with different alphabets within the same storage day (lower case) or same treatment group at different storage days (upper case) are significantly different (Tukey's HSD Test, $p \leq 0.05$).

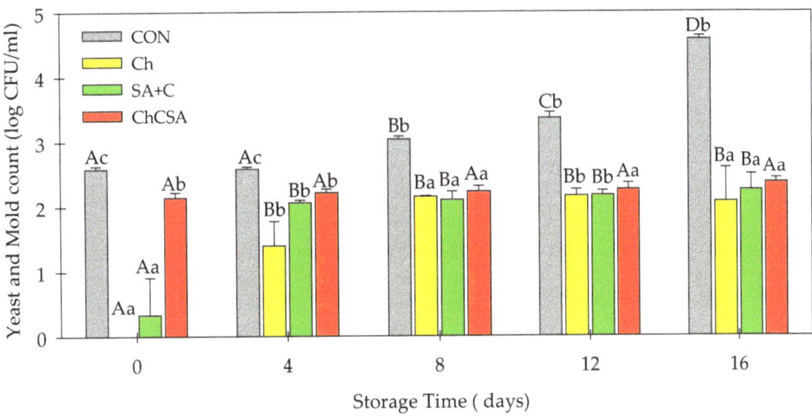

Figure 5. The effect of single-layer and gel coatings on the yeast and mold on fresh-cut purple sweet potatoes. Vertical bars represent means and standard deviation. Bars with different alphabets within the same storage day (lower case) or same treatment group at different storage days (upper case) are significantly different (Tukey's HSD Test, $p \leq 0.05$).

2.5. Effect of Coatings on CO_2 Production

The thin film layer formed by coatings on the food surface controls gas permeability, and provides a delicate balance between inhibiting over-ripening and preventing senescence. In addition, it regulates normal gas exchange to avoid the buildup of CO_2, which promotes anaerobic conditions that lead to off flavors [37]. High rate of respiration is one of the problems for fresh-cut products [11]. The composition of CO_2 in the headspace gas was used to explain the rate of respiration in packaged samples (Figure 6). There were no noticeable differences in CO_2 production during the first 8 days of storage. Thereafter, CO_2 concentration slowly increased in all samples until the end of storage. Notably, a sharp increase in CO_2 production was observed in uncoated fresh-cuts compared to the coated samples. Gel coatings (SA + C and ChCSA) showed better effectiveness in retarding CO_2 production. High CO_2 production in fruits corresponds to high oxygen consumption. Thus, the low oxygen permeability of coated samples resulted in tissue respiration, and subsequently, low CO_2 production [38].

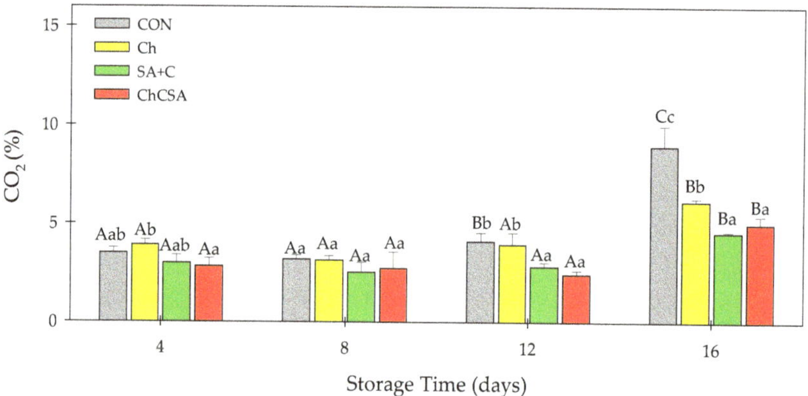

Figure 6. The effect of layer-by-layer and single-layer coatings on the percentage of carbon dioxide gas emission on fresh-cut purple sweet potatoes. Vertical bars represent means and standard deviation. Bars with different alphabets within the same storage day (lower case) or same treatment group at different storage days (upper case) are significantly different (Tukey's HSD Test, $p \leq 0.05$).

By modifying the gas atmosphere around the fruit tissue, polysaccharide coatings with semipermeable properties on the surface of fruits impede the rate of respiration and ripening during storage, thus retaining the quality attributes of products [39]. Similar gaseous barrier effects of polysaccharide-based coatings on fresh-cut products have been reported [33,40].

2.6. *Effect of Coatings on Soluble Solid Concentration and pH*

As shown in Figure 7, the total soluble solid (TSS), measured as °Brix value, increased in all samples during the storage period. The increase in TSS is due to the conversion of starch and non-starch polysaccharides to simple sugar by hydrolytic processes [27]. After 16 days of storage, TSS was highest in CON (21.3 °Brix) and lowest in ChCSA (14.3 °Brix). Chitosan- and alginate-based coatings were observed to inhibit metabolic and hydrolytic reactions associated with TSS increase in various fruits, including Chinese winter jujube, longan, and fig fruits [27,40–42].Thus, low TSS in ChCSA-coated samples could be attributed to the effective combination of Ch and SA, which reduced metabolic reactions and retarded polysaccharides breakdown processes [13].

Noticeable changes in the pH value of samples occurred after 16 days of storage (Figure 8). CON sample showed a sharp decrease (6.5 to 4.9), while coated samples showed marginal pH changes after the storage period. During post-harvest storage, a decrease in pH is typical and attributed to the production of organic acids by respiratory metabolism [34]. Low pH in CON may be related to the utilization of polysaccharide substrates by microorganisms, which led to the increased production of acidic metabolites [6]. Similar marginal changes in pH value were observed for coated nectarine slices [25] and fresh-cut watermelon [43].

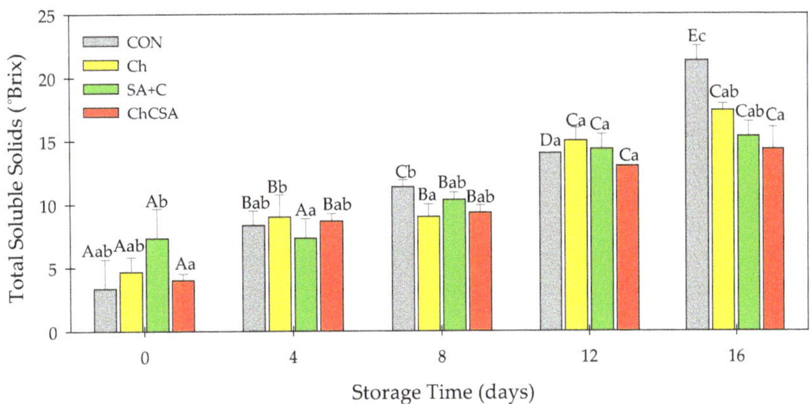

Figure 7. The effect of layer-by-layer and single-layer coatings on the percentage of °Brix of fresh-cut purple sweet potatoes. Vertical bars represent means and standard deviation. Bars with different alphabets within the same storage day (lower case) or same treatment group at different storage days (upper case) are significantly different (Tukey's HSD Test, $p \leq 0.05$).

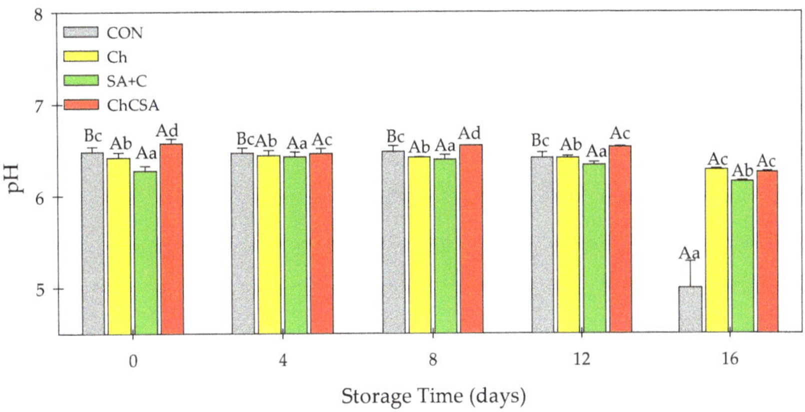

Figure 8. The effect of single-layer and gel coatings on the pH on fresh-cut purple sweet potato potatoes. Vertical bars represent means and standard deviation. Bars with different alphabets within the same storage day (lower case) or same treatment group at different storage days (upper case) are significantly different (Tukey's HSD Test, $p \leq 0.05$).

2.7. *Effect of Coatings on Total Anthocyanin Content and Total Phenolic Content*

Variations in the total anthocyanin content (TAC) were observed in samples during storage, with a more pronounced decrease in CON from 11.1 to 8.4 mg cyanidin-3-glucoside/g after 16 days of storage (Figure 9). These data are consistent with previous studies which showed that anthocyanin content was influenced by the storage time as well as the coating treatment [24]. Moreover, edible coatings have been reported to be beneficial in inhibiting the degradation pathways of anthocyanins in various anthocyanin-rich produce [44–46]. Moreover, variations in TAC during storage according to different edible coatings have been previously observed [27,47].

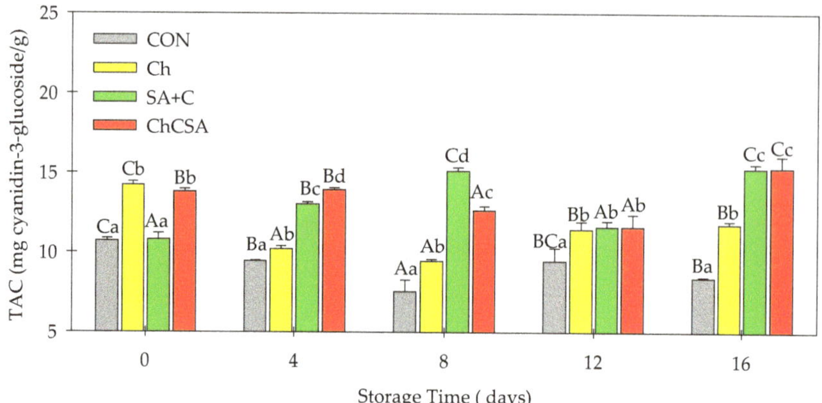

Figure 9. The effect of single-layer and gel coatings on the anthocyanin content on fresh-cut purple sweet potatoes. Vertical bars represent means and standard deviation. Bars with different alphabets within the same storage day (lower case) or same treatment group at different storage days (upper case) are significantly different (Tukey's HSD Test, $p \leq 0.05$).

Similar trends were observed for TPC, in which coated samples prevented phenolic compounds oxidation and degradation, having higher TPC values (2.27–3.56 mg GAE/g) compared to uncoated samples (1.41 mg GAE/g) throughout the storage period (Figure 10). Connor et al. [48] reported that several causes of physiological stress could promote the enzymatic oxidation of phenolic compounds during storage. However, coatings, especially composite coatings, could be beneficial in alleviating these oxidation processes [49]. Similar to the reports of Kou et al. [27], composite ChCSA-coated samples maintained a higher phenolic content throughout 16 days of storage. Moreover, the increase in the phenolic contents could be explained by the effect of Ch/SA coating in promoting phenylalanine ammonia-lyase (PAL) activity which led to the accumulation of phenolic compounds [27].

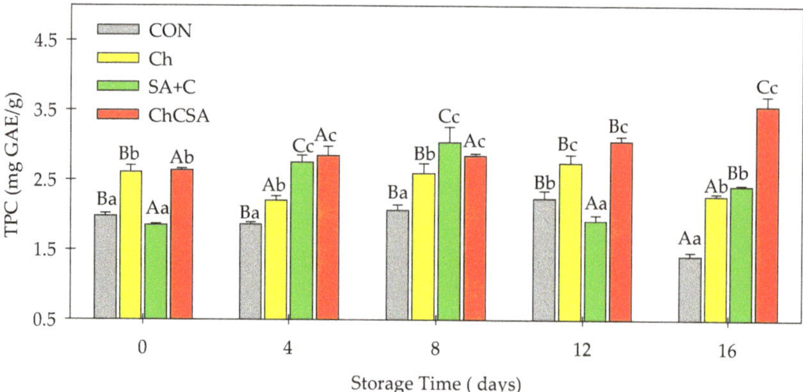

Figure 10. The effect of single-layer and gel coatings on the total phenolic content on fresh-cut purple sweet potatoes. Vertical bars represent means and standard deviation. Bars with different alphabets within the same storage day (lower case) or same treatment group at different storage days (upper case) are significantly different (Tukey's HSD Test, $p \leq 0.05$).

3. Conclusions

This study examined the effect of chitosan-, sodium alginate-, and their composite gel-coatings on the post harvest quality and shelf life of fresh-cut purple sweet potatoes. During

16 days of storage, various physiological and biochemical reactions associated with quality deterioration were effectively controlled in coated samples. For instance, improved quality retention and microbial inhibitions were observed in samples preserved with gel coating formed by Ch and SA multilayer solutions in the presence of $CaCl_2$, as a cross linking agent. The observed effects were attributed to enhanced barrier properties and antimicrobial properties, which regulated quality losses by transpiration, respiration, oxidation, and cellular degradation. In summary, ChCSA gel coating achieved the best preservative effect on the post harvest quality and shelf life of fresh-cut purple sweet potatoes, indicating the superiority of layer-by-layer coating over single-layer coating.

4. Materials and Methods

4.1. Materials

The experiments were performed with mature purple flesh sweet potato (PFSP) from a farm in Haenam-gun in Korea. The samples were stored at 5 °C. In addition, sodium alginate (32–250 kDa, Duksan Chemicals, Ansan-si, Republic of Korea), high molecular weight chitosan (≥75% deacetylation, Sigma Aldrich, USA), calcium chloride, glacial acetic acid, and Tween-80 were obtained from Sigma Aldrich (St. Louis, MO, USA).

4.2. Coating Solutions Preparation

The chitosan solution was prepared according to the method described by [50]. Chitosan powder was mixed with distilled water containing glacial acetic acid (0.5% v/v) at 70 °C under stirring until fully dissolved to produce a 2% chitosan solution. Finally, the pH of the solution was adjusted to 5.6 with 1 N NaOH.

The sodium alginate solution was prepared according to the method described by [11], with some modifications. Sodium alginate powder was dissolved in 100 mL of distilled water to obtain a 2% concentration. Then, the solution was stirred in a 70 °C water bath for 2 h to dissolve completely. Finally, the sodium alginate solution was cooled at room temperature.

Calcium chloride was used as the cross linking agent to produce gel coatings via layer-by-layer treatment. Calcium chloride was weighed and dissolved in 100 mL of distilled water to obtain a 2% solution. Then, the solution was shaken in the incubator to become dissolved entirely.

4.3. Sample Preparation

Purple flesh sweet potatoes without mechanical injuries or fungal infections were selected and washed in running water. Then, they were peeled and diced to get 1 cm pieces for flesh-cut coating.

4.4. Coating Application on the Samples

The coating procedure is illustrated in Figure 11. For a single-layer Ch coating, approximately 2 kg of flesh-cut purple sweet potatoes were dipped in 5 L of Ch solution for 2 min and dried at room temperature for 30 min. For SA gel coating, flesh-cuts were dipped in SA solution, rinsed for 30 s to remove the residual solution, and thereafter immersed in calcium chloride solution and dried. For the multilayer gel coating (ChCSA), the fresh-cuts were dipped in Ch solution for 2 min, rinsed for 30 s to remove the residual solution, immersed in calcium chloride for 2 min, then rinsed for 30 s and finally dipped in SA solution for 2 min before air-drying. Lastly, distilled water was used as an immersion solution for uncoated samples.

For each coating treatment, approximately 200 g of coated fresh-cuts were weighed and stored in triplicate in Ziploc bags (5 °C). Stored samples were removed at 4-days interval during a 16-day storage period and analyzed for quality parameters.

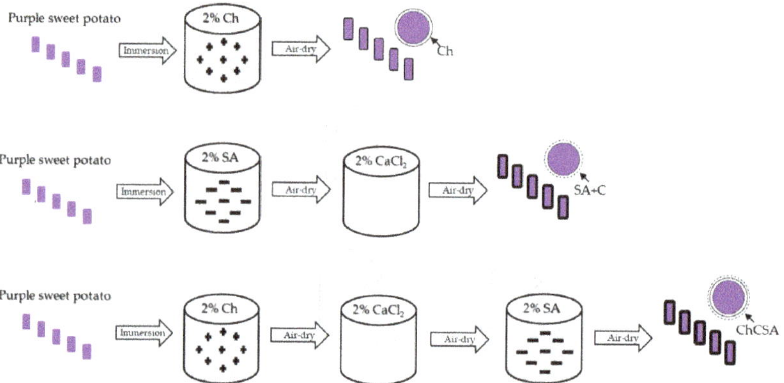

Figure 11. Illustration of coating procedure for fresh-cut purple flesh sweet potatoes.

4.5. Color Measurement

The surface color of the samples was determined by randomly selecting 3 samples and taking 3 readings for each treatment using a chromameter (CR-300, Minolta Co., Osaka, Japan). The L*, a*, b* value (CIE L a b) system was numerically specified in a three-dimensional spherical space defined by the three perpendicular axes: the L-axis (brightness) ranged from 0 (black) to 100% (white); the a-axis ranged from − a (green) to + a (red); and the b-axis ranged from − b (blue) to + b (yellow). Total color difference (ΔE) was calculated using L, a, and b values with the following equation [51]:

$$\Delta E = \sqrt{(L_2 - L_1)^2 + (a_2 - a_1)^2 + (b_2 - b_1)^2} \quad (1)$$

where subscripts 1 and 2 represent the final and initial readings, respectively at a particular storage interval.

4.6. Weight Loss Measurement

The coated and control fresh-cut sweet purple potatoes were individually weighed using a digital laboratory scale (Mettler Toledo, CH/PL 3002) at each data collection interval. The weight loss was calculated as follows:

$$\text{Weight loss } (\%) = \left[\frac{w_{in} - w_{fin}}{w_{in}}\right] \times 100 \quad (2)$$

where w_{in} and w_{fin} represent the initial and the final weight, respectively, measured at a particular storage interval.

4.7. Firmness Measurement

The firmness of the fresh-cut purple sweet potato samples was measured using a texture analyzer (Compac-100, Scientific Co., Tokyo, Japan) equipped with a 3 mm cylinder probe was used to assess the hardness of a fresh-cut purple sweet potato. A puncture test was carried out on a horizontally-positioned sample over a 13 mm hole at the speed of 60 mm/min and a travel distance of 20 mm [3]. The maximum force required to penetrate the sample was recorded for seven randomly selected fresh-cut PFSP per treatment group.

4.8. Microbial Analysis

The microbial growth in samples during storage was evaluated by counting the total number of aerobic bacteria and total fungi (yeast and mold). Ten g of the sample was taken aseptically from each treatment and transferred into sterile plastic bags with 30 mL of 0.1% peptone in water. The materials were homogenized in a Stomacher blender (Thomas

Scientific, Swedesboro, NJ, USA) and filtered to obtain the sample stock for microbial analysis. Dilutions were done using 0.1% peptone water prior to plating.

Total aerobic bacteria counts were determined by inoculating 100 µL of the diluted extract on the surface of plate count agar (PCA; Becton Dickinson, NJ, USA). The plates were incubated at 37 °C for 24 h. Total fungi counts were determined using the surface inoculation of potato dextrose agar (PDA; Becton Dickinson, NJ, USA), supplemented with ampicillin to control bacterial growth. The plates were also incubated at 30 °C for 48 h [11]. Afterward, the colonies were enumerated, and the results were expressed as the logarithm of colony-forming unit per mL (Log CFU g/mL) of sweet potato.

4.9. Carbon Dioxide Production

The samples were analyzed using a digital gas analyzer (Quantek Gas Analyzer Model 902D, Quantek Instruments, Grafton, MA, USA) by inserting the device's needle probe into the packaging film, enclosing the sample to determine the CO_2 concentration. The CO_2 concentration was displayed on the device screen and was computed as the % CO_2 produced using the following equation [3]:

$$CO_2 \text{ produced } (\%) = CO_{2fin} - CO_{2in}$$

where CO_{2in} and CO_{2fin} are CO_2 concentration on the first day, and at each storage interval.

4.10. Soluble Solid Concentration and pH

The soluble solid concentration (TSS) and pH of a sample were measured using the juice extracted from 1 g of the treated sample blended with 20 mL of distilled water in a tissue homogenizer. Soluble solid concentration was determined using a digital refractometer (Atago refractometer model PAL-1, Co., Ltd., Saitama, Japan), and the results were described as °Brix. The pH was measured using a pH meter (METTLER TOLEDO AG8608, Schwerzenbach, Switzerland).

4.11. Sample Extraction for Total Anthocyanin Content and Total Phenolic Content

Before analyzing the TAC and TPC of fresh-cut purple sweet potatoes, the sample pieces were frozen (−80 °C) and freeze-dried (FDS8518, Ilsin BioBase Co. Ltd., Dongducheon-si, Republic of Korea) for 7 days. Finally, the freeze-dried samples were ground, and their powder was stored in a freezer at −20 °C using extraction. 0.5 g of dried powder was weighed into a centrifuge tube and dissolved in 10 mL of 50% ethanol. Next, the sample solutions were homogenized for 30 min using an ultrasonic cleaner (JAC-3010; KODO, Hwaseong, Republic of Korea). The tube was placed in a centrifuge (45,000 rpm for 15 min), and finally, clear supernatant was obtained after filtration.

4.11.1. Total Anthocyanin Content

Total anthocyanin content was determined using the pH differential method [52]. Anthocyanin content was measured at the absorbance of 530 and 700 nm at pH 1.0 and 4.5. The results were described as mg of cyanidin-3-glucoside/g (Cy3G/g) of fresh purple sweet potato.

4.11.2. Total Phenolic Content

Total Phenolic Content (TPC) was analyzed using the Folin-Ciocalteu reagent as described in our previous study [9]. TPC values were presented in mg gallic acid equivalents (GAE) per g of fresh-cut purple sweet potato.

4.12. Statistical Analysis

The data were analyzed using IBM SPSS (V.20, SPSS Inc., Chicago, IL, USA); The Tukey's HSP test (honest significant differences) was used to determine the significance of the differences among the treatment means. The results were expressed as the mean ± standard deviation (Turkey's HSD Test, $p \leq 0.05$).

Author Contributions: Conceptualization, I.F.O., C.-S.C. and W.Y.L.; methodology, C.-S.C. and I.F.O.; software, C.-S.C.; validation, I.F.O. and J.J.P.; formal analysis, C.-S.C., I.F.O. and J.J.P.; investigation, C.-S.C. and I.F.O.; resources, W.Y.L.; writing—original draft preparation, I.F.O. and C.-S.C.; writing—review and editing, I.F.O., and W.Y.L.; visualization, C.-S.C. and I.F.O.; supervision, I.F.O. and W.Y.L. All authors have read and agreed to the published version of the manuscript.

Funding: This research received no external funding.

Institutional Review Board Statement: Not applicable.

Informed Consent Statement: Not applicable.

Data Availability Statement: Not applicable.

Acknowledgments: The co-first author Olawuyi I.F acknowledges the scholarship support received from Samsung Dream Scholarship Foundation, Republic of Korea.

Conflicts of Interest: The authors declare no conflict of interest.

References

1. Hu, Y.; Deng, L.; Chen, J.; Zhou, S.; Liu, S.; Fu, Y.; Yang, C.; Liao, Z.; Chen, M. An analytical pipeline to compare and characterise the anthocyanin antioxidant activities of purple sweet potato cultivars. *Food Chem.* **2016**, *194*, 46. [CrossRef] [PubMed]
2. Fossen, T.; Andersen, Ø. Spectroscopic techniques applied to flavonoids. *Flavonoids Chem. Biochem. Appl.* **2006**, 37–142.
3. Olawuyi, I.F.; Park, J.J.; Lee, J.J.; Lee, W.Y. Combined effect of chitosan coating and modified atmosphere packaging on fresh-cut cucumber. *Food Sci. Nutr.* **2019**, *7*, 1043. [CrossRef] [PubMed]
4. Zhang, H.; Han, M.; Xie, Y.; Wang, M.; Cao, C. Application of ethylene-regulating packaging in post-harvest fruits and vegetables storage: A review. *Packag. Technol. Sci.* **2022**, *35*, 461. [CrossRef]
5. Arnon-Rips, H.; Poverenov, E. Improving food products' quality and storability by using Layer by Layer edible coatings. *Trends Food Sci. Technol.* **2018**, *75*, 81. [CrossRef]
6. Olawuyi, I.F.; Lee, W. Influence of chitosan coating and packaging materials on the quality characteristics of fresh-cut cucumber. *Korean J. Food Preserv.* **2019**, *26*, 371. [CrossRef]
7. Aayush, K.; McClements, D.J.; Sharma, S.; Sharma, R.; Singh, G.P.; Sharma, K.; Oberoi, K. Innovations in the development and application of edible coatings for fresh and minimally processed Apple. *Food Control* **2022**, *141*, 109188. [CrossRef]
8. Olawuyi, I.F.; Kim, S.R.; Lee, W.Y. Application of plant mucilage polysaccharides and their techno-functional properties' modification for fresh produce preservation. *Carbohydr. Polym.* **2021**, *272*, 118371. [CrossRef]
9. Park, D.H.; Park, J.J.; Olawuyi, I.F.; Lee, W.Y. Quality of White mushroom (*Agaricus bisporus*) under argon-and nitrogen-based controlled atmosphere storage. *Sci. Hortic.* **2020**, *265*, 109229. [CrossRef]
10. Dhall, R. Advances in edible coatings for fresh fruits and vegetables: A review. *Crit. Rev. Food Sci. Nutr.* **2013**, *53*, 435. [CrossRef]
11. Poverenov, E.; Danino, S.; Horev, B.; Granit, R.; Vinokur, Y.; Rodov, V. Layer-by-layer electrostatic deposition of edible coating on fresh cut melon model: Anticipated and unexpected effects of alginate–chitosan combination. *Food Bioprocess Technol.* **2014**, *7*, 1424. [CrossRef]
12. Bourtoom, T. Edible films and coatings: Characteristics and properties. *Int. Food Res. J.* **2008**, *15*, 237.
13. Medeiros, B.G.d.S.; Pinheiro, A.C.; Carneiro-da-Cunha, M.G.; Vicente, A.A. Development and characterization of a nanomultilayer coating of pectin and chitosan–Evaluation of its gas barrier properties and application on 'Tommy Atkins' mangoes. *J. Food Eng.* **2012**, *110*, 457. [CrossRef]
14. Brasil, I.; Gomes, C.; Puerta-Gomez, A.; Castell-Perez, M.; Moreira, R. Polysaccharide-based multilayered antimicrobial edible coating enhances quality of fresh-cut papaya. *LWT-Food Sci. Technol.* **2012**, *47*, 39. [CrossRef]
15. Vargas, M.; Pastor, C.; Chiralt, A.; McClements, D.J.; Gonzalez-Martinez, C. Recent advances in edible coatings for fresh and minimally processed fruits. *Crit. Rev. Food Sci. Nutr.* **2008**, *48*, 496. [CrossRef]
16. Acevedo, C.A.; López, D.A.; Tapia, M.J.; Enrione, J.; Skurtys, O.; Pedreschi, F.; Brown, D.I.; Creixell, W.; Osorio, F. Using RGB image processing for designing an alginate edible film. *Food Bioprocess Technol.* **2012**, *5*, 1511. [CrossRef]
17. Zapata, P.J.; Guillén, F.; Martínez-Romero, D.; Castillo, S.; Valero, D.; Serrano, M. Use of alginate or zein as edible coatings to delay postharvest ripening process and to maintain tomato (*Solanum lycopersicon* Mill) quality. *J. Sci. Food Agric.* **2008**, *88*, 1287. [CrossRef]
18. Maftoonazad, N.; Ramaswamy, H.S.; Marcotte, M. Shelf-life extension of peaches through sodium alginate and methyl cellulose edible coatings. *Int. J. Food Sci. Technol.* **2008**, *43*, 951. [CrossRef]
19. Nehchiri, N.; Amiri, S.; Radi, M. Improving the water barrier properties of alginate packaging films by submicron coating with drying linseed oil. *Packag. Technol. Sci.* **2021**, *34*, 283. [CrossRef]
20. Ouattara, B.; Simard, R.; Piette, G.; Begin, A.; Holley, R. Diffusion of acetic and propionic acids from chitosan-based antimicrobial packaging films. *J. Food Sci.* **2000**, *65*, 768. [CrossRef]
21. Devlieghere, F.; Vermeulen, A.; Debevere, J. Chitosan: Antimicrobial activity, interactions with food components and applicability as a coating on fruit and vegetables. *Food Microbiol.* **2004**, *21*, 703. [CrossRef]

22. Han, C.; Lederer, C.; McDaniel, M.; Zhao, Y. Sensory evaluation of fresh strawberries (*Fragaria ananassa*) coated with chitosan-based edible coatings. *J. Food Sci.* **2005**, *70*, S172. [CrossRef]
23. Chien, P.-J.; Sheu, F.; Yang, F.-H. Effects of edible chitosan coating on quality and shelf life of sliced mango fruit. *J. Food Eng.* **2007**, *78*, 225. [CrossRef]
24. Chiabrando, V.; Giacalone, G. Quality evaluation of blueberries coated with chitosan and sodium alginate during postharvest storage. *Int. Food Res. J.* **2017**, *24*, 1553–1561.
25. Chiabrando, V.; Giacalone, G. Effect of different coatings in preventing deterioration and preserving the quality of fresh-cut nectarines (cv Big Top). *CyTA-J. Food* **2013**, *11*, 285. [CrossRef]
26. Soison, B.; Jangchud, K.; Jangchud, A.; Harnsilawat, T.; Piyachomkwan, K. Characterization of starch in relation to flesh colors of sweet potato varieties. *Int. Food Res. J.* **2015**, *22*, 2302.
27. Kou, X.; He, Y.; Li, Y.; Chen, X.; Feng, Y.; Xue, Z. Effect of abscisic acid (ABA) and chitosan/nano-silica/sodium alginate composite film on the color development and quality of postharvest Chinese winter jujube (*Zizyphus jujuba* Mill. cv. Dongzao). *Food Chem.* **2019**, *270*, 385. [CrossRef]
28. Aider, M. Chitosan application for active bio-based films production and potential in the food industry. *LWT Food Sci. Technol.* **2010**, *43*, 837. [CrossRef]
29. Tapia, M.; Rojas-Graü, M.; Carmona, A.; Rodríguez, F.; Soliva-Fortuny, R.; Martin-Belloso, O. Use of alginate-and gellan-based coatings for improving barrier, texture and nutritional properties of fresh-cut papaya. *Food Hydrocoll.* **2008**, *22*, 1493. [CrossRef]
30. Rojas-Graü, M.A.; Tapia, M.S.; Martín-Belloso, O. Using polysaccharide-based edible coatings to maintain quality of fresh-cut Fuji apples. *LWT-Food Sci. Technol.* **2008**, *41*, 139. [CrossRef]
31. Oms-Oliu, G.; Soliva-Fortuny, R.; Martín-Belloso, O. Using polysaccharide-based edible coatings to enhance quality and antioxidant properties of fresh-cut melon. *LWT-Food Sci. Technol.* **2008**, *41*, 1862. [CrossRef]
32. Azarakhsh, N.; Osman, A.; Ghazali, H.M.; Tan, C.P.; Adzahan, N.M. Lemongrass essential oil incorporated into alginate-based edible coating for shelf-life extension and quality retention of fresh-cut pineapple. *Postharvest Biol. Technol.* **2014**, *88*, 1. [CrossRef]
33. Souza, M.P.; Vaz, A.F.; Cerqueira, M.A.; Texeira, J.A.; Vicente, A.A.; Carneiro-da-Cunha, M.G. Effect of an edible nanomultilayer coating by electrostatic self-assembly on the shelf life of fresh-cut mangoes. *Food Bioprocess Technol.* **2015**, *8*, 647. [CrossRef]
34. Chiabrando, V.; Giacalone, G. Anthocyanins, phenolics and antioxidant capacity after fresh storage of blueberry treated with edible coatings. *Int. J. Food Sci. Nutr.* **2015**, *66*, 248. [CrossRef]
35. Embuscado, M.E.; Huber, K.C. *Edible Films and Coatings for Food Applications*; Springer: Berlin/Heidelberg, Germany, 2009; Volume 9.
36. Baldwin, E.A.; Hagenmaier, R.; Bai, J. *Edible Coatings and Films to Improve Food Quality*; CRC Press: Boca Raton, FL, USA, 2011.
37. Mishra, B.; Khatkar, B.; Garg, M.; Wilson, L. Permeability of edible coatings. *J. Food Sci. Technol.* **2010**, *47*, 109. [CrossRef]
38. Ghidelli, C.; Mateos, M.; Rojas-Argudo, C.; Pérez-Gago, M.B. Extending the shelf life of fresh-cut eggplant with a soy protein–cysteine based edible coating and modified atmosphere packaging. *Postharvest Biol. Technol.* **2014**, *95*, 81. [CrossRef]
39. Xing, Y.; Li, X.; Xu, Q.; Jiang, Y.; Yun, J.; Li, W. Effects of chitosan-based coating and modified atmosphere packaging (MAP) on browning and shelf life of fresh-cut lotus root (Nelumbo nucifera *Gaerth*). *Innov. Food Sci. Emerg. Technol.* **2010**, *11*, 684. [CrossRef]
40. Shi, S.; Wang, W.; Liu, L.; Wu, S.; Wei, Y.; Li, W. Effect of chitosan/nano-silica coating on the physicochemical characteristics of longan fruit under ambient temperature. *J. Food Eng.* **2013**, *118*, 125. [CrossRef]
41. Vieira, T.M.; Moldão-Martins, M.; Alves, V.D. Composite coatings of chitosan and alginate emulsions with olive oil to enhance postharvest quality and shelf life of fresh figs (*Ficus carica* L. cv.'Pingo De Mel'). *Foods* **2021**, *10*, 718. [CrossRef]
42. Al-Hilifi, S.A.; Al-Ali, R.M.; Al-Ibresam, O.T.; Kumar, N.; Paidari, S.; Trajkovska Petkoska, A.; Agarwal, V. Physicochemical, Morphological and Functional Characterization of Edible Anthocyanin-Enriched Aloe Vera Coatings on Fresh Figs (*Ficus carica* L.). *Gels* **2022**, *8*, 645. [CrossRef]
43. Sipahi, R.; Castell-Perez, M.; Moreira, R.; Gomes, C.; Castillo, A. Improved multilayered antimicrobial alginate-based edible coating extends the shelf life of fresh-cut watermelon (*Citrullus lanatus*). *LWT-Food Sci. Technol.* **2013**, *51*, 9. [CrossRef]
44. Zam, W. Effect of alginate and chitosan edible coating enriched with olive leaves extract on the shelf life of sweet cherries (*Prunus avium* L.). *J. Food Qual.* **2019**, *2019*, 8192964. [CrossRef]
45. Qamar, J.; Ejaz, S.; Anjum, M.A.; Nawaz, A.; Hussain, S.; Ali, S.; Saleem, S. Effect of Aloe vera gel, chitosan and sodium alginate based edible coatings on postharvest quality of refrigerated strawberry fruits of cv. Chandler. *J. Hortic. Sci. Technol.* **2018**, *1*, 8. [CrossRef]
46. Valero, D.; Díaz-Mula, H.M.; Zapata, P.J.; Guillén, F.; Martínez-Romero, D.; Castillo, S.; Serrano, M. Effects of alginate edible coating on preserving fruit quality in four plum cultivars during postharvest storage. *Postharvest Biol. Technol.* **2013**, *77*, 1. [CrossRef]
47. Nair, M.S.; Saxena, A.; Kaur, C. Effect of chitosan and alginate based coatings enriched with pomegranate peel extract to extend the postharvest quality of guava (*Psidium guajava* L.). *Food Chem.* **2018**, *240*, 245. [CrossRef]
48. Connor, A.M.; Luby, J.J.; Hancock, J.F.; Berkheimer, S.; Hanson, E.J. Changes in fruit antioxidant activity among blueberry cultivars during cold-temperature storage. *J. Agric. Food Chem.* **2002**, *50*, 893. [CrossRef]
49. Song, H.; Yuan, W.; Jin, P.; Wang, W.; Wang, X.; Yang, L.; Zhang, Y. Effects of chitosan/nano-silica on postharvest quality and antioxidant capacity of loquat fruit during cold storage. *Postharvest Biol. Technol.* **2016**, *119*, 41. [CrossRef]

50. Ali, A.; Muhammad, M.T.M.; Sijam, K.; Siddiqui, Y. Effect of chitosan coatings on the physicochemical characteristics of Eksotika II papaya (*Carica papaya* L.) fruit during cold storage. *Food Chem.* **2011**, *124*, 620. [CrossRef]
51. Arroyo, B.J.; Bezerra, A.C.; Oliveira, L.L.; Arroyo, S.J.; de Melo, E.A.; Santos, A.M.P. Antimicrobial active edible coating of alginate and chitosan add ZnO nanoparticles applied in guavas (*Psidium guajava* L.). *Food Chem.* **2020**, *309*, 125566. [CrossRef]
52. Lee, J.; Durst, R.W.; Wrolstad, R.E.; Eisele, T.; Giusti, M.M.; Hach, J.; Hofsommer, H.; Koswig, S.; Krueger, D.A.; Kupina, S.; et al. Determination of total monomeric anthocyanin pigment content of fruit juices, beverages, natural colorants, and wines by the pH differential method: Collaborative study. *J. AOAC Int.* **2005**, *88*, 1269. [CrossRef]

 gels

MDPI

Article

Physicochemical, Morphological, and Functional Characterization of Edible Anthocyanin-Enriched *Aloe vera* Coatings on Fresh Figs (*Ficus carica* L.)

Sawsan Ali Al-Hilifi [1,*], Rawdah Mahmood Al-Ali [1], Orass T. Al-Ibresam [1], Nishant Kumar [2], Saeed Paidari [3], Anka Trajkovska Petkoska [4] and Vipul Agarwal [5]

1 Department of Food Science, College of Agriculture, University of Basrah, Basrah 61004, Iraq
2 Department of Food Science and Technology, National Institute of Food Technology Entrepreneurship and Management, Sonepat 131028, India
3 Department of Food Science and Technology, Isfahan (Khorasgan) Branch, Islamic Azad University, Isfahan 81551-39998, Iran
4 Faculty of Technology and Technical Social Sciences, St. Kliment Ohridski University-Bitola, Dimitar Vlahov, 1400 Veles, North Macedonia
5 Cluster for Advanced Macromolecular Design (CAMD), School of Chemical Engineering, University of New South Wales, Sydney, NSW 2052, Australia
* Correspondence: sawsan.hameed@uobasrah.edu.iq

Abstract: In the present investigation, *Aloe vera* gel (AVG)-based edible coatings enriched with anthocyanin were prepared. We investigated the effect of different formulations of aloe-vera-based edible coatings, such as neat AVG (T1), AVG with glycerol (T2), *Aloe vera* with 0.2% anthocyanin + glycerol (T3), and AVG with 0.5% anthocyanin + glycerol (T4), on the postharvest quality of fig (*Ficus carica* L.) fruits under refrigerated conditions (4 °C) for up to 12 days of storage with 2-day examination intervals. The results of the present study revealed that the T4 treatment was the most effective for reducing the weight loss in fig fruits throughout the storage period (~4%), followed by T3, T2, and T1. The minimum weight loss after 12 days of storage (3.76%) was recorded for the T4 treatment, followed by T3 (4.34%), which was significantly higher than that of uncoated fruit (~11%). The best quality attributes, such as the total soluble solids (TSS), titratable acidity (TA), and pH, were also demonstrated by the T3 and T4 treatments. The T4 coating caused a marginal change of 0.16 in the fruit titratable acidity, compared to the change of 0.33 in the untreated fruit control after 12 days of storage at 4 °C. Similarly, the total soluble solids in the T4-coated fruits increased marginally (0.43 °Brix) compared to the uncoated control fruits (>2 °Brix) after 12 days of storage at 4 °C. The results revealed that the incorporation of anthocyanin content into AVG is a promising technology for the development of active edible coatings to extend the shelf life of fig fruits.

Keywords: coating; anthocyanin; fruit decay; postharvest shelf life; *Aloe vera*; antioxidant; fig

Citation: Al-Hilifi, S.A.; Al-Ali, R.M.; Al-Ibresam, O.T.; Kumar, N.; Paidari, S.; Trajkovska Petkoska, A.; Agarwal, V. Physicochemical, Morphological, and Functional Characterization of Edible Anthocyanin-Enriched *Aloe vera* Coatings on Fresh Figs (*Ficus carica* L.). *Gels* **2022**, *8*, 645. https://doi.org/10.3390/gels8100645

Academic Editors: Aris E. Giannakas, Constantinos Salmas and Charalampos Proestos

Received: 20 September 2022
Accepted: 6 October 2022
Published: 11 October 2022

1. Introduction

In the past decade, there has been remarkable growth in the development of bio-based and active edible packaging to ensure food safety and quality and reduce postharvest losses of fruits and vegetables [1–5]. Edible packaging is an ecofriendly and sustainable approach to maintaining the quality attributes of fruits and vegetables during storage by (i) minimizing lipid peroxidation, (ii) altering the respiration rate, (iii) reducing weight loss, and (iv) maintaining other quality attributes [6–8]. Furthermore, edible coatings also increase fruit microbiological safety and protect them from the effects of external environmental conditions, hence extending their shelf lives [9]. Moreover, the application of functional agents/compounds such as antioxidants, antimicrobials, and nutraceuticals in edible coatings improves the quality attributes and postharvest characteristics of fruits and vegetables [10–13].

Figs (*Ficus carica.* L.) are a popular fruit due their health benefits and functional properties, including boosting immunity and being rich in fiber and antioxidant compounds. However, they are highly perishable due to their susceptibility to deterioration, oxidation, and microbial growth [14–18]. These fruits are also sensitive to microbial contamination under cold storage conditions, resulting in an unpleasant taste and aroma that affect consumer acceptability [19]. The postharvest shelf life of fig fruits can be extended using various technologies, i.e., cold storage, modified atmosphere packaging, controlled atmospheres, vacuum application, and edible packaging (coating) [20–23].

Edible coating is an alternative postharvest management strategy to extend the shelf life of fig fruits by preserving their quality attributes [17,24]. Different types of edible coatings based on polysaccharides, proteins, lipids, or composites have been used for fruit and vegetable preservation. The advantages of these edible coatings include their eco-friendliness, biodegradability in nature, and ability to extend the shelf life of fruits and vegetables [6,7]. *Aloe vera* is a succulent plant of the *Asphodelaceae* family of the genus Aloe that has been used as a medicinal plant for millennia. *Aloe vera* gel (AVG) is the pulp produced by *Aloe vera* plants, and its gelatinous matrix can be used to produce natural edible coatings. Its applicability in the preservation of certain fruits has been reported previously [25]. The preservation effectiveness of AVG coatings has been attributed to its ability to (i) reduce phenolic oxidation, (ii) inhibit peroxidase and polyphenol oxidase enzyme activity, and (iii) reduce the browning of fruits while preserving their quality from the detrimental effects of weight loss, enzymatic browning, electrolytic leakage, respiration rate degradation, and chlorophyll degradation [26]. Fruit preservation quality can be assessed in terms of firmness, visual appeal, nutrient content, and freshness [27,28]. AVG is considered an excellent source of nutritional and phytochemical compounds such phenols, flavonoids, terpenoids, lectins, and fatty acids. AVG also contains vitamins and polysaccharides, which function as natural antioxidants [29]. AVG has shown antiviral, antibacterial, laxative, antioxidant, anti-inflammation, anticancer, antidiabetic, antiallergic, immunostimulatory, and UV-protective properties due to presence of high amounts of bioactive compounds [30]. However, neat AVG is highly hydrophilic, which limits its functional duration as a fruit protective coating. To this end, AVG is supplemented with other constituents to improve its properties and longevity. While a range of different coatings have been developed and tested on various fruits [31,32], to the best of our knowledge, no study has explored the potential of anthocyanin (as a natural antioxidant) and AVG supplemented with anthocyanin in natural edible coatings.

The present study aimed to investigate the physicochemical, morphological, and functional characteristics of AVG-based edible coatings enriched with anthocyanin extracted from onion peel. We assessed the effects of different treatments based on the developed AVG edible coating with or without anthocyanin on the shelf life of *Ficus carica* fruit during 12 days of storage at 4 °C.

2. Results and Discussion

2.1. Physicochemical Properties of Aloe vera Gel (AVG)

Table 1 depicts the physicochemical observations for the AVG. The results showed that the AVG had a moisture content of 97.52%, a pH of 5.29 with 0.06% acidity, a carbohydrate content of 0.65%, a viscosity of 4.66, a refractive index of 1.33, and TSS of 3.10 °Bx. Our results were in agreement with previous studies [33–35]. The obtained refractive index of 1.33 was in line with the International Aloe Science Council's recommendation (IASC). A gel's refractive index is a physical measure that defines its purity when compared to double-distilled water. The optimum treatment for the coating process is a gel with the lowest refractive index. Impurities in the extracted gel are indicated by a higher refractive index [36].

Table 1. Physicochemical characteristics of AVG.

Characteristic Parameter	Mean± Standard Deviation
Moisture	97.52 ± 0.83
pH	5.29 ± 0.037
Acidity	0.06 ± 0.00
Carbohydrates	0.65 ± 0.01
Viscosity	4.66 ± 0.00
Refractive index	1.33 ± 0.00
Total soluble solids	3.10 ± 0.00

2.2. Microstructure Analysis

The microstructure analysis was conducted using SEM (Figure 1). The AVG image showed a more organized and smoother structure compared to the others, with an intact parenchymal cell of a well-rounded shape and characteristic diameter, whose dimensions were in the range of 28–50 nm due to the presence of a high water content. Similarly, the close contact between the walls of adjacent cells was in line with previous reports [37,38]. Compared to the control samples, all coating samples exhibited decreased intracellular integrity and cell shape regularity. The results showed that the incorporation of glycerol and anthocyanin into the AVG introduced ruptured cell structures. The microstructure of the AVG with anthocyanin content was more shrinkable and less porous compared to the others. The pore sizes of the AVG + glycerol + anthocyanin coating (T3 and T4) samples (Figure 1c,d) were smaller than those of the neat AVG and AVG + glycerol coatings (Figure 1a,b). The observed differences were attributed to the weaker hydrogen bonds between the carboxylic group of the AVG and anthocyanin. An increased concentration of anthocyanin in the AVG resulted in a more complex and rougher film microstructure. The obtained results were similar to those of a previous report on aloe-vera–gelatin–glycerol edible films enriched with *Pimenta dioica* L. Merrill essential oil [39]. In this study, it was reported that the incorporation of the active ingredient (*Pimenta dioica* L. Merrill essential oil) affected the microstructure of the AV–gelatin–glycerol-based films by increasing the roughness and the flocculation rate on the surface.

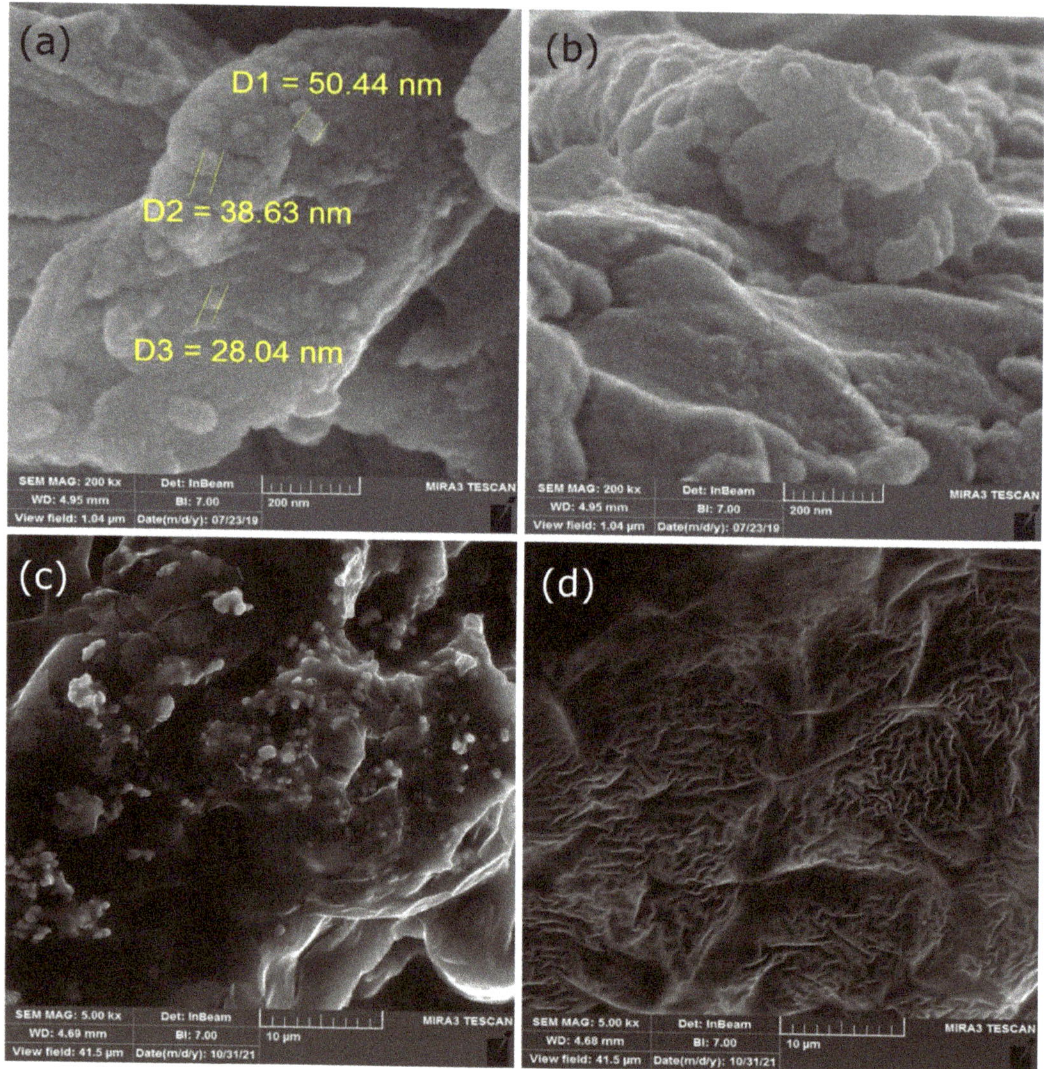

Figure 1. SEM images showing the microstructure of AVG-based edible coatings. (**a**) AVG (T1), (**b**) AVG + glycerol (T2), (**c**) AVG + glycerol + anthocyanin + 0.2% (T3), (**d**) AVG + glycerol + anthocyanin + 0.5% (T4). Scale bar: (**a**,**b**) = 200 nm, (**c**,**d**) = 10 μm. Magnifications: (**a**,**b**) 200,000×, (**b**,**c**) 5000×.

2.3. *Effect of AVG-Based Edible Coatings on Postharvest Quality of Fig Fruits*

2.3.1. Weight Loss

Figs are highly susceptible to weight loss due to their thin peels, which allow for rapid water loss and tissue deterioration. Figure 2 presents the weight loss (%) of the coated and uncoated figs. We observed significantly reduced weight loss ($p < 0.05$) in the coated fruits after 10 days of storage (<2.94%) compared to the uncoated figs (maximum weight loss of 9.68%). The combination of anthocyanin with AVG + glycerol effectively contributed to reducing weight loss. The observed reduction in weight loss in the coated fruits was attributed to the formation of a semi-permeable barrier that prevented water loss [25,40]. Furthermore, the superior performance of the AVG + glycerol + anthocyanin coating could have been caused by a reduction in the water loss of the fruits due the crosslinking of anthocyanin, glycerol, and AVG [41,42]. In the presence of plant extracts (AVG), the properties of biopolymeric films can be changed as a result of interactions between the biopolymer and polyphenolic compounds (anthocyanin) [43]. The weight loss measured for the fruits coated using the developed (AVG + glycerol-based) edible coatings was either similar to or significantly better than the results obtained by previous studies [17,44,45]. For example, contrary to our study, the application of zein containing cystein (0.2%), ascorbic acid (0.2%), and jamun leaf extract (0.2%) coatings on jamun fruit resulted in significantly higher weight loss as the storage period progressed [46]. In another study, an *Aloe vera* and gum tragacanth coating applied to button mushrooms was shown to cause a significant weight loss of ~40–50% over the 13-day storage period, compared to the figure of ~2–5% obtained for our AVG + glycerol + anthocyanin-coated figs over the 12-day storage period [47]. The difference between the abovementioned work [47] and our study could be attributed to a combination of (i) the superior performance of our coatings and (ii) the higher water content of button mushrooms compared to fig fruits. Furthermore, the significant difference in weight loss between the uncoated and coated fruits observed in our study was in disagreement with a previous report on the application of quinoa protein/chitosan coatings containing a thymol nanoemulsion on refrigerated strawberries, which showed no difference between the coated and uncoated fruit [48]. The observed disagreement could be attributed to the significant differences in the (fruit) surface adhesion of the two coatings, i.e., chitosan + quinoa protein + thymol nanoemulsion and AVG + glycerol + anthocyanin.

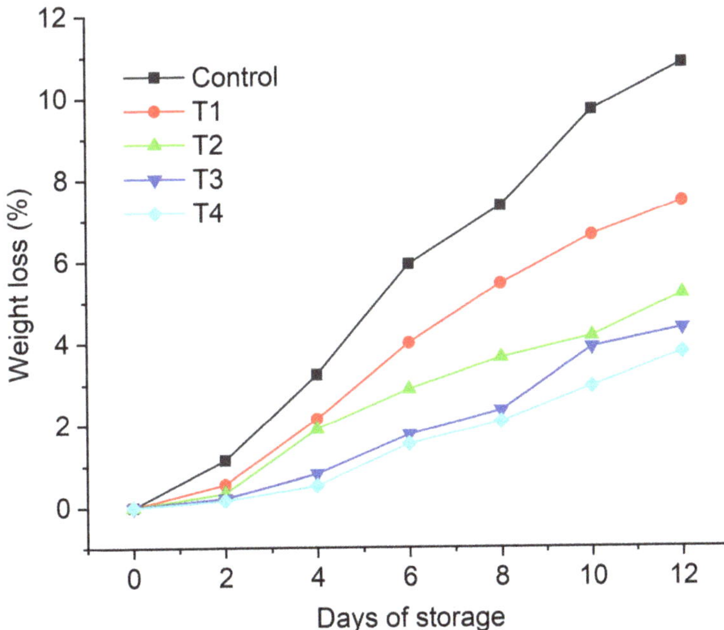

Figure 2. Effect of edible coatings on weight loss of *Ficus carica* fruits: uncoated control, AVG (T1), AVG + glycerol (T2), AVG + glycerol + anthocyanin + 0.2% (T3), AVG + glycerol + anthocyanin + 0.5% (T4). Data are presented as mean ± SD, (n = 3) (error bars are significantly smaller than the data points).

2.3.2. pH

The change in pH is an important indicator of fruit properties, with an increase in pH indicating the ripening and oxidation of the fruit over time. To prolong a fruit's shelf life, it is important that the change in pH is marginal over the storage period. The pH analysis of the juice extracted from the coated and uncoated fruits revealed a gradual increase as the storage period progressed (Figure 3). Compared to the control fruits (T0—water-washed fruits), the coated fruits exhibited a lower increase in pH values over time. We observed a maximum increase in pH for the control fruits (from ~4.3 at day 0 to ~4.7 at day 12) as the storage period progressed. The lowest pH increase was observed for the figs coated with T4 (AVG + glycerol + anthocyanin 0.5%) over the storage period. Coatings are known to reduce the respiratory and metabolic rates of fruit, thereby limiting the utilization of organic acids and restricting the pH change over the storage period [49]. The addition of active compounds such as anthocyanin promotes coatings' functional performance, enhancing the stability; quality (reducing biochemical deterioration, enzymatic browning, and the development of off-flavors); and safety of foods [50,51]. Similar results in terms of marginal changes in pH value have been reported previously for an AVG-based edible coating on freshly cut papaya [28], a nanostructured lipid carriers + cinnamon essential oil coating on tangerine [52], and a pectin + candelilla wax + aloe mucilage + glycerol + polyphenol Larrea leaf extract coating on avocados [53].

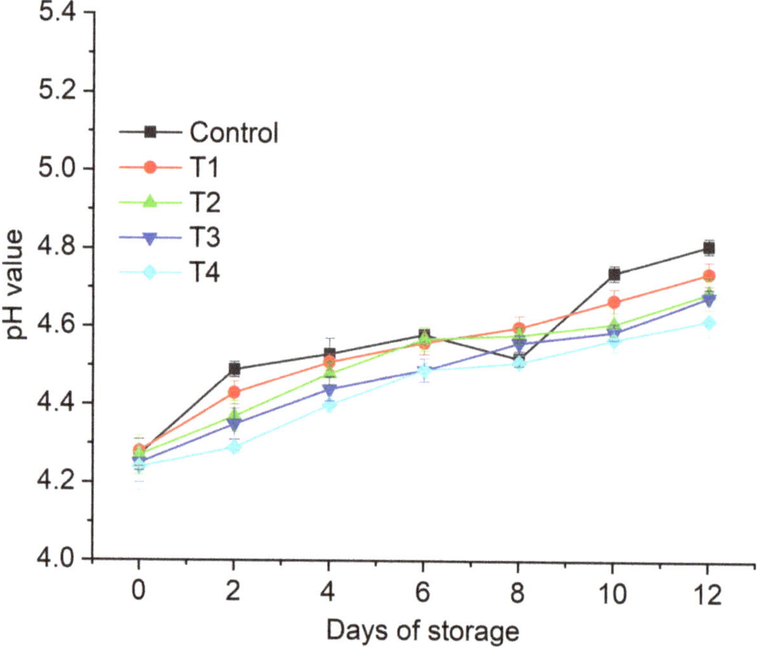

Figure 3. Effect of edible coatings on pH of *Ficus carica* fruits: uncoated control, AVG (T1), AVG + glycerol (T2), AVG + glycerol + anthocyanin + 0.2% (T3), and AVG + glycerol + anthocyanin + 0.5% (T4). Data are presented as mean ± SD, ($n = 3$).

2.3.3. Titratable Acidity (TA)

The TA of the treated and untreated figs gradually reduced with increasing storage time (Figure 4). The fig fruits treated with T2 (AVG + glycerol), T3 (AVG + glycerol + anthocyanin 0.2%), and T4 (AVG + glycerol + anthocyanin 0.5%) showed a slower reduction in TA compared to the control (water-washed) and T1 (AVG)-treated fruits. Between T3 and T4, a lower reduction in TA was observed in the figs coated with T4 compared to T3. The significantly lower reduction in the coated figs could be attributed to the restriction of the respiration rate and water loss in the fruits [54]. The TSS were significantly reduced in the coated fruits compared to the uncoated control fruits after 10 days of storage due to the increase in the respiration rate and fruit maturity in the uncoated fruits compared to the coated fruits. Based on our findings, we postulated that the anthocyanin-containing coatings (T3 and T4) decreased the oxidation and fruit cellular senescence, resulting in a higher TA compared to the other coatings over the 12-day storage period. These results supported the changes observed in the pH values, with the coated samples showing significantly lower pH changes over the storage period compared to the uncoated (T0) fruits.

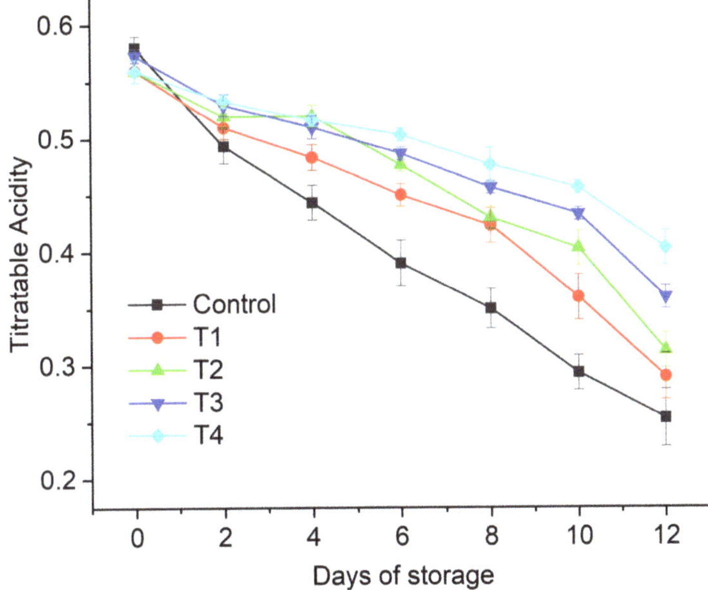

Figure 4. Effect of edible coatings on titratable acidity of *Ficus carica* fruit: uncoated control (water-washed), AVG (T1), AVG + glycerol (T2), AVG + glycerol + anthocyanin + 0.2% (T3), AVG + glycerol + anthocyanin + 0.5% (T4). Data are presented as mean ± SD, (n = 3).

2.3.4. Total Soluble Solids (TSS)

The TSS are a very important characteristic of fruits and vegetables, being indicative of their freshness and sweetness. We observed an increase in the TSS of all the treated and control fruits with increasing storage time (Figure 5). However, we observed a marginal increase in the coated figs compared to the uncoated control fruits, which exhibited the highest TSS value (16.34 ± 0.02 °Brix at day 12). Furthermore, the lowest increase in TSS was observed in the figs coated with T3 (AVG + glycerol + anthocyanin 0.2%) and T4 (AVG + glycerol + anthocyanin 0.5%). The superior performance of the T3 and T4 coatings could be ascribed to the reduction in water loss and the minimized oxidation of the fruits due to the presence of anthocyanin [21]. On the 12th day of storage, the higher TSS (14.93 ± 0.01 °Brix) value was observed for the fruits coated with T3 compared to the T4-coated fruits (14.70 ± 0.01 °Brix). Similar trends in results for fig fruits have been reported previously [55]. In study [55], a chitosan and alginate emulsion coating enriched with olive-oil was shown to inhibit a TSS increase in coated figs. The presence of anthocyanins as an active agent in an AVG-based coating has been previously shown to maintain the TSS of fruits throughout the storage period due to natural fruit ripening processes [56,57]. Furthermore, an increase in TSS during storage could be linked to the transformation of pectic compounds, starch hydrolysis, and the solubilization of polyuronides and hemicelluloses in fruit cell walls, as well as the hydrolysis of insoluble polysaccharides into simple sugars [58,59]. The incorporation of polysaccharides to support bioactive compounds from plant sources could be a potential way to extend the shelf life of fresh fruit during postharvest storage [47]. In addition, the performance of our AVG + glycerol + anthocyanin coatings in maintaining fruit TSS was similar to that of other types of coatings reported previously, including a alginate + black cumin extract coating on guava fruit [44], a sodium alginate + cinnamaldehyde-loaded nanostructured lipid carrier coating on date palm fruit [60], and a chitosan coating on sweet cherry cultivars [61]. Our AVG + glycerol + anthocyanin coatings performed better in maintaining TSS levels over the storage period than other coatings reported in the literature, including a chitosan + quinoa protein + thymol nanoemulsion coating on refrigerated strawberries [48] and a starch + mango peel powder coating on apple slices [62].

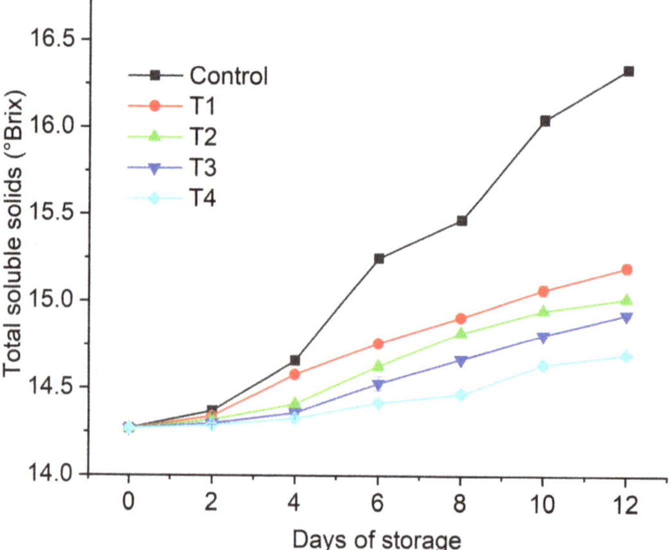

Figure 5. Effect of edible coatings on total soluble solids of *Ficus carica* fruits: uncoated control, AVG (T1), AVG + glycerol (T2), AVG + glycerol + anthocyanin 0.2% (T3), AVG + glycerol + anthocyanin 0.5% (T4). Data are presented as mean ± SD, (n = 3) (error bars are significantly smaller than the data points).

3. Conclusions

Edible coatings continue to attract significant attention as means to extend the shelf life of perishable foods. In this study, we explored the potential of aloe-vera-based edible coatings. *Aloe vera* gel (AVG) enriched with anthocyanin was developed as an edible coating to improve the shelf and storage life of fig fruits. Different edible coatings were tested on fig fruits over a 12-day storage period under regular refrigerated conditions. We observed significant improvements in coating performance and the maintenance of fruit quality with the inclusion of anthocyanin in the AVG coatings, as judged by the weight of the coated fruits and the changes in their pH, total soluble solids, and titratable acidity. The AVG + anthocyanin (0.5%) (T4) coating extended the fruit shelf life by limiting weight loss (~4%) compared to the uncoated control fruit (which lost >10% weight) after 12 days of storage at 4 °C, due to the loss of water from the fruits. The T4 coating also preserved fruit acidity (20% reduction in acidity) and total soluble solids (0.25 °Brix) compared to the uncoated control fruits (with a ~60% reduction in acidity and 2.5 °Brix total soluble solids). The coating performance improved significantly with an increasing amount of anthocyanin, with the T4 coatings exhibiting better performance in preserving the innate properties of coated fig fruits compared to the T3 (AVG + anthocyanin (0.2%)) and neat AVG (T1) coatings, as judged by the changes in weight loss, titratable acidity, and total soluble solids. Overall, this study found that the inclusion of anthocyanin in natural edible nanocomposite coatings can significantly improve the coating performance and shelf life of coated fruits. In the future, we hope to carry out a taste test of fruits coated using AVG + anthocyanin coatings. The developed strategy and coatings are envisaged to inspire further research exploring the commercial use of naturally occurring anthocyanin in edible coatings.

4. Material and Methods

4.1. Materials

Fresh and homogeneously sized figs (*Ficus carica*), free from physical and microbial damage, were procured from a local market in Basrah, Iraq in July 2020. The fruits were chosen based on their size, maturity stage, color, and the absence of visible defects, and they were transported in refrigerated containers to the laboratories of the Department of Food Sciences of the College of Agriculture at the University of Basrah. Fresh *Aloe vera* (*Aloe barbadensis* Miller) leaves were collected from a local orchard in Basrah city, Iraq.

4.2. Experimental Methods

Preparation of *Aloe vera* Gel (AVG)

For the preparation of AVG, the leaves were washed using chlorinated water and dried to remove dirt and contamination. The margin of the leaves was mechanically cut to remove the external epidermis and extract the gel. We then added 1% (*w*/*w*) vitamin C to the obtained gel and stirred for 30 min at 50 °C to prevent browning, and the mixture was stored in airtight, opaque glass containers for further analysis and use.

4.3. Physiochemical Analysis of AVG

4.3.1. pH

The pH of the AVG was analyzed by AOAC (2010) standard methods. The pH of the samples was assessing using a pH meter (pH-EMCO-256071, Japan). After the homogenization of the samples, pH was measured by the direct immersion of the electrode.

4.3.2. Moisture Content

AOAC (2012) gravimetric techniques were used to determine moisture content. Ten grams of each sample was weighed and dried in an oven for 24 h at 120 °C (Heraeus, Hanau, Germany). The change in the weight was determined using the standard gravimetry method and presented in terms of sample mass loss (percentage).

4.3.3. Viscosity

An Ostwald viscometer size D was used to measure the viscosity of the AVG at 21 °C, which was calculated using the following equation:

$$r_1 = \frac{r_2 \times p_1 \times t_1}{p_2 \times t_2}$$

where r_1 = viscosity of the gel, p_1 = density of the gel, r_2 = viscosity of water, t_1 = gel descent time in seconds, p_2 = density of water, and t_2 = water descent time in seconds.

4.3.4. Refractive Index

The refractive index of the AVG was measured using an Abbe Refractometer (A87117, Bellingham, UK). At 22 °C, the refractive indices ranged between 1.3000 and 1.7000, with an accuracy of 0.0003. Instrument calibration was conducted using distilled water. Data are presented as the average ± standard deviation of five measurements.

4.4. Anthocyanin Pigment Extraction from Red Onion Peel

For the extraction of anthocyanin from red onion peel, 10 g sections of fresh red onion (*Allium cepa*) peel were washed with tap water to remove impurities. Clean wet peels were then dried in an oven at 50 °C for 24 h. Dried peels were then grounded to obtain a powder. To extract anthocyanin, 10 g of clean, dried onion peel powder was dispersed in 40 mL of acidified ethanol and 1% ascorbic acid, and the mixture was stirred for 1 min using a magnetic hot plate stirrer (GFL, Korea). The solution was incubated overnight at 30 °C in an oven (Binder) for the extraction and recovery of anthocyanin pigments. The obtained solution was filtered using Whatman filter paper (540), after which the solvent was evaporated using a rotary evaporator (Franklin Electric, UK) at 40 °C to obtain dry powdered anthocyanin. The obtained anthocyanin pigment powder was stored in a sterile container for further use.

4.5. Preparation of AVG-Based Coating

AVG was diluted with distilled water to 40% (*v*/*v*) as a coating solution. We added 2% glycerol as a plasticizer, and the solution was homogenized using a magnetic stirrer at 1500 rpm for 6 h. Anthocyanin (0.2 mg/100 mL and 0.5 mg/100 mL) was added to the coating solution as an antioxidant agent. The pH of the coating mixture was maintained at 4 using citric acid (4.5–4.6 g/L). We placed 15 mL of each AVG solution into petri dishes to dry at 45 °C in a sterilized oven to obtain different coatings.

4.6. Microstructure Analysis

The surface morphology of the AVG-based edible coatings enriched with anthocyanin was analyzed by a scanning electron microscope (SEM) (Supra 55 VP, Carl Zeiss, Germany) with a voltage of 1 KV and a magnification power of 1000–15,000×. The results are expressed in terms of the appearance of the scattered electrons at different magnification powers.

4.7. Application of AVG-Based Coatings Enriched with Anthocyanin

Fig fruits was carefully washed with chlorinated water to remove any foreign debris, such as dust and dirt. Maximum effort was put into selecting fruits that were uniform in size, high in quality, and free from injury or disease. The different formulations of edible coatings, i.e., AVG (T1), AVG + glycerol (T2), AVG + glycerol + anthocyanin + 0.2% (T3), and AVG + glycerol + anthocyanin + 0.5% (T4), were applied to the fruits. The deposition of coating materials on the fruits was carried out via a dipping procedure for 5 min, and fruits were dried using an oven at 45 °C for 30 min. The fruits were stored under refrigerated conditions at 4 °C throughout the storage period of 12 days. Deionized water used as a control (T0) to treat fig fruits. The postharvest quality evaluation was performed at two-day intervals (i.e., on days 2, 4, 6, 8, and 12).

4.8. Scanning Electron Microscopy (SEM) Imaging of Edible Coatings

We conducted SEM imaging of coating samples, which were prepared by dropcasting 0.5 mL of each coating solution onto a copper stub and drying under ambient conditions. Dried coatings were gold-coated prior to imaging using an SEM (Zeiss SUPRA 55VP, Germany) at a 10 kV accelerating voltage.

4.9. Physicochemical Analyses of Ficus carica Fruit

To prepare samples for physicochemical analysis, 10 g of uncoated and coated fruits at different time points (2, 4, 6, 8, 12 days) were homogenized with 80 mL distilled water using a kitchen blender (MX-KM5070, Panasonic, Malaysia). The obtained homogenized fruit was squeezed through a muslin cloth to obtain fruit juice, which was used for further analysis.

4.9.1. Weight Loss

The weight loss of control and treated fig fruits was calculated using the mass difference technique, as reported previously [12]. The fruits were weighed every 2 days over the storage period of 12 days using an analytical weighing balance. Weight loss was calculated by taking the difference between fruit weights at specific time points and day 0. Three repeats were performed under all conditions at each time point, and data are presented as an average ± standard deviation.

4.9.2. pH

The pH of the extracted fruit juice was determined using the standard AOAC (2010) method. A 10 mL sample of juice from control (water-washed fruit) and treated fig fruits was placed in a beaker, before an electrode of the digital pH meter (pH-EMCO-256071, Japan) was dipped inside the beaker containing the sample and left for 10 min. This technique was carried out a minimum of three times for all samples, and data are presented as average ± standard deviation. Before use, the pH meter was calibrated with buffer solutions of pH 4 and 7.

4.9.3. Titratable Acidity (TA)

The TA of control and AVG-coated fruits was measured using the titration method reported previously, with minor modifications [32]. A 25 mL sample of extracted fruit juice prepared using 10 g of fruit pulp in 40 mL distilled water was titrated with 0.1 N NaOH solution. Phenolphthalein was used as an indicator for marking the end point. TA was measured as malic acid (%) and calculated using the equation below:

$$\text{Titratable acidity } (\%) = \text{volume of NaOH} \times \text{miliequivalent weight of acid} \times \text{normality of NaOH} \times \text{volume of sample}$$

4.9.4. Total Soluble Solids (TSS)

The TSS of the control and treated fruits were estimated using a refractometer (A87117, Bellingham, UK) with a precision of ±0.1%. Distilled water was used to calibrate the refractometer. Two drops of fruit juice were placed on the refractometer prism, and the measurements were taken again. This procedure was repeated three times for each fruit sample, and the prism was cleaned with ethanol after each measurement.

Author Contributions: S.A.A.-H.: conceptualization, methodology, validation, investigation, and writing—original draft, review, and editing. R.M.A.-A., O.T.A.-I., and N.K.: formal analysis, conceptualization, investigation, software, and data curation. V.A., S.P., and A.T.P.: writing—editing and reviewing. All authors have read and agreed to the published version of the manuscript.

Funding: This research received no external funding.

Institutional Review Board Statement: Not applicable.

Informed Consent Statement: Not applicable.

Data Availability Statement: Data are available on request from the authors.

Acknowledgments: The authors are grateful to the Agriculture college of Basrah University for their infrastructural support.

Conflicts of Interest: The authors declare no conflict of interest.

References

1. Galus, S.; Arik Kibar, E.A.; Gniewosz, M.; Kraśniewska, K. Novel Materials in the Preparation of Edible Films and Coatings—A Review. *Coatings* **2020**, *10*, 674. [CrossRef]
2. Wang, Q.; Chen, W.; Zhu, W.; McClements, D.J.; Liu, X.; Liu, F. A review of multilayer and composite films and coatings for active biodegradable packaging. *Npj Sci. Food* **2022**, *6*, 18. [CrossRef] [PubMed]
3. Siddiqui, S.A.; Zannou, O.; Bahmid, N.A.; Fidan, H.; Alamou, A.-F.; Nagdalian, A.A.; Hassoun, A.; Fernando, I.; Ibrahim, S.A.; Arsyad, M. Consumer behavior toward nanopackaging—A new trend in the food industry. *Future Foods* **2022**, *6*, 100191. [CrossRef]

4. Amin, U.; Khan, M.U.; Majeed, Y.; Rebezov, M.; Khayrullin, M.; Bobkova, E.; Shariati, M.A.; Chung, I.M.; Thiruvengadam, M. Potentials of polysaccharides, lipids and proteins in biodegradable food packaging applications. *Int. J. Biol. Macromol.* **2021**, *183*, 2184–2198. [CrossRef] [PubMed]
5. Gvozdenko, A.A.; Siddiqui, S.A.; Blinov, A.V.; Golik, A.B.; Nagdalian, A.A.; Maglakelidze, D.G.; Statsenko, E.N.; Pirogov, M.A.; Blinova, A.A.; Sizonenko, M.N.; et al. Synthesis of CuO nanoparticles stabilized with gelatin for potential use in food packaging applications. *Sci. Rep.* **2022**, *12*, 12843. [CrossRef]
6. Kumar, N.; Neeraj. Polysaccharide-based component and their relevance in edible film/coating: A review. *Nutr. Food Sci.* **2019**, *49*, 793–823. [CrossRef]
7. Suhag, R.; Kumar, N.; Petkoska, A.T.; Upadhyay, A. Film formation and deposition methods of edible coating on food products: A review. *Food Res. Int.* **2020**, *136*, 109582. [CrossRef] [PubMed]
8. Kumar, N.; Pratibha; Neeraj; Petkoska, A.T.; AL-Hilifi, S.A.; Fawole, O.A. Effect of Chitosan–Pullulan Composite Edible Coating Functionalized with Pomegranate Peel Extract on the Shelf Life of Mango (*Mangifera indica*). *Coatings* **2021**, *11*, 764. [CrossRef]
9. Al-Ali, R.M.; Al-Hilifi, S.A.; Rashed, M.M.A. Fabrication, characterization, and anti-free radical performance of edible packaging-chitosan film synthesized from shrimp shell incorporated with ginger essential oil. *J. Food Meas. Charact.* **2021**, *15*, 2951–2962. [CrossRef]
10. Lin, D.; Zhao, Y. Innovations in the Development and Application of Edible Coatings for Fresh and Minimally Processed Fruits and Vegetables. *Compr. Rev. Food Sci. Food Saf.* **2007**, *6*, 60–75. [CrossRef]
11. Salgado, P.R.; Ortiz, C.M.; Musso, Y.S.; Di Giorgio, L.; Mauri, A.N. Edible films and coatings containing bioactives. *Curr. Opin. Food Sci.* **2015**, *5*, 86–92. [CrossRef]
12. Kumar, N.; Pratibha; Neeraj; Ojha, A.; Upadhyay, A.; Singh, R.; Kumar, S. Effect of active chitosan-pullulan composite edible coating enrich with pomegranate peel extract on the storage quality of green bell pepper. *LWT* **2021**, *138*, 110435. [CrossRef]
13. Al-Hilifi, S.A.; Al-Ali, R.M.; Petkoska, A.T. Ginger Essential Oil as an Active Addition to Composite Chitosan Films: Development and Characterization. *Gels* **2022**, *8*, 327. [CrossRef] [PubMed]
14. Venditti, T.; Molinu, M.G.; Dore, A.; D'Hallewin, G.; Fiori, P.; Tedde, M.; Agabbio, M. Treatments with gras compounds to keep fig fruit (*Ficus carica* L.) quality during cold storage. *Commun. Agric. Appl. Biol. Sci.* **2005**, *70*, 339–343. [PubMed]
15. Doster, M.A.; Michailides, T.J. Fungal Decay of First-Crop and Main-Crop Figs. *Plant Dis.* **2007**, *91*, 1657–1662. [CrossRef]
16. Crisosto, H.; Ferguson, L.; Bremer, V.; Stover, E.; Colelli, G. 7-Fig (*Ficus carica* L.). In *Postharvest Biology and Technology of Tropical and Subtropical Fruits*; Yahia, E.M., Ed.; Woodhead Publishing: Sawston, UK, 2011; pp. 134–160e.
17. Reyes-Avalos, M.C.; Femenia, A.; Minjares-Fuentes, R.; Contreras-Esquivel, J.C.; Aguilar-González, C.N.; Esparza-Rivera, J.R.; Meza-Velázquez, J.A. Improvement of the Quality and the Shelf Life of Figs (*Ficus carica*) Using an Alginate–Chitosan Edible Film. *Food Bioprocess Technol.* **2016**, *9*, 2114–2124. [CrossRef]
18. Arvaniti, O.S.; Samaras, Y.; Gatidou, G.; Thomaidis, N.S.; Stasinakis, A.S. Review on fresh and dried figs: Chemical analysis and occurrence of phytochemical compounds, antioxidant capacity and health effects. *Food Res. Int.* **2019**, *119*, 244–267. [CrossRef]
19. Paolucci, M.; Di Stasio, M.; Sorrentino, A.; La Cara, F.; Volpe, M.G. Active Edible Polysaccharide-Based Coating for Preservation of Fresh Figs (*Ficus carica* L.). *Foods* **2020**, *9*, 1793. [CrossRef]
20. Villalobos, M.d.C.; Serradilla, M.J.; Martín, A.; Ruiz-Moyano, S.; Pereira, C.; Córdoba, M.d.G. Synergism of defatted soybean meal extract and modified atmosphere packaging to preserve the quality of figs (*Ficus carica* L.). *Postharvest Biol. Technol.* **2016**, *111*, 264–273. [CrossRef]
21. Allegra, A.; Sortino, G.; Inglese, P.; Settanni, L.; Todaro, A.; Gallotta, A. The effectiveness of Opuntia ficus-indica mucilage edible coating on post-harvest maintenance of 'Dottato' fig (*Ficus carica* L.) fruit. *Food Packag. Shelf Life* **2017**, *12*, 135–141. [CrossRef]
22. Opara, U.L.; Caleb, O.J.; Belay, Z.A. 7—Modified atmosphere packaging for food preservation. In *Food Quality and Shelf Life*; Galanakis, C.M., Ed.; Academic Press: Cambridge, MA, USA, 2019; pp. 235–259.
23. Song, C.; Li, A.; Chai, Y.; Li, Q.; Lin, Q.; Duan, Y. Effects of 1-Methylcyclopropene Combined with Modified Atmosphere on Quality of Fig (*Ficus carica* L.) during Postharvest Storage. *J. Food Qual.* **2019**, *2019*, 2134924. [CrossRef]
24. Adiletta, G.; Zampella, L.; Coletta, C.; Petriccione, M. Chitosan Coating to Preserve the Qualitative Traits and Improve Antioxidant System in Fresh Figs (*Ficus carica* L.). *Agriculture* **2019**, *9*, 84. [CrossRef]
25. Sogvar, O.B.; Koushesh Saba, M.; Emamifar, A. *Aloe vera* and ascorbic acid coatings maintain postharvest quality and reduce microbial load of strawberry fruit. *Postharvest Biol. Technol.* **2016**, *114*, 29–35. [CrossRef]
26. Ali, S.; Khan, A.S.; Nawaz, A.; Anjum, M.A.; Naz, S.; Ejaz, S.; Hussain, S. *Aloe vera* gel coating delays postharvest browning and maintains quality of harvested litchi fruit. *Postharvest Biol. Technol.* **2019**, *157*, 110960. [CrossRef]
27. Mubarak, A.; Engakanah, T. *Aloe vera* edible coating retains the bioactive compounds of wax apples (*Syzygium samarangense*). *Malays. Appl. Biol.* **2017**, *46*, 141–148.
28. Farina, V.; Passafiume, R.; Tinebra, I.; Scuderi, D.; Saletta, F.; Gugliuzza, G.; Gallotta, A.; Sortino, G. Postharvest Application of *Aloe vera* Gel-Based Edible Coating to Improve the Quality and Storage Stability of Fresh-Cut Papaya. *J. Food Qual.* **2020**, *2020*, 8303140. [CrossRef]
29. Rizwana, N.; Agarwal, V.; Nune, M. Antioxidant for Neurological Diseases and Neurotrauma and Bioengineering Approaches. *Antioxidants* **2022**, *11*, 72. [CrossRef]
30. Ray, A.; Dutta Gupta, S. A panoptic study of antioxidant potential of foliar gel at different harvesting regimens of *Aloe vera* L. *Ind. Crops Prod.* **2013**, *51*, 130–137. [CrossRef]

31. Ribeiro, A.M.; Estevinho, B.N.; Rocha, F. Preparation and Incorporation of Functional Ingredients in Edible Films and Coatings. *Food Bioprocess Technol.* **2021**, *14*, 209–231. [CrossRef]
32. Kumar, N.; Neeraj; Pratibha; Trajkovska Petkoska, A. Improved Shelf Life and Quality of Tomato (*Solanum lycopersicum* L.) by Using Chitosan-Pullulan Composite Edible Coating Enriched with Pomegranate Peel Extract. *ACS Food Sci. Technol.* **2021**, *1*, 500–510. [CrossRef]
33. Scala, K.D.; Vega-Gálvez, A.; Ah-Hen, K.; Nuñez-Mancilla, Y.; Tabilo-Munizaga, G.; Pérez-Won, M.; Giovagnoli, C. Chemical and physical properties of *Aloe vera* (*Aloe barbadensis* Miller) gel stored after high hydrostatic pressure processing. *Food Sci. Technol.* **2013**, *33*, 52–59. [CrossRef]
34. Eshun, K.; He, Q. *Aloe vera*: A Valuable Ingredient for the Food, Pharmaceutical and Cosmetic Industries—A Review. *Crit. Rev. Food Sci. Nutr.* **2004**, *44*, 91–96. [CrossRef] [PubMed]
35. Hamman, J.H. Composition and Applications of *Aloe vera* Leaf Gel. *Molecules* **2008**, *13*, 1599–1616. [CrossRef]
36. Chandegara, D.V.; Varshney, A. *Aloe vera: Development of Gel Extraction Process for Aloe vera Leaves*; Lambert Academic Publication: Saarbrücken, Germany, 2012; ISBN 978-659-21648-0.
37. Miranda, M.; Maureira, H.; Rodríguez, K.; Vega-Gálvez, A. Influence of temperature on the drying kinetics, physicochemical properties, and antioxidant capacity of *Aloe vera* (*Aloe barbadensis* Miller) gel. *J. Food Eng.* **2009**, *91*, 297–304. [CrossRef]
38. Vega-Gálvez, A.; Uribe, E.; Perez, M.; Tabilo-Munizaga, G.; Vergara, J.; Garcia-Segovia, P.; Lara, E.; Di Scala, K. Effect of high hydrostatic pressure pretreatment on drying kinetics, antioxidant activity, firmness and microstructure of *Aloe vera* (*Aloe barbadensis Miller*) gel. *LWT-Food Sci. Technol.* **2011**, *44*, 384–391. [CrossRef]
39. Flores-Martínez, N.; Valdez-Fragoso, A.; Jiménez-Islas, H.; Perez-Perez, C. Physical, barrier, mechanical and microstructural properties of *Aloe vera*-gelatin-glycerol edible films incorporated with *Pimenta dioica* L. Merrill essential oil. *Rev. Mex. De Ing. Quim.* **2017**, *16*, 109–119. [CrossRef]
40. Riaz, A.; Aadil, R.M.; Amoussa, A.M.O.; Bashari, M.; Abid, M.; Hashim, M.M. Application of chitosan-based apple peel polyphenols edible coating on the preservation of strawberry (*Fragaria ananassa* cv Hongyan) fruit. *J. Food Process. Preserv.* **2021**, *45*, e15018. [CrossRef]
41. Balaguer, M.P.; Gómez-Estaca, J.; Gavara, R.; Hernandez-Munoz, P. Functional Properties of Bioplastics Made from Wheat Gliadins Modified with Cinnamaldehyde. *J. Agric. Food Chem.* **2011**, *59*, 6689–6695. [CrossRef]
42. Iñiguez-Moreno, M.; Ragazzo-Sánchez, J.A.; Calderón-Santoyo, M. An Extensive Review of Natural Polymers Used as Coatings for Postharvest Shelf-Life Extension: Trends and Challenges. *Polymers* **2021**, *13*, 3271. [CrossRef]
43. Staroszczyk, H.; Kusznierewicz, B.; Malinowska-Pańczyk, E.; Sinkiewicz, I.; Gottfried, K.; Kołodziejska, I. Fish gelatin films containing aqueous extracts from phenolic-rich fruit pomace. *LWT* **2020**, *117*, 108613. [CrossRef]
44. Hasan, K.; Islam, R.; Hasan, M.; Sarker, S.H.; Biswas, M.H. Effect of Alginate Edible Coatings Enriched with Black Cumin Extract for Improving Postharvest Quality Characteristics of Guava (*Psidium guajava* L.) Fruit. *Food Bioprocess Technol.* **2022**, *15*, 2050–2064. [CrossRef]
45. Yang, Z.; Zou, X.; Li, Z.; Huang, X.; Zhai, X.; Zhang, W.; Shi, J.; Tahir, H.E. Improved Postharvest Quality of Cold Stored Blueberry by Edible Coating Based on Composite Gum Arabic/Roselle Extract. *Food Bioprocess Technol.* **2019**, *12*, 1537–1547. [CrossRef]
46. Baraiya, N.S.; Rao, T.V.R.; Thakkar, V.R. Improvement of Postharvest Quality and Storability of Jamun Fruit (*Syzygium cumini* L. var. Paras) by Zein Coating Enriched with Antioxidants. *Food Bioprocess Technol.* **2015**, *8*, 2225–2234. [CrossRef]
47. Mohebbi, M.; Ansarifar, E.; Hasanpour, N.; Amiryousefi, M.R. Suitability of *Aloe vera* and Gum Tragacanth as Edible Coatings for Extending the Shelf Life of Button Mushroom. *Food Bioprocess Technol.* **2012**, *5*, 3193–3202. [CrossRef]
48. Robledo, N.; López, L.; Bunger, A.; Tapia, C.; Abugoch, L. Effects of Antimicrobial Edible Coating of Thymol Nanoemulsion/Quinoa Protein/Chitosan on the Safety, Sensorial Properties, and Quality of Refrigerated Strawberries (*Fragaria* × *ananassa*) Under Commercial Storage Environment. *Food Bioprocess Technol.* **2018**, *11*, 1566–1574. [CrossRef]
49. Pinzon, M.I.; Sanchez, L.T.; Garcia, O.R.; Gutierrez, R.; Luna, J.C.; Villa, C.C. Increasing shelf life of strawberries (*Fragaria* ssp) by using a banana starch-chitosan—*Aloe vera* gel composite edible coating. *Int. J. Food Sci. Technol.* **2020**, *55*, 92–98. [CrossRef]
50. Valdés, A.; Burgos, N.; Jiménez, A.; Garrigós, M.C. Natural Pectin Polysaccharides as Edible Coatings. *Coatings* **2015**, *5*, 865–886. [CrossRef]
51. Marín, A.; Atarés, L.; Chiralt, A. Improving function of biocontrol agents incorporated in antifungal fruit coatings: A review. *Biocontrol Sci. Technol.* **2017**, *27*, 1220–1241. [CrossRef]
52. Radi, M.; Ahmadi, H.; Amiri, S. Effect of Cinnamon Essential Oil-Loaded Nanostructured Lipid Carriers (NLC) Against *Penicillium citrinum* and *Penicillium expansum* Involved in Tangerine Decay. *Food Bioprocess Technol.* **2022**, *15*, 306–318. [CrossRef]
53. Aguirre-Joya, J.A.; Cerqueira, M.A.; Ventura-Sobrevilla, J.; Aguilar-Gonzalez, M.A.; Carbó-Argibay, E.; Castro, L.P.; Aguilar, C.N. Candelilla Wax-Based Coatings and Films: Functional and Physicochemical Characterization. *Food Bioprocess Technol.* **2019**, *12*, 1787–1797. [CrossRef]
54. Kumar, N.; Neeraj; Pratibha; Singla, M. Enhancement of Storage Life and Quality Maintenance of Litchi (*Litchi chinensis* Sonn.) Fruit Using Chitosan: Pullulan Blend Antimicrobial Edible Coating. *Int. J. Fruit Sci.* **2020**, *20*, S1662–S1680. [CrossRef]
55. Vieira, T.M.; Moldão-Martins, M.; Alves, V.D. Composite Coatings of Chitosan and Alginate Emulsions with Olive Oil to Enhance Postharvest Quality and Shelf Life of Fresh Figs (*Ficus carica* L. cv. 'Pingo De Mel'). *Foods* **2021**, *10*, 718. [CrossRef] [PubMed]
56. Tapia, M.S.; Rojas-Graü, M.A.; Rodríguez, F.J.; Ramírez, J.; Carmona, A.; Martin-Belloso, O. Alginate- and Gellan-Based Edible Films for Probiotic Coatings on Fresh-Cut Fruits. *J. Food Sci.* **2007**, *72*, E190–E196. [CrossRef] [PubMed]

57. Rojas-Graü, M.A.; Tapia, M.S.; Martín-Belloso, O. Using polysaccharide-based edible coatings to maintain quality of fresh-cut Fuji apples. *LWT-Food Sci. Technol.* **2008**, *41*, 139–147. [CrossRef]
58. Tanada-Palmu, P.S.; Grosso, C.R.F. Effect of edible wheat gluten-based films and coatings on refrigerated strawberry (*Fragaria ananassa*) quality. *Postharvest Biol. Technol.* **2005**, *36*, 199–208. [CrossRef]
59. Passafiume, R.; Gugliuzza, G.; Gaglio, R.; Busetta, G.; Tinebra, I.; Sortino, G.; Farina, V. Aloe-Based Edible Coating to Maintain Quality of Fresh-Cut Italian Pears (*Pyrus communis* L.) during Cold Storage. *Horticulturae* **2021**, *7*, 581. [CrossRef]
60. Akhavan, H.-R.; Hosseini, F.-S.; Amiri, S.; Radi, M. Cinnamaldehyde-Loaded Nanostructured Lipid Carriers Extend the Shelf Life of Date Palm Fruit. *Food Bioprocess Technol.* **2021**, *14*, 1478–1489. [CrossRef]
61. Petriccione, M.; De Sanctis, F.; Pasquariello, M.S.; Mastrobuoni, F.; Rega, P.; Scortichini, M.; Mencarelli, F. The Effect of Chitosan Coating on the Quality and Nutraceutical Traits of Sweet Cherry During Postharvest Life. *Food Bioprocess Technol.* **2015**, *8*, 394–408. [CrossRef]
62. Rojas-Bravo, M.; Rojas-Zenteno, E.G.; Hernández-Carranza, P.; Ávila-Sosa, R.; Aguilar-Sánchez, R.; Ruiz-López, I.I.; Ochoa-Velasco, C.E. A Potential Application of Mango (*Mangifera indica* L. cv Manila) Peel Powder to Increase the Total Phenolic Compounds and Antioxidant Capacity of Edible Films and Coatings. *Food Bioprocess Technol.* **2019**, *12*, 1584–1592. [CrossRef]

 gels

Article

Edible Xanthan/Propolis Coating and Its Effect on Physicochemical, Microbial, and Sensory Quality Indices in Mackerel Tuna (*Euthynnus affinis*) Fillets during Chilled Storage

Aly Farag El Sheikha [1,2,3,4,5,*,†], Ayman Younes Allam [5,†], Emel Oz [6], Mohammad Rizwan Khan [7], Charalampos Proestos [8,*] and Fatih Oz [6]

1 College of Bioscience and Bioengineering, Jiangxi Agricultural University, 1101 Zhimin Road, Nanchang 330045, China
2 School of Nutrition Sciences, Faculty of Health Sciences, University of Ottawa, 25 University Private, Ottawa, ON K1N 6N5, Canada
3 Bioengineering and Technological Research Centre for Edible and Medicinal Fungi, Jiangxi Agricultural University, 1101 Zhimin Road, Nanchang 330045, China
4 Jiangxi Key Laboratory for Conservation and Utilization of Fungal Resources, Jiangxi Agricultural University, 1101 Zhimin Road, Nanchang 330045, China
5 Department of Food Science and Technology, Faculty of Agriculture, Minufiya University, Shibin El Kom 32511, Egypt; ayman_alaam@yahoo.com
6 Department of Food Engineering, Faculty of Agriculture, Ataturk University, Erzurum 25240, Türkiye; emel.oz@atauni.edu.tr (E.O.); fatihoz@atauni.edu.tr (F.O.)
7 Department of Chemistry, College of Science, King Saud University, Riyadh 11451, Saudi Arabia; mrkhan@ksu.edu.sa
8 Laboratory of Food Chemistry, Department of Chemistry, School of Sciences, National and Kapodistrian University of Athens, 15772 Athens, Greece
* Correspondence: elsheikha_aly@yahoo.com (A.F.E.S.); harpro@chem.uoa.gr (C.P.)
† These authors are considered the first author.

Citation: Sheikha, A.F.E.; Allam, A.Y.; Oz, E.; Khan, M.R.; Proestos, C.; Oz, F. Edible Xanthan/Propolis Coating and Its Effect on Physicochemical, Microbial, and Sensory Quality Indices in Mackerel Tuna (*Euthynnus affinis*) Fillets during Chilled Storage. *Gels* **2022**, *8*, 405. https://doi.org/10.3390/gels8070405

Academic Editor: Miguel A. Cerqueira

Received: 3 May 2022
Accepted: 21 June 2022
Published: 25 June 2022

Abstract: Worldwide aquaculture production is increasing, but with this increase comes quality and safety related problems. Hence, there is an urgent need to develop potent technologies to extend the shelf life of fish. Xanthan gum is commonly used in the food industry because of its high-water solubility, stability of its aqueous solutions in a wide pH range, and high viscosity. One of its modern food applications is its use as a gelling agent in edible coatings building. Therefore, in this study, the effect of xanthan coating containing various concentrations (0, 1, 2%; *w*/*v*) of ethanolic extract of propolis (EEP) on physicochemical, microbial, and sensory quality indices in mackerel fillets stored at 2 °C for 20 days was evaluated. The pH, peroxide value, K-value, TVB-N, TBARS, microbiological and sensory characteristics were determined every 5 days over the storage period (20 days). Samples treated with xanthan (XAN) coatings containing 1 and 2% of EEP were shown to have the highest level of physicochemical protection and maximum level of microbial inhibition ($p < 0.05$) compared to uncoated samples (control) over the storage period. Furthermore, the addition of EEP to XAN was more effective in notably preserving ($p < 0.05$) the taste and odor of coated samples compared to control.

Keywords: bioactive packaging; xanthan; propolis; physicochemical properties; microbiological analyses; sensory evaluation

1. Introduction

With the continuous increase in the world population, which may reach over 9 billion by the year 2050, global food production needs to increase by an estimated 50% at least to keep pace with this population increase and meet its nutritional needs [1]. Fish and fishery products can meet a significant proportion of the world's food needs by 2050 [2]. In this context, it was noted that the world per capita fish consumption increased from an average of 9.9 kg in the 1960s to 20.5 kg in 2018 [3]. Although fish has high commercial value, it is an extremely perishable product. Mackerel tuna (*Euthynnus affinis*) is a commercially important fish in high demand. As with other fish, Mackerel tuna is easily and rapidly damaged (highly perishable), so it is very susceptible to quality degradation. The major factor causing quality degradation in fish, including mackerel tuna, is microbial activity, even though the first changes are caused by the endogenous enzymes of fish, which ultimately shortens their shelf life [4–6]. Microbial deterioration proceeds fast because of the presence of large amounts of low-molecular-weight compounds, high water activity, and high post-mortem (pH > 6) in fish muscles. Hence, cooling is necessary to prolong the shelf life of fish and is often combined with vacuum packaging to prevent the growth of aerobic microflora that cause spoilage [5,7]. Thus, the application of appropriate packaging and/or processing techniques will be the best solution to extend the shelf life of fish and fish products [8]. In addition to the short shelf life, another challenge facing fresh fish consumption is that seafood products take a long time to prepare as meals [9]. Despite freezing and cold storage being significant and frequent methods of preserving fish and fish products, they cannot completely prevent chemical and oxidation reactions in fish and fish products [10]. The reason for the occurrence of these reactions may be related to the presence of polyunsaturated fatty acids in fish and fish products, which oxidize rapidly in the presence of oxygen [11]. Therefore, the use of preservatives, especially natural ones, has become an urgent need to extend the shelf life of perishable foods such as fish [12].

Propolis (bee glue) is a balsamic product obtained from exotic Africanized bees *Apis mellifera* L. [13]. Propolis extract is well-known for its functional properties, such as anti-inflammatory, pharmacological, antiviral, anticancer, antioxidant, antifungal, and antibacterial activities [14]. Natural preservatives benefits have recently been enhanced by incorporating them into various edible coatings and films on food products [15,16]. Due to the edible coatings' simplicity and eco-friendly nature. Several characteristics distinguish the edible coating. It acts as a carrier for bioactive components and is a semi-permeable barrier to moisture loss, gas exchange, and oxidative reactions [17,18]. Several studies have been conducted showing the possibility of using natural gums such as xanthan in formulating edible coatings and improving their characteristics [19–22].

Xanthan gum is a polysaccharide produced by *Xanthomonas campestris*, and a food additive that is commonly added to foods as a thickener or stabilizer [23,24]. Xanthan is featured because of its ability to enhance food flavor, consistency, texture, shelf life, and appearance [24]. Xanthan has multiple technological advantages that make it a rich raw material with various applications, especially for food. The features of xanthan could be listed as follows [23,25]: (1) high viscosity at low concentrations: for example, a solution with a concentration of 1% appears almost gel-like at rest, yet pours readily and has a very low resistance to mixing and pumping; (2) high resistance to a wide pH range (2–12) makes xanthan well-suited to foods; (3) high thermal stability; the viscosity is not affected by temperatures in the range of (0–100 °C), and it has excellent freeze-thaw ability; (4) high solubility of xanthan gum renders it appropriate for many applications, including foods; (5) high compatibility with most of the commercially available thickeners.

Furthermore, one of the new food applications of xanthan is its use as a gelling agent in edible coating building [26].

Bioactive edible coatings or films from natural preservatives with antioxidant and antibacterial properties, prolong the shelf life of fish and fish products [27]. The main advantage is that the edible film helps in the reduction of environmental pollution [28,29]. In this sense, there is just one study that indicates the effect of using edible coatings on extending the shelf life of mackerel tuna fish (*Euthynnus affinis*) fillets [30]. Kumar and others' [30] study aimed to develop a bioactive edible coating from gelatin and chitosan, incorporated with different concentrations of clove oil as a natural preservative, and evaluate their effect on the shelf life of mackerel fillets under refrigerated conditions (4 °C). This edible coating demonstrated its potential as a natural antibacterial agent which can be used for packaging tuna and other fishery products.

Due to the features of propolis as a natural preservative (antioxidant and antimicrobial agent), it could be integrated with xanthan gum in formulating gel-based edible coating [31] to extend the shelf life of fish and fishery products. There is no published data on the application of a xanthan/propolis composite coating to preserve fish and fish products' quality and shelf life. Therefore, to our knowledge, this is the first paper to study the effects of using a composite edible coating from xanthan containing various levels (0, 1, and 2%) of ethanolic extract of propolis (EEP) on the physicochemical, microbiological, and sensory quality parameters of mackerel tuna fish (*Euthynnus affinis*) fillets during chilled storage (2 °C) for 20 days.

2. Results and Discussion

2.1. Test Probabilities for Physicochemical, Microbiological, and Sensory Criteria of Mackerel Tuna Fillets—Multi-Aspect Variance Analysis, including Interactions

The storage time and coating treatment of mackerel tuna fillets can have a significant impact on their physicochemical, microbiological, and sensory quality indices. The data presented in Table 1 show a significant effect ($p < 0.001$) of storage period on all measured parameters except taste, in which the significant effect was $p < 0.01$. Furthermore, Table 1 illustrates a significant effect ($p < 0.001$) of coating treatment on all measured parameters except TBARS, TVC, PTC, and Enterobacteriaceae, in which the significant effect was $p < 0.01$. In addition, the coating treatment had a smaller effect on K-value ($p < 0.05$). The interactions between the storage time and the coating treatment were also indicated for all the tested parameters. The storage time and coating treatment had a significant effect ($p < 0.001$) on all measured parameters.

Table 1. Test probabilities for physicochemical, microbiological, and sensory quality indices of mackerel tuna fillets—multi-aspect variance analysis, including interactions.

Quality Indices		Effect		Interaction T × ST
		Treatment (T)	Storage Time (ST)	
Physicochemical properties	pH	XXX [1]	XXX	XXX
	Peroxide value	XXX	XXX	XXX
	TBARS	XX [2]	XXX	XXX
	TVB-N	XXX	XXX	XXX
	K-value	X [3]	XXX	XXX
Microbiological analyses	TVC	XX	XXX	XXX
	PTC	XX	XXX	XXX
	Enterobacteriaceae	XX	XXX	XXX
	E. coli	XXX	XXX	XXX
	Pseudomonas fluorescens	XXX	XXX	XXX
	Lactic acid bacteria	XXX	XXX	XXX
	Yeasts/molds	XXX	XXX	XXX
Sensory evaluation	Taste	XXX	XX	XXX
	Odor	XXX	XXX	XXX
	Overall acceptability	XXX	XXX	XXX

[1] XXX: significant effect ($p < 0.001$); [2] XX: significant effect ($p < 0.01$); [3] X: significant effect ($p < 0.05$).

2.2. Physicochemical Analyses of Mackerel Tuna Fillets

2.2.1. pH Values of Mackerel Fillets

Figure 1 shows the pH value changes in mackerel tuna fish fillets stored at 2 °C for 20 days. The primary pH values of fresh mackerel tuna fillets (pH 5.93–5.98) were consistent with previous studies [32,33]. In our study, the pH value of control samples increased from 5.95 to 7.21 after 20 days of cold storage, while the pH values for XAN-EEP 0%, XAN-EEP 1%, and XAN-EEP 2% samples after 20 days of storage were 6.81, 6.60, and 6.35, respectively. These results exhibited the protective effect of XAN edible coating against spoilage, which was significantly ($p < 0.05$) increased by propolis, especially in the higher dose group.

The lower pH value of the other treatments (XAN-EEP 0%, XAN-EEP 1%, and XAN-EEP 2%) could have prevented exogenous (microbial) and endogenous proteases from acting in treated mackerel tuna fillets through the storage period. Propolis' antimicrobial and antioxidant properties may be responsible for the observed pH changes in stored fish fillets, preventing changes in proteolysis and microbiological development [34].

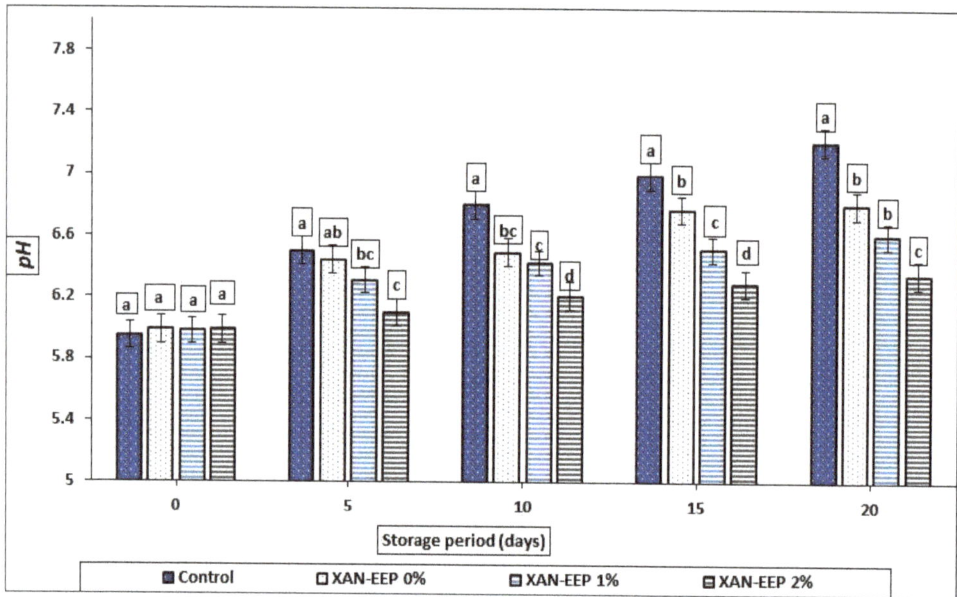

Figure 1. The influence of coating treatments on pH values in mackerel tuna fillet samples during storage at 2 °C for 20 days. Control: Uncoated mackerel tuna fillet samples (soaked samples in sterile distilled water). XAN-EEP 0%: Coated samples with xanthan containing (0%) ethanolic extract of propolis. XAN-EEP 1%: Coated samples with xanthan containing (1%) ethanolic extract of propolis. XAN-EEP 2%: Coated samples with xanthan containing (2%) ethanolic extract of propolis. $^{a–d}$: Within a column, different superscripts indicate significant differences ($p < 0.05$).

Furthermore, the pH values of coated and uncoated mackerel tuna fillets increased as the storage period increased. At the end of the storage period, the increase in pH values of the uncoated samples (controls) was more pronounced. This can happen as a result of the accumulation of ammonia and amino acid degradation products, which causes the pH to rise [33]. An increase in the pH values of stored fish may be linked to the production of peptides, amino acids, and ammonia due to increased protease activity or microbial development [35,36].

2.2.2. Oxidative Stability of Mackerel Tuna Fillets

The deterioration of the quality of fish and its products during storage is mainly due to the oxidation of lipid [11]. Lipid peroxidation is the reaction of oxygen with unsaturated lipids; hence, one of the methods that delays or prevents oxidation processes is the use of edible coatings for fish fillets [37] because the edible coatings can guarantee performance as a low oxygen barrier [24].

The peroxide value (POV) is a substantial indicator of fat rancidity, but how does fat rancidity happen? Rancidity happens through the process of lipid oxidation, which is accompanied by the production of free radicals, which in turn leads to the formation of aldehydes and ketones, all of which, of course, negatively affect the quality of fish [38]. As the storage period progressed, the peroxide values in coated and uncoated mackerel tuna fillets increased, with the control (uncoated) samples having the highest ($p < 0.05$) peroxide value at each interval storage period (Figure 2). The peroxide value of the control (uncoated fillet samples) increased from 2.22 to 17.32 meq peroxides/kg lipid, while during this time, the peroxide value of XAN-EEP 0%, XAN-EEP 1%, and XAN-EEP 2% increased from 2.23 to 14.55, 2.18 to 9.89, and 2.25 to 8.44 meq peroxides/kg lipid, respectively, after 20 days of chilled storage. In all treatments (coated fillet samples), the values were significantly reduced ($p < 0.05$) compared to the control samples. In this context, Roy et al. [39] found that the composite coating based on propolis could reduce the peroxide index in coated meat products over the storage period compared to the control (uncoated samples). The XAN-EEP 2% treatment resulted in a maximal decrease in the peroxide formation, followed by XAN-EEP 1% and XAN-EEP 0%; this condition may be due to the potent antioxidant activity of XAN-EEP 2%. These results may be in line with what was mentioned by Shavisi et al. [40]. They observed that a polylactic acid (PLA) film containing ethanolic extract of propolis (EEP) reduced the peroxide value of minced beef more than the control samples that were stored in the refrigerator.

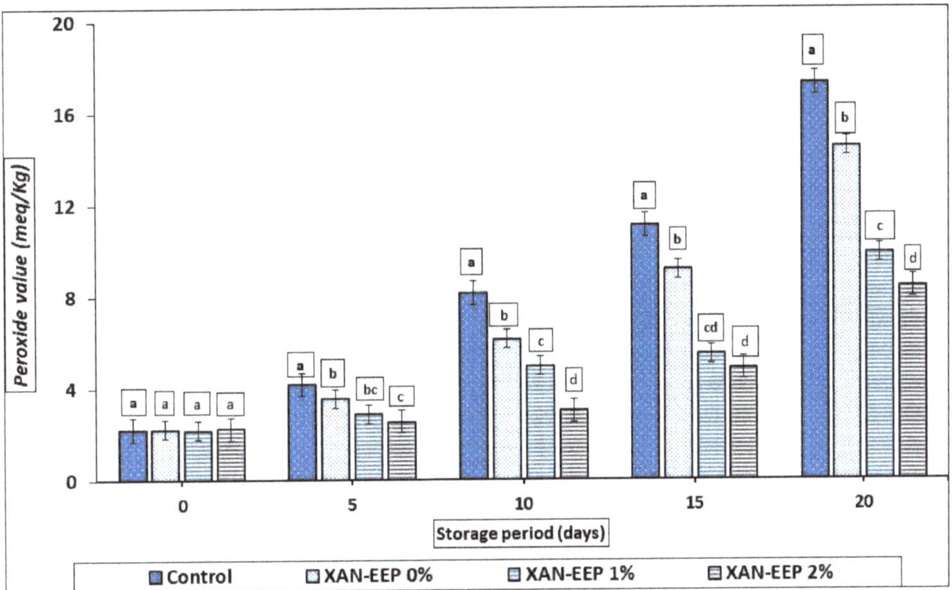

Figure 2. The influence of coating treatments on peroxide values (meq/kg) in mackerel tuna fillet samples during storage at 2 °C for 20 days. Control: Uncoated mackerel tuna fillet samples (soaked samples in sterile distilled water). XAN-EEP 0%: Coated samples with xanthan containing (0%) ethanolic extract of propolis. XAN-EEP 1%: Coated samples with xanthan containing (1%) ethanolic extract of propolis. XAN-EEP 2%: Coated samples with xanthan containing (2%) ethanolic extract of propolis. $^{a–d}$: Within a column, different superscripts indicate significant differences ($p < 0.05$).

The results for TBARS which is an indicator of lipid oxidation [41] of mackerel tuna fillets coated in (Figure 3) showed a significant effect of the coating on the oxidation of mackerel fillets. During refrigeration, the TBARS values of XAN-EEP 0%, XAN-EEP 1%, and XAN-EEP 2% were significantly ($p < 0.05$) lower than the control. After 20 days of cold storage, the TBARS values of the XAN-EEP 0%, XAN-EEP 1%, and XAN-EEP 2% treatments were 2.25, 1.98, and 1.31 mg MDA/kg, respectively. The malondialdehyde (MDA) levels in the treated fillets (XAN, XAN-EEP 1%, and XAN-EEP 2%) were significantly lower ($p < 0.05$) than in the control sample (2.9 mg MDA/kg). Connell [42] also mentioned that the acceptable limit for the value of TBARS in a fish sample is in the range of 1 to 2 mg

MDA/kg, and if the value exceeds this limit, an unpleasant smell of fish begins to develop. All tested samples for all treatments in our study exceeded the TBARS value limit after 20 days of storage.

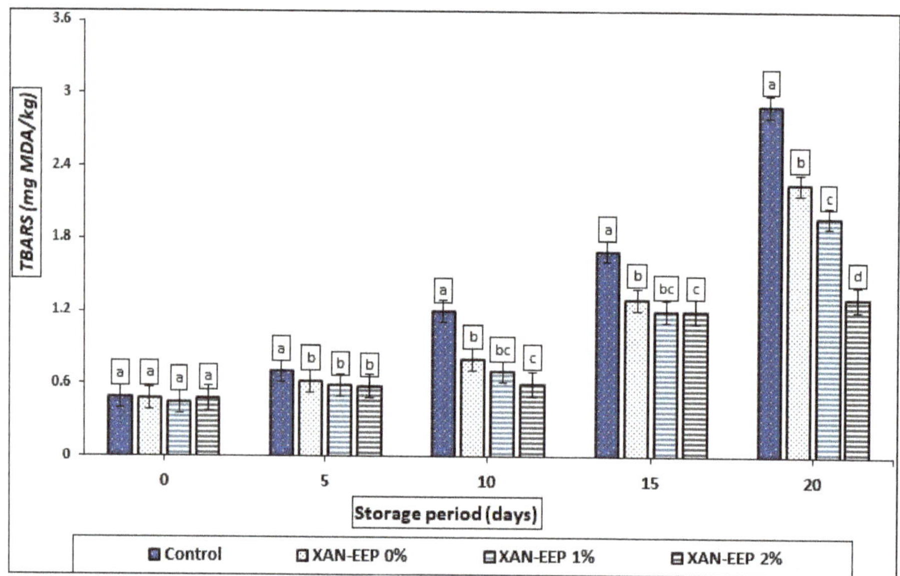

Figure 3. The influence of coating treatments on TBARS values (MDA mg/kg) in mackerel tuna fillet samples during storage at 2 °C for 20 days. Control: Uncoated mackerel tuna fillet samples (soaked samples in sterile distilled water). XAN-EEP 0%: Coated samples with xanthan containing (0%) ethanolic extract of propolis. XAN-EEP 1%: Coated samples with xanthan containing (1%) ethanolic extract of propolis. XAN-EEP 2%: Coated samples with xanthan containing (2%) ethanolic extract of propolis. $^{a-d}$: Within a column, different superscripts indicate significant differences ($p < 0.05$).

TBARS values were significantly lower in the EEP-containing coated mackerel tuna fillet samples than in others, most likely due to the antioxidants present (EEP) [40]. Additionally, the highest effects were noted with XAN-EEP at a concentration of 2%.

2.2.3. Total Volatile Basic Nitrogen (TVB-N) of Mackerel Tuna Fillet Samples

Amongst the important indicators of spoilage is TVB-N, which results from the degradation of proteins and non-protein nitrogen compounds as a response to bacterial activity as well as the presence of endogenous enzymes [43]. Figure 4 shows the changes in TVB-N values for all chip processors during cryogenic storage. At the start of storage (zero-time), TVB-N content ranged from 8.12 to 8.20 mg N/100 g for all mackerel tuna fillet samples. Over time TVB-N values increased for all samples, which was, of course, consistent with increases in pH values during later stages of storage. The results of our study are in line with those obtained by Yu et al. [44]. On the 20th day, TVB-N values for the control, XAN-EEP 0%, XAN-EEP 1%, and XAN-EEP 2% were 50.19, 42.15, 27.14, and 22.87 mg N/100 g, respectively. Thirty-five to forty milligrams of nitrogen per one hundred grams is the acceptable limit for TVB-N values in fresh fish, as reported by Connell [42]. Grigorakis et al. [45] suggested that the acceptable limit for TVB-N values in chilled sea bass is 19–20 mg N/100 g. Twenty-five to thirty-five of nitrogen per one hundred grams was considered a limit for mackerel tuna fillet damage in our study. The edible films may extend the shelf life of the fish fillet by reducing gas permeability and penetration, especially oxygen permeability, thereby limiting bacterial growth and activity [46].

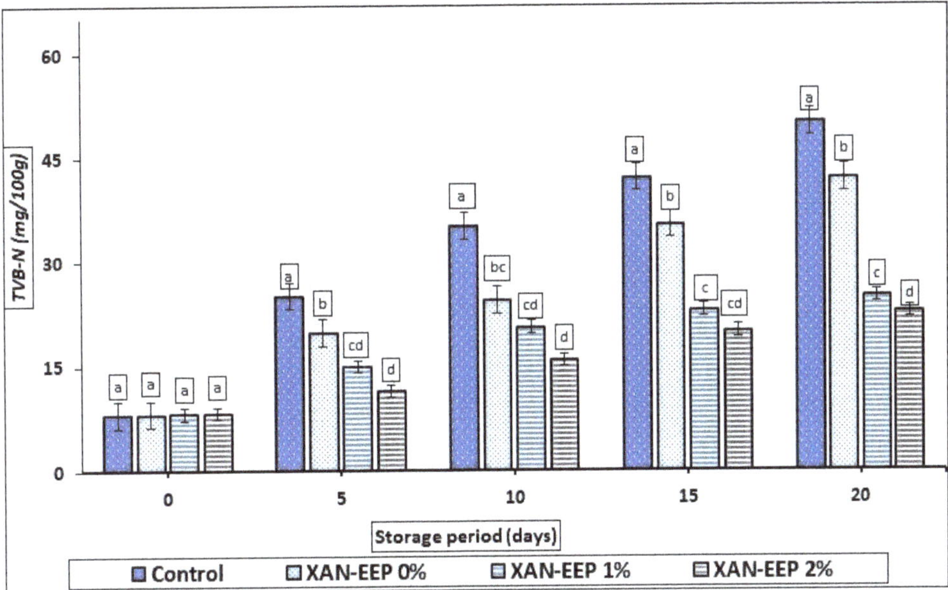

Figure 4. The influence of coating treatments on TVB-N values (mg/100 g) in mackerel tuna fillet samples during storage at 2 °C for 20 days. Control: Uncoated mackerel tuna fillet samples (soaked samples in sterile distilled water). XAN-EEP 0%: Coated samples with xanthan containing (0%) ethanolic extract of propolis. XAN-EEP 1%: Coated samples with xanthan containing (1%) ethanolic extract of propolis. XAN-EEP 2%: Coated samples with xanthan containing (2%) ethanolic extract of propolis. $^{a-d}$: Within a column, different superscripts indicate significant differences ($p < 0.05$).

In comparison to the other treatments, the propolis-treated mackerel tuna fillets had the lowest TVB-N values. Similar observations were obtained by Bazargani-Gilani et al. [47]. These results can be interpreted based on the ability of propolis to inhibit microbial activity, including the inhibition of bacteria responsible for the deamination reaction of non-protein nitrogen (NPN) components [46].

2.2.4. *K*-Value of Mackerel Tuna Fillet Samples

Endogenous biochemical changes occur in fish muscle during postmortem fish storage, among which is nucleotide degradation [48]. Calculation of the contents of ATP and its associated degradation products is an effective indicator for monitoring the freshness of fish fillets [47]. Changes in the K value during cryogenic storage of mackerel tuna fillets are shown in Figure 5. The initial K-values of the control and treated mackerel tuna samples ranged from 15.31 to 16.82%. K-values of uncoated (control) and coated mackerel tuna fillet samples increased significantly ($p < 0.05$) with storage time. Additionally, the treatments illustrated significantly lower K-values ($p < 0.05$) than the control sample. According to previous studies, the rejection level of the K-value was close to 60% [49]. Control exceeded this limit on the 10th day (68.14%), while XAN-EEP 0% exceeded this limit on the 15th day (72.19%), and both XAN-EEP 1% and XAN-EEP 2% exceeded this limit on the 20th day (68.04, 59.99%).

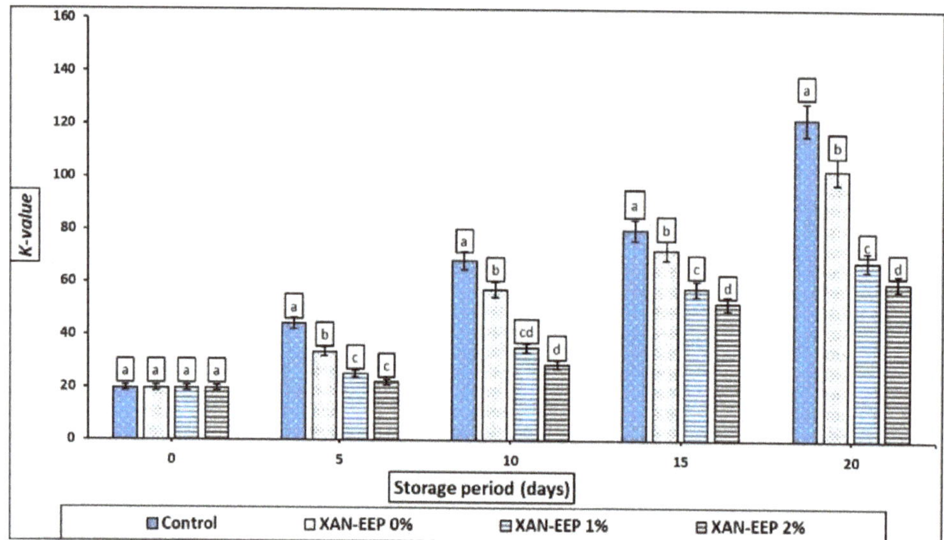

Figure 5. The influence of coating treatments on K values (%) in mackerel tuna fillet samples during storage at 2 °C for 20 days. Control: Uncoated mackerel tuna fillet samples (soaked samples in sterile distilled water). XAN-EEP 0%: Coated samples with xanthan containing (0%) ethanolic extract of propolis. XAN-EEP 1%: Coated samples with xanthan containing (1%) ethanolic extract of propolis. XAN-EEP 2%: Coated samples with xanthan containing (2%) ethanolic extract of propolis. $^{a-d}$: Within a column, different superscripts indicate significant differences ($p < 0.05$).

2.3. *Microbiological Analyses of Mackerel Tuna Fillets*

Generally, propolis' antimicrobial characteristics may be responsible for the inhibition of microbial growth in stored fish fillets [50]. Specifically, the following sections will focus on the changes in each microbial group in mackerel tuna fillet samples.

2.3.1. Total Viable Count (TVC)

Changes in total viable count (TVC) of mackerel tuna fillet samples during refrigerated storage are shown in Figure 6A. The initial TVC ($\log_{10}$ CFU/g) of all samples, including the control and treatments, ranged from 2.5 to 3.0. Compared to the values reported by Yu et al. [44] for grass carp fillets (4.90 $\log_{10}$ CFU/g), the values obtained in our study were lower. The reason for this may be attributed to individual differences or the handling of the fish during processing. The lower initial TVC for coated mackerel tuna fillets indicated that XAN-EPP coating reduced the microbial population. According to the International Commission on Microbiological Specifications for Foods (ICMSF) [51], the maximum allowable TVC is 7.0 $\log_{10}$ CFU/g. Based on that and looking at the results of our study, it was found that over the storage period, there was a noticeable increase ($p < 0.05$) in TVC for untreated samples compared to treated samples, until the untreated samples (control group) exceeded the permissible limit of TVC after 11 days. XAN-EEP 0% samples have exceeded the TVC limit after 16 days. For mackerel tuna fillets treated with 1% and 2% ethanolic extract of propolis (EEP), TVC was below the limit level during the whole storage period.

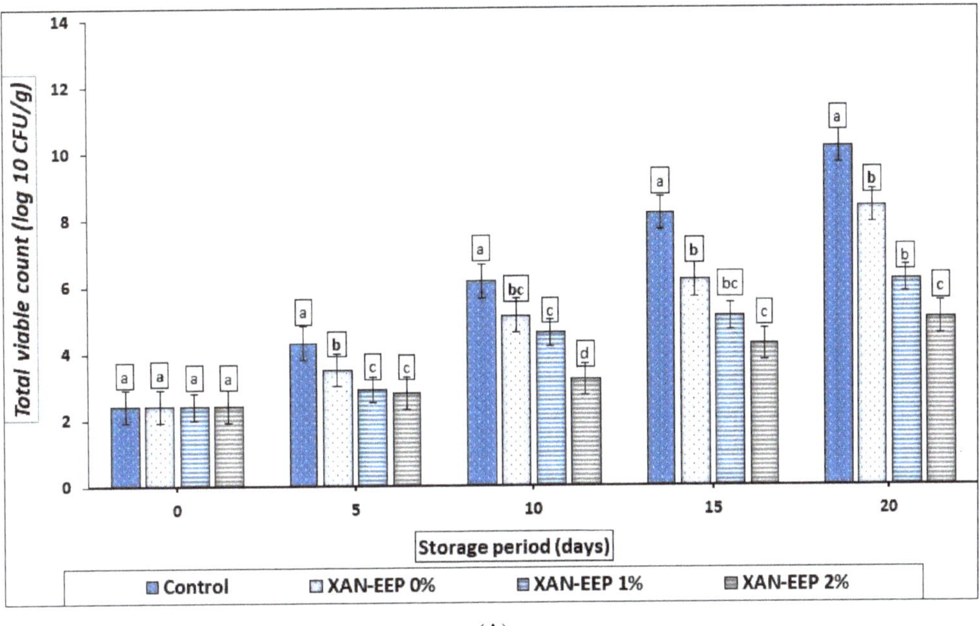

(A)

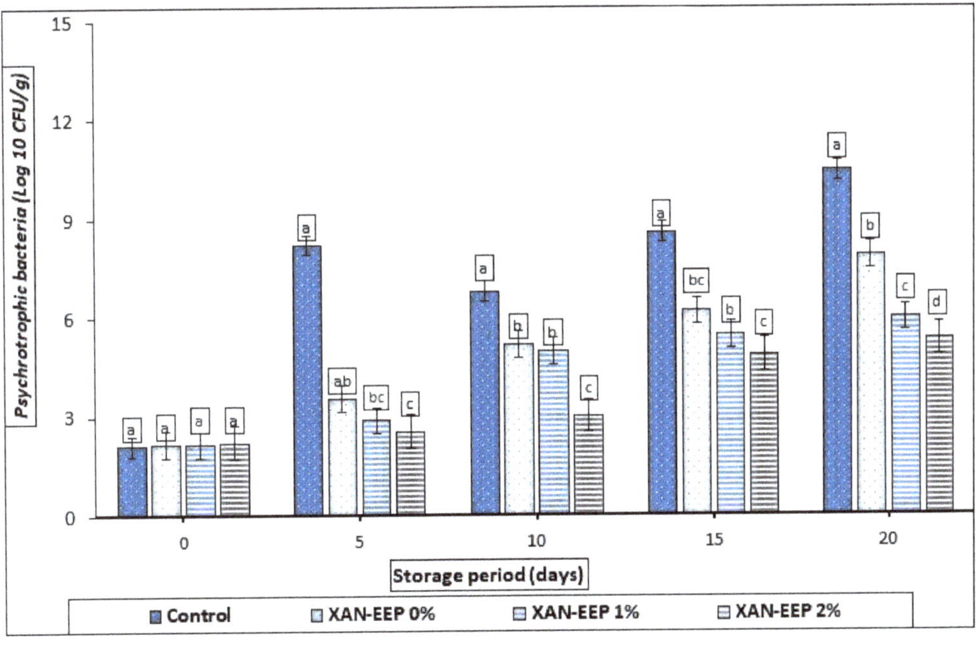

(B)

Figure 6. *Cont.*

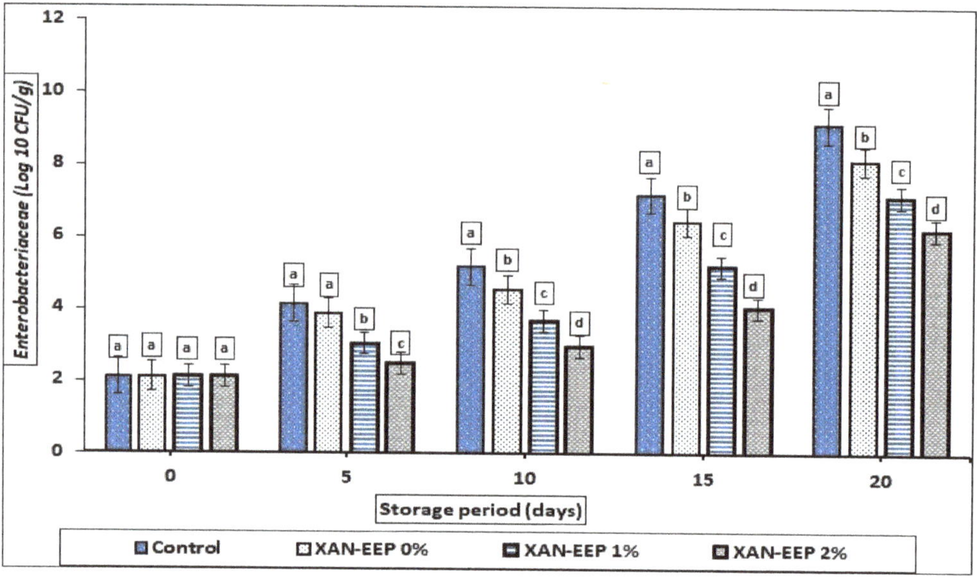

(C)

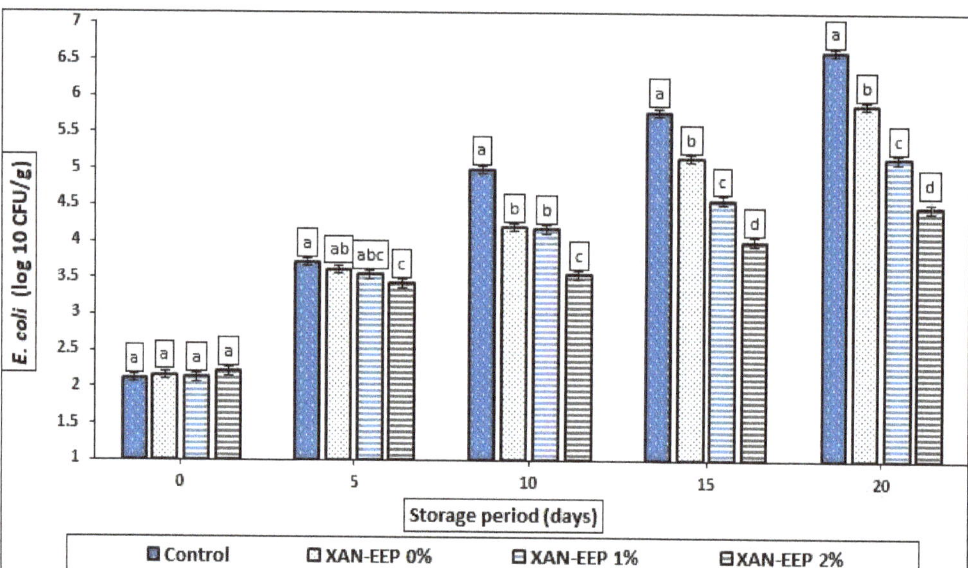

(D)

Figure 6. *Cont.*

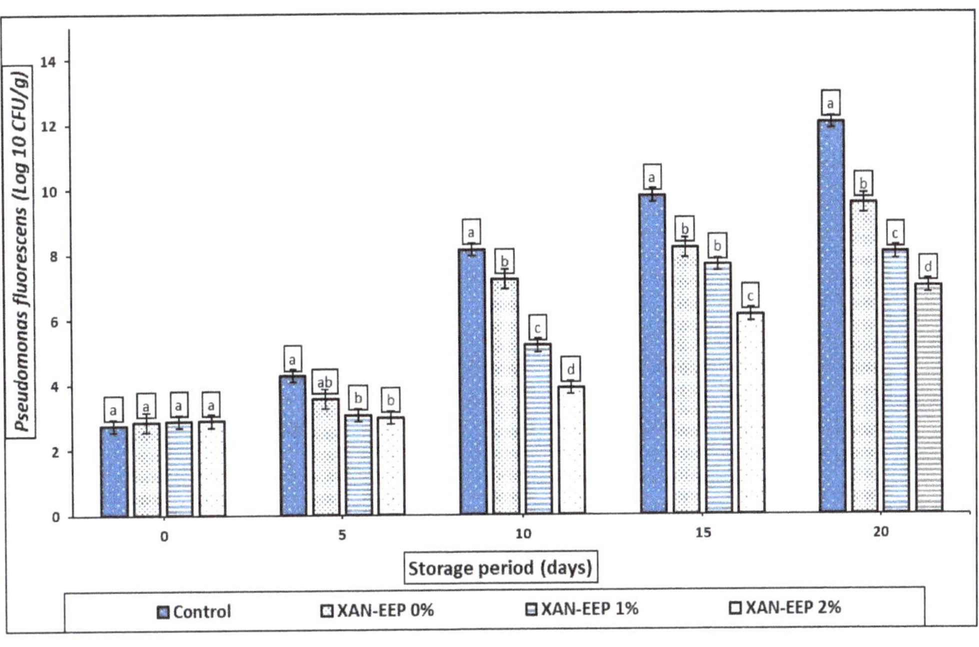

(E)

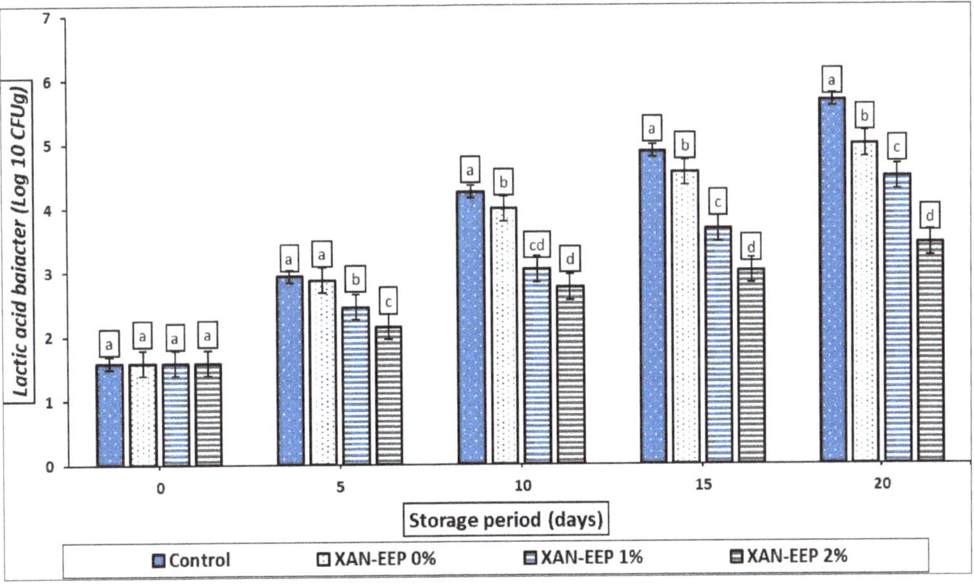

(F)

Figure 6. *Cont.*

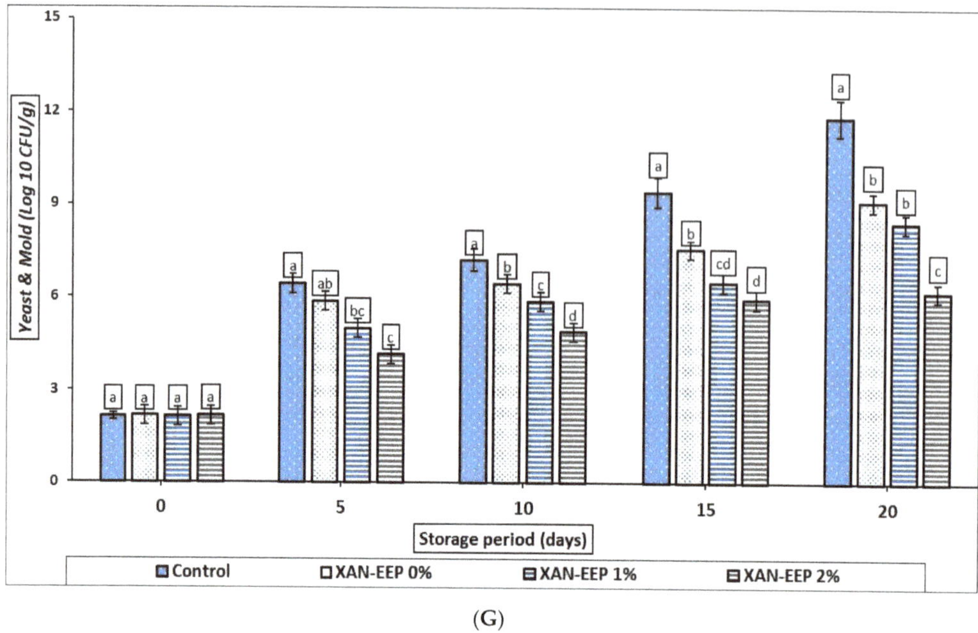

(G)

Figure 6. The influence of coating treatments on (**A**) total viable count (TVC) (log_{10} CFU/g), (**B**) psychotropic count (PTC) (log_{10} CFU/g), (**C**) Enterobacteriaceae (log_{10} CFU/g), (**D**) E. coli (log10 CFU/g), (**E**) *Pseudomonas fluorescens* (log10 CFU/g), (**F**) lactic acid bacteria (LAB) (log_{10} CFU/g), and (**G**) yeasts and molds (log_{10} CFU/g) in mackerel tuna fillet samples during storage at 2 °C for 20 days. Control: Uncoated mackerel tuna fillet samples (soaked samples in sterile distilled water). XAN-EEP 0%: Coated samples with xanthan containing (0%) ethanolic extract of propolis. XAN-EEP 1%: Coated samples with xanthan containing (1%) ethanolic extract of propolis. XAN-EEP 2%: Coated samples with xanthan containing (2%) ethanolic extract of propolis. $^{a–d}$: Within a column, different superscripts indicate significant differences ($p < 0.05$).

2.3.2. Psychotropic Count (PTC)

Among the major pathogens of microbial spoilage of refrigerated fish fillets are psychotropic bacteria [45]. Changes in the PTC of fish fillets are shown in Figure 6B. PTC of mackerel tuna fillets increased progressively ($p < 0.05$) from initial values of 2.11, 2.17, 2.15, and 2.22 log_{10} CFU/g to final values of 10.44, 7.88, 5.99, and 5.34 log_{10} CFU/g for control, XAN-EEP 0%, XAN-EEP 1%, and XAN-EEP 2%, respectively. The results depicted that the composite edible coating formed from xanthan and ethanolic extracted propolis significantly ($p < 0.05$) inhibited the growth of the total psychotropic bacteria.

2.3.3. Enterobacteriaceae

Based on previous studies, Enterobacteriaceae were found to be among the main spoilers in rainbow trout fillets stored at 4 °C [52]. The initial count of Enterobacteriaceae was 2.12 log_{10} CFU/g on trout fillets coated under fridge temperature [47,53]. After 15 storage days, Enterobacteriaceae counts reached 4.04, 5.19, 6.44, and 7.18 log_{10} CFU/g in the XAN-EEP 2%, XAN-EEP 1%, XAN-EEP 0%, and control fillets, respectively (Figure 6C). Moreover, according to the studies done by Volpe et al. [53], Bazargani-Gilani et al. [44], steady growth in Enterobacteriaceae was observed for refrigerator-stored trout chips. The findings of these studies are consistent with what was revealed by our study of the ability of coating with XAN and EEP to reduce the growth rate of Enterobacteriaceae in fillets ($p < 0.05$) compared to the uncoated samples (control samples) during cold storage. The lowest number of

Enterobacteriaceae communities was found in XAN-EEP 2% fillet samples, followed by XAN-EEP 1% and XAN-EEP 0%.

As mentioned by Jalali et al. [54], *Escherichia coli* O157:H7 is the major member of Enterobacteriaceae found in the chilled silver carp flesh. The initial count of *E. coli* O157:H7 ranged from 2.11 to 2.21 $\log_{10}$ CFU/g. After 20 storage days, *E. coli* O157:H7 counts reached 4.48, 5.15, 5.88, and 6.61 $\log_{10}$ CFU/g in the XAN-EEP 2%, XAN-EEP 1%, XAN-EEP 0%, and control fillets, respectively (Figure 6D). A previous study showed that propolis extracts can be considered natural preservatives. Their efficacy has been proven to inhibit *Escherichia coli* bacteria in vitro due to the polyphenol compounds that propolis extracts contain, which are known for their antimicrobial effect. Among these phenolic compounds is gallic acid, known for its antibacterial activity [55]. Phenolic compounds act on the bacterial cell membrane, interfere with nucleic acid synthesis, inhibit bacterial metabolism, coagulate cytoplasmic proteins, and interfere with biofilm formation [56].

2.3.4. Pseudomonas Fluorescens

According to the primary count of about 2.74–2.89 $\log_{10}$ CFU/g for *Pseudomonas fluorescens* (day 0) of the mackerel tuna fillet samples (Figure 6E), similar initial numeration (day 0) related to the rainbow trout was also found by other studies [44]. During the storage time, the *P. fluorescens* count rose to the final numeration of 12.03 $\log_{10}$ CFU/g (control fillets), while the counts of XAN-EEP 0%, XAN-EEP 1%, and XAN-EEP 2% reached 9.55, 8.04, and 6.99 $\log_{10}$ CFU/g at the last interval, being less than the control fillets. The *P. fluorescens* count in all groups was significantly ($p < 0.05$) less than the control, showing that EEP-containing treatments were the strongest concerning the inhibition treatments of *P. fluorescens*. The use of propolis extracts can be effective in inhibiting the activity of *P. aeruginosa* that causes chronic putrefaction as shown by studies conducted by Mohammadzadeh et al. [57], and De Marco et al. [58].

2.3.5. Lactic Acid Bacteria

Lactic acid bacteria (LAB) as facultative anaerobic bacteria are part of the original microflora of mackerel tuna flesh; hence, the number of these bacteria can increase under both aerobic and anaerobic conditions [59]. As shown in Figure 6F, it is clear that the initial number of LAB was 1.59 $\log_{10}$ CFU/g and did not exceed 4.88 $\log_{10}$ CFU/g in control fillets until the 15th day of the storage period. It was also observed that the LAB counts of the XAN and XAN-EEP fillet samples were significantly lower ($p < 0.05$) compared to the control samples (uncoated fillets) during the refrigerated storage period.

The best treatment in terms of inhibiting LAB proliferation in mackerel tuna fillet samples among the other tested groups was XAN-EEP 2% compared to the other tested groups, and this could be attributed to the synergistic antimicrobial effect of EEP. It has been known that LAB bacteria are the most resistant Gram-positive bacteria to antimicrobial agents [60]. The results of our study confirmed this as LAB bacteria were more resistant compared with other spoilage bacteria versus XAN combined with EEP.

This conclusion regarding the synergistic antimicrobial effect of EEP may agree with what was indicated by Duman and Özpolat [61] concerning the effect of aqueous extract of propolis during storage of shibuta (*Barbus grypus*) fillets at 4 °C. It was observed that 0.5% aqueous extract of propolis significantly reduced the number of shibuta's lactic acid bacteria at all storage times compared to the control samples. Duman and Özpolat [61] attributed this effect to the phenolic content of propolis.

In another study conducted in Greece and Cyprus on evaluating the antibacterial activities of propolis ethanolic extracts (PEs), the results concluded that the minimum inhibitory concentration (MIC) of all studied propolis ethanolic extracts was higher for lactic acid bacteria compared to the other tested bacterial species (*Listeria monocytogenes*, *Staphylococcus aureus*, and *Bacillus cereus*) [62].

2.3.6. Yeasts/Molds

The yeast and mold species are common agents of microbial spoilage in refrigerated fish [63]. A in prior investigations, the primary count (day 0) of yeast/mold of mackerel tuna fillets was 2.13–2.16 $\log_{10}$ CFU/g (Figure 6G) [44,64]. It was shown from the results of all treatments (XAN-EEP 0%, XAN-EEP 1%, and XAN-EEP 2%) in the present study that they had a significant ability ($p < 0.05$) to reduce the number of yeasts/molds compared to the untreated fillet samples (control) under cooling conditions (Figure 6G). These results were in agreement with Duman and Özpolat [61], confirming the antifungal activity of propolis from refrigerated shibuta fillets.

Other studies have highlighted the role of propolis extracts against *Candida tropicalis* and *Candida albicans* [62], and also against *Aspergillus niger* and *Candida albicans* [57].

2.4. Sensory Evaluation

It is worth noting that the use of xanthan and ethanolic extract of propolis as food-grade components in coating-forming makes it safe for consumers [65,66]. The freshness of mackerel tuna fillets during storage was assessed sensorially by taste, odor, and overall acceptability. At the beginning of the storage period, all groups of fillets were characterized by a smell of fresh fish and a distinctive shiny surface, but with the continuation of the cold storage process, the sensory properties of the samples deteriorated, but the rate of deterioration was significantly faster ($p < 0.05$) in the uncoated fillet samples (control) compared to the samples coated with xanthan/ethanolic extract of propolis (Figure 7A–C; Supplementary Table S1). According to Bazargani-Gilani and Pajohi-Alamoti [67], the permissible sensory level must be higher than 4 for the samples of fish fillets to be fit for consumption. Fishy and putrid odors increased gradually in control after 10 days of storage. Microbial damage and the consequent accumulation of receptors, such as trimethylamine (TMA) and biogenic amines, are the cause of unpleasant odors [68]. The results obtained in our study from the panelists in terms of the general acceptance of all the fillet samples under examination showed that: (a) uncoated fillet samples (control) had a shelf life of fewer than 11 days; (b) treatment with XAN-EEP 0% had a viability of more than 11 days; (c) treatment with XAN-EEP 1%, viable for 15 days; (d) treatment with XAN-EEP 2% had a viability of 20 days.

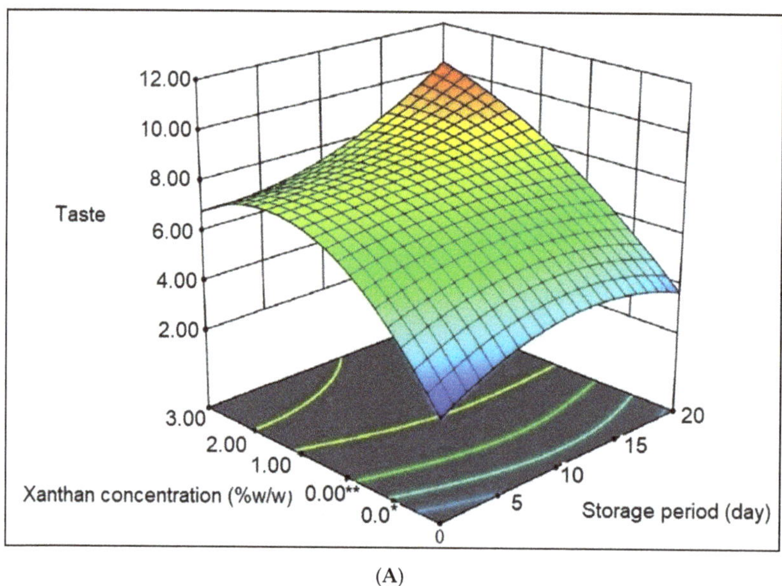

(A)

Figure 7. *Cont.*

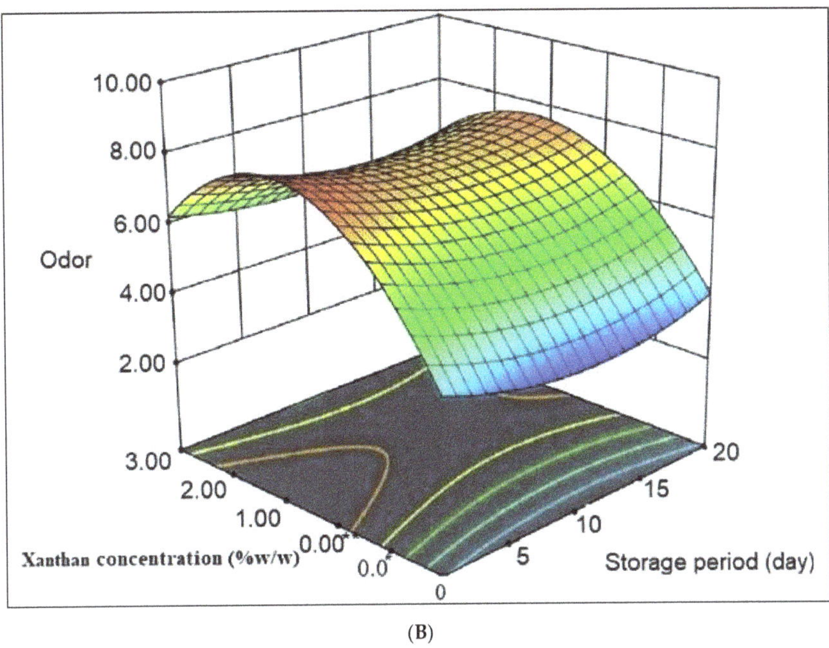

(B)

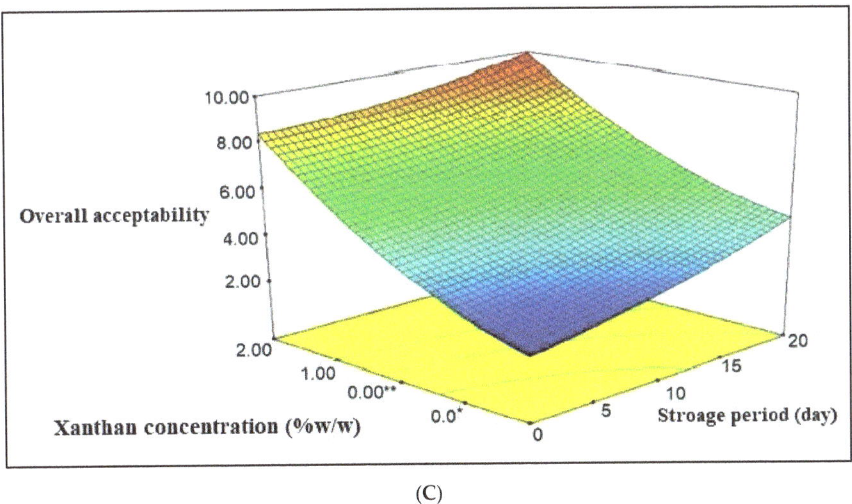

(C)

Figure 7. The response surface plot of coating treatments on the (**A**) taste, (**B**) odor, and (**C**) overall acceptability of mackerel tuna fillet samples during storage at 2 °C for 20 days. 0.0 *: Control samples [Uncoated mackerel tuna fillet samples (soaked samples in sterile distilled water)]. 0.00 **: XAN-EEP 0% [Coated samples with xanthan containing (0%) ethanolic extract of propolis]. 1.00: XAN-EEP 1% [Coated samples with xanthan containing (1%) ethanolic extract of propolis]. 2.00: XAN-EEP 2% [Coated samples with xanthan containing (2%) ethanolic extract of propolis].

These results can be attributed to the fact that the incorporation of EEP into the XAN coating significantly ($p < 0.05$) preserved the general acceptability scores and the fresh organoleptic characteristics of taste and aroma in trout meat until the last period of time.

These results are also in agreement with what was reported by Duman and Özpolat [61] about shibuta fillets.

The results of general acceptance of the studied treatments can be linked to the antimicrobial effect of EEP associated with its content of phenolic compounds [47].

3. Conclusions

Through the results of physical, chemical, microbiological, and sensory analyses conducted in our study, it can be concluded that the composite coating of xanthan and ethanolic extract of propolis preserves the properties of mackerel tuna fillets for a longer time when stored under refrigerated conditions. For example, the shelf life of mackerel tuna steaks extended by about 4 and 7 days for XAN-EEP 1% and XAN-EEP 2%, respectively, compared to the uncoated mackerel tuna fillet samples (control). Xanthan gum is a natural gelling agent that has advantages over synthetic ones owing to its safer, biodegradable nature. Moreover, the orientation of the food coating industry toward these naturally derived gelling agents has led to increasing efforts to discover, extract, and purify such compounds from the natural origin.

This work presented an edible coating containing propolis as a potent, natural, safer, and cost-effective alternative to synthetic preservatives to produce active packaging coatings that might be applicable for other food types. However, further studies are required to identify the effect of propolis on other fish species and fishery products. Furthermore, more applications of EEP-containing packages or coatings on new food models should be tested in future works.

4. Materials and Methods

4.1. Materials

Propolis was taken from an apiary around Shibin El-Kom City, Menoufia, Egypt, and kept frozen (−18 °C) until used. A total number of 40 samples of mackerel tuna fish (*Euthynnus affinis*) with an average weight of 100–150 g were obtained from a local aquacultural farm (Sadat City, Egypt). Mackerel tuna fish samples were transported to the laboratory in insulated boxes containing ice within one hour of fishing. All experiments were performed in April 2021 at the Department of Food Science and Technology, Faculty of Agriculture, Menofiya University, Shebin El-Kom City, Egypt.

4.2. Chemicals and Reagents

Xanthan gum with a molecular weight of around 500 kDa, glycerol, ethanol, methanol, potato dextrose agar, peptone, nutrient agar, plate count agar, and chemical reagents were obtained from Sigma-Aldrich (St. Louis, MO, USA). Trichloroacetic acid (TCA) and thiobarbituric acid (TBA) were procured from Sigma-Aldrich Chemie GmbH (Eschenstr, Taufkirchen, Germany). Stomacher was obtained from Lab Blender 400 (London, UK). Violet red bile glucose agar (VRBGA) was obtained from Trafalgar Scientific Ltd. (Leicester, UK). Potato dextrose agar (PDA) was purchased from Biokar diagnostics (Allonne, France). Man rogosa sharpe agar (MRS) was purchased from Oxoid Ltd. (Basingstoke, UK). Sorbitol-MacConkey agar (SMAC) was obtained from Merck (Darmstadt, Germany).

4.3. Preparation of Ethanolic Extract of Propolis (EEP)

The frozen crude propolis kept at −18 °C was grinded in a mortar until a powder was obtained, then blended with ethanol at a ratio of 25:100 g/mL, stirred at 500 rpm for 24–30 h using an Earlene shaker. The resulting solution was then filtered and evaporated at 40–45 °C using a rotary evaporator (Rotavapor RE121, Büchi, Fawil, Switzerland). Lastly, the concentrated extracts were dried at 50–55 °C in a vacuum oven. The final ethanolic extract of propolis (EEP) was stored at −18 °C until used.

4.4. Preparation of Coating Formulas

The coating solution was prepared using sterile distilled water and 1.5% xanthan (*w*/*v*). As a plasticizer agent, glycerol 0.5% (*v*/*v*) was utilized in the xanthan (XAN) coating solution, followed by stirring for 30 min on magnetic stirrer. The gel formed after heating the coating solution at 85–90 °C for 3–5 min, then the solution was cooled at room temperature. Different percentages (0, 1, and 2%) of ethanolic extract of propolis (EEP) were added to xanthan to prepare the composite coating formulas (XAN-EEP 0%, XAN -EEP 1%, and XAN-EEP 2%).

4.5. Dip Coating Procedure

There were 48 fish divided into four groups (each 12 fish). The mackerel fillet samples were skinned and washed with sterile water. Then the fillets were cut into portions of approximately 3.5 cm × 2.5 cm × 1.6 cm (10–15 g). The mackerel fillet samples were divided into 4 groups. The first group was soaked in sterile distilled water to prepare the control samples (uncoated). The other three groups were soaked in the XAN/EEP composite coating solution for 2 min to prepare the three coated treatments (XAN-EEP 0%, XAN-EEP 1%, and XAN-EEP 2%). The mackerel fillet samples were dipped in respective coating solutions (XAN-EEP 0%, XAN-EEP 1%, and XAN-EEP 2%) in a ratio of 1:2 (*w*/*v*) for 5 min and were then air dried on filter paper for 15–20 min. After that, each sample for each treatment was placed individually in a sterile zip plastic stomach bag under aerobic conditions and all packages were stored in the refrigerator at a temperature of 2 °C. Physical, chemical, microbial, and sensory analyses were carried out every 5 days, starting from 0 to over 20 days from the storage process.

4.6. Physicochemical Quality Criteria Analyses

4.6.1. Analysis of pH Value

Ten grams of each sample of fish fillets from all treatments were placed in 100 mL of distilled water and then the homogenization process was carried out for about 30 s; the pH-meter (350 Jenway pH meter, Fisher Scientific, Leicestershire, UK) was calibrated with the pH 7 buffer solution and pH 4 buffer solution, then the electrode was rinsed with distilled water and wipe with a lint-free tissue. After that, the electrode was submerged into each prepared sample to measure its pH value [69].

4.6.2. Oxidative Stability of Mackerel Fillets

Peroxide and TBARS values were used to determine the oxidative stability of mackerel fillets.

1. Determination of peroxide value

By adopting the procedure described by Shon and Chin [70], the peroxide value of mackerel fillets was calculated. Five grams of fillet sample was heated in a water bath at 60 °C for 3 min, followed by the addition of 30 mL of a solution of acetic acid-chloroform (3:2 *v*/*v*) accompanied by thorough mixing by stirring to ensure homogeneity of the sample and also to dissolve the fat. Then a filtration process was carried out, followed by the addition of 0.5 mL of saturated potassium iodide solution to the filtrate. Titration was carried out with a standard solution of sodium thiosulfate (25 g/L) in the presence of a starch solution as an indicator. The value of peroxide was expressed in peroxide equivalent units, in milliequivalent peroxides per kilogram of lipid, which was calculated by the following equation:

$$POV\left(\text{meq.kg}^{-1}\right) = \frac{S \times N}{W} \times 1000 \quad (1)$$

where S is the volume of titration (mL), N is the normality of the sodium thiosulfate solution, and W is sample weight (kg).

2. Determination of TBARS (Thiobarbituric Acid-Reactive Substances)

Each fish fillet sample (5 g) was dispersed in 20 mL of thiobarbituric acid solution (0.375% thiobarbituric acid, 15% trichloroacetic acid, and 0.25 mol/L HCl). The mixture was heated in boiling water for 10 min, cooled with water, and centrifuged at 3600× g for 20 min at room temperature. Then, the TBARS value of the coated fish fillets was assessed spectrophotometrically at 531 nm by spectrophotometer (Model UV-VIS- 2802PC, USA) according to the method described by Song et al. [49]. TBARS values were expressed in milligrams of malondialdehyde (MDA) per kilogram of fish fillet.

4.6.3. Total Volatile Basic Nitrogen (TVB-N) Measurement

Total volatile basic nitrogen (TVB-N) was measured in a mackerel meat sample (10 g) according to the method described by Sallam et al. [71], where the sample was dispersed in 100 mL of distilled water by stirring for 30 min and then filtered. Then, to 5 mL of the filtrate, 5 mL of a MgO solution (1%) was added and a Kjeldahl apparatus was used to distill the sample. The results were calculated as milligrams of nitrogen (N) per 100 g of fish fillets.

4.6.4. *K*-Value Determination

The *K*-value was measured, according to the method described by Choi et al. [72] and by using high-performance liquid chromatography (HPLC) (1100 series; Agilent Technologies, Palo Alto, CA, USA). The nucleic acid-related compounds (NARCs) (adenosine triphosphate (*ATP*), adenosine diphosphate (*ADP*), adenosine monophosphate (*AMP*), inosine monophosphate (*IMP*), inosine, (*HXR*), and hypoxanthine (*HX*)) were analyzed under the following conditions: UV detection at 254 nm, the absorbed dose range (AUF) was 0.5, μBondapak column C18 (3.9 mm × 300 mm; water, Milford, MA, USA), and the column oven temperature was 40 °C, the flow rate was 2.0 mL/min, and the mobile phase was 1% triethylamine (pH 6.5) modified with 10% H_3PO_4. The K-value was calculated using the following equation:

$$K - value\ (\%) = \frac{[(HXR) + (HX)]}{[(ATP) + (ADP) + (AMP) + (IMP) + (HXR) + (HX)]} \times 100 \quad (2)$$

4.7. Microbiological Analysis

The total viable count (TVC) and psychotropic count (PTC) as well as yeasts and molds were determined by the method described by Yu et al. [44], under aseptic conditions 10 g of each sample was naturalized with 90 mL of sterile normal saline (0.85%). A series of dilutions were then prepared from each sample and an aliquot (1 mL) of the diluent was poured into a petri dish and mixed with platelet agar medium. The inoculated plates were incubated at 30 °C for 2 days to measure TVC. The inoculated plates were incubated at 10 °C for a week for PTC. The inoculated plates were incubated at 25 °C for 5 days to count the yeasts and molds. Using the overlay casting method using violet red bile glucose agar (VRBGA), Enterobacteriaceae were enumerated as the corresponding plates were incubated for 24 h at 37 °C [73].

Escherichia coli O157:H7 were enumerated using sorbitol-McConkey agar (SMAC). Dilutions were coated on SMAC using the casting plate technique and then the plates were incubated for 24 h at 37 °C [74].

According to the procedure described by Tang et al. [75] *Pseudomonas fluorescens* were enumerated on king agar medium, and the inoculated plates were then incubated for 24 h at 30 °C.

Using de man rogosa Sharpe agar (MRS) and under anaerobic conditions lactic acid bacteria (LAB) were enumerated at 30 °C for 24 h [76].

All counts were expressed as log10 colony-forming units (CFU) g^{-1}.

4.8. Sensory Evaluation

A sensory evaluation of cooked mackerel fillet samples (microwave oven for six min at 60% of maximum power (Mw60)) was conducted by 25 trained panelists of staff members (aged 21–40 years) of the Department of Food Science and Technology, Faculty of Agriculture, Menofiya University, according to the method described by Allam et al. [74,77].

Panelists were selected based on their interests and availability. The panelists were asked to rate the cooked samples' color and odor using a scale point ranging from 0 to 10, where 10 = excellent; 9 = very good; 8 = good; 7 = acceptable; 6 = poor. The product was defined as unacceptable after the onset of a bad odor or unpleasant taste. The fresh mackerel fillet was used as a reference. The sensory analysis was done in three independent sessions.

4.9. Statistical Analysis

The study was replicated three times. Data were analyzed using the SPSS software (IBM SPSS statistics 21). Mean values of different parameters were used to compare chemical and microbiological indices. Sensory attributes data were analyzed by analysis of variance (ANOVA). The mean values ± standard deviation (SD) of the analyses were calculated. When a significant main effect was detected, the means were separated with the least significant difference (LSD) procedure. A two-way analysis of variance was used for multiple variable comparisons. Using analysis of variance (ANOVA), Tukey's test, and independent sample *t*-test for all data interpretation at $p < 0.05$.

Supplementary Materials: The following supporting information can be downloaded at: https://www.mdpi.com/article/10.3390/gels8070405/s1, Table S1: Effect of storage conditions (2 °C for 20 days) and treatments on sensory evaluation during storage of mackerel tuna fillets.

Author Contributions: Conceptualization, A.F.E.S.; methodology, A.Y.A.; software, A.Y.A.; validation, A.F.E.S. and F.O.; formal analysis, A.Y.A.; data curation, A.F.E.S.; writing—review and editing, A.F.E.S., E.O., M.R.K., C.P. and F.O.; supervision, A.F.E.S.; funding acquisition, A.F.E.S. and M.R.K. All authors have read and agreed to the published version of the manuscript.

Funding: This research was supported by the Researchers Supporting Project No. (RSP-2021/138) King Saud University, Riyadh, Saudi Arabia. This research was also supported by the Research Start-Up Fund from Jiangxi Agricultural University, China (grant number: 9232307245) and SPECIAL ACCOUNT FOR RESEARCH GRANTS of the National and Kapodistrian University of Athens, Greece (grant number: 15363).

Institutional Review Board Statement: Not applicable.

Informed Consent Statement: Not applicable.

Data Availability Statement: The data presented in this study are available on request from the corresponding author.

Acknowledgments: The authors thank to the Jiangxi Agricultural University (China), Minufiya University (Egypt), Ataturk University (Turkey), King Saud University (Saudi Arabia), and National and Kapodistrian University of Athens (Greece), UoA/S.A.R.G. The authors would like also to thank all staff in these universities for their intellectual and research support.

Conflicts of Interest: The authors declare no conflict of interest.

References

1. Diana, J.S.; Egna, H.S.; Chopin, T.; Peterson, M.S.; Cao, L.; Pomeroy, R.; Verdegem, M.; Slack, W.T.; Bondad-Reantaso, M.G.; Cabello, F. Responsible Aquaculture in 2050: Valuing Local Conditions and Human Innovations Will Be Key to Success. *BioScience* **2013**, *63*, 255–262. [CrossRef]
2. El Sheikha, A.F.; Xu, J. Traceability as a Key of Seafood Safety: Reassessment and Possible Applications. *Rev. Fish. Sci. Aquac.* **2017**, *25*, 158–170. [CrossRef]
3. Food and Agriculture Organization of the United Nations (FAO). The State of World Fisheries and Aquaculture (SOFIA). Sustainability in Action. Available online: https://www.fao.org/3/ca9229en/online/ca9229en.html (accessed on 16 January 2022).
4. Mikš-Krajnik, M.; Yoon, Y.-J.; Ukuku, D.O.; Yuk, H.-G. Volatile chemical spoilage indexes of raw Atlantic salmon (Salmo salar) stored under aerobic condition in relation to microbiological and sensory shelf lives. *Food Microbiol.* **2016**, *53*, 182–191. [CrossRef] [PubMed]
5. Jääskeläinen, E.; Jakobsen, L.M.A.; Hultman, J.; Eggers, N.; Bertram, H.C.; Björkroth, J. Metabolomics and bacterial diversity of packaged yellowfin tuna (*Thunnus albacares* L.) and salmon (*Salmo salar* L.) show fish species-specific spoilage development during chilled storage. *Int. J. Food Microbiol.* **2019**, *293*, 44–52. [CrossRef]

6. Sulfiana, S.; Nursinah, A.; Arni, M.; Metusalach, P. The Quality of Fresh Mackerel Tuna (*Euthynnus affinis* L.) Preserved with Different Icing Methods. *Int. J. Environ. Agric. Biotechnol.* **2022**, *7*, 054–060. [CrossRef]
7. Silbande, A.; Adenet, S.; Smith-Ravin, J.; Joffraud, J.-J.; Rochefort, K.; Leroi, F. Quality assessment of ice-stored tropical yellowfin tuna (*Thunnus albacares* L.) and influence of vacuum and modified atmosphere packaging. *Food Microbiol.* **2016**, *60*, 62–72. [CrossRef] [PubMed]
8. Tsironi, T.N.; Taoukis, P.S. Current Practice and Innovations in Fish Packaging. *J. Aquat. Food Prod. Technol.* **2018**, *27*, 1024–1047. [CrossRef]
9. Corbo, M.R.; Speranza, B.; Filippone, A.; Conte, A.; Sinigaglia, M.; Del Nobile, M.A. Natural compounds to preserve fresh fish burgers. *Int. J. Food Sci. Technol.* **2009**, *44*, 2021–2027. [CrossRef]
10. Tavares, J.; Martins, A.; Fidalgo, L.G.; Lima, V.; Amaral, R.A.; Pinto, C.A.; Silva, A.M.; Saraiva, J.A. Fresh Fish Degradation and Advances in Preservation Using Physical Emerging Technologies. *Foods* **2021**, *10*, 780. [CrossRef]
11. Secci, G.; Parisi, G. From farm to fork: Lipid oxidation in fish products. A review. *Ital. J. Anim. Sci.* **2016**, *15*, 124–136. [CrossRef]
12. Mei, J.; Ma, X.; Xie, J. Review on Natural Preservatives for Extending Fish Shelf Life. *Foods* **2019**, *8*, 490. [CrossRef] [PubMed]
13. de Oliveira, M.S.; Cruz, J.N.; Ferreira, O.O.; Pereira, D.S.; Pereira, N.S.; Oliveira, M.E.C.; Venturieri, G.C.; Guilhon, G.M.S.P.; Souza Filho, A.P.d.S.; Andrade, E.H.d.A. Chemical Composition of Volatile Compounds in Apis mellifera Propolis from the Northeast Region of Pará State, Brazil. *Molecules* **2021**, *26*, 3462. [CrossRef] [PubMed]
14. Anjum, S.I.; Ullah, A.; Khan, K.A.; Attaullah, M.; Khan, H.; Ali, H.; Bashir, M.A.; Tahir, M.; Ansari, M.J.; Ghramh, H.A.; et al. Composition and functional properties of propolis (bee glue): A review. *Saudi J. Biol. Sci.* **2019**, *26*, 1695–1703. [CrossRef] [PubMed]
15. Ju, J.; Xie, Y.; Guo, Y.; Cheng, Y.; Qian, H.; Yao, W. Application of edible coating with essential oil in food preservation. *Crit. Rev. Food Sci. Nutr.* **2019**, *59*, 2467–2480. [CrossRef] [PubMed]
16. Chawla, R.; Sivakumar, S.; Kaur, H. Antimicrobial edible films in food packaging: Current scenario and recent nanotechnological advancements—A review. *Carbohydr. Polym. Technol. Appl.* **2021**, *2*, 100024. [CrossRef]
17. Sharma, S.; Rao, T.V.R. Xanthan gum based edible coating enriched with cinnamic acid prevents browning and extends the shelf-life of fresh-cut pears. *LWT—Food Sci. Technol.* **2015**, *62*, 791–800. [CrossRef]
18. Fan, W.; Sun, J.; Chen, Y.; Qiu, J.; Zhang, Y.; Chi, Y. Effects of chitosan coating on quality and shelf life of silver carp during frozen storage. *Food Chem.* **2009**, *115*, 66–70. [CrossRef]
19. Atieno, L.; Owino, W.; Ateka, E.M.; Ambuko, J. Influence of Coating Application Methods on the Postharvest Quality of Cassava. *Int. J. Food Sci.* **2019**, *2019*, 2148914. [CrossRef]
20. Salehi, F.; Satorabi, M. Effect of Basil Seed and Xanthan Gums Coating on Colour and Surface Change Kinetics of Peach Slices During Infrared Drying. *Acta Technol. Agric.* **2021**, *24*, 150–156. [CrossRef]
21. Quoc, L.; Hoa, D.; Ngoc, H.; Phi, T. Effect of Xanthan gum Solution on the Preservation of Acerola (*Malpighia glabra* L.). *Cercet. Agron. Mold.* **2015**, *48*, 89–97. [CrossRef]
22. Soleimani-Rambod, A.; Zomorodi, S.; Naghizadeh Raeisi, S.; Khosrowshahi Asl, A.; Shahidi, S.-A. The Effect of Xanthan Gum and Flaxseed Mucilage as Edible Coatings in Cheddar Cheese during Ripening. *Coatings* **2018**, *8*, 80. [CrossRef]
23. Hidalgo, M.E.; Armendariz, M.; Wagner, J.R.; Risso, P.H. Effect of Xanthan Gum on the Rheological Behavior and Microstructure of Sodium Caseinate Acid Gels. *Gels* **2016**, *2*, 23. [CrossRef] [PubMed]
24. Pullen, C. Xanthan Gum—Is This Food Additive Healthy or Harmful? Available online: https://www.healthline.com/nutrition/xanthan-gum (accessed on 23 April 2022).
25. Gowthaman, M.K.; Prasad, M.S.; Karanth, N.G. Fermentation (Industrial) | Production of Xanthan Gum. In *Encyclopedia of Food Microbiology*; Robinson, R.K., Ed.; Elsevier: Oxford, UK, 1999; pp. 699–705.
26. Energias Market Research. Gelling Agent-Polymer Performing Multiwork, To Witness a CAGR of 5.1% during 2017–2023. Available online: https://www.globenewswire.com/news-release/2018/01/25/1304730/0/en/Gelling-Agent-Polymer-Performing-Multiwork-To-Witness-a-CAGR-of-5-1-During-2017-2023.html (accessed on 23 April 2022).
27. Lu, F.; Ding, Y.; Ye, X.; Liu, D. Cinnamon and nisin in alginate–calcium coating maintain quality of fresh northern snakehead fish fillets. *LWT—Food Sci. Technol.* **2010**, *43*, 1331–1335. [CrossRef]
28. Díaz-Montes, E.; Castro-Muñoz, R. Edible films and coatings as food-quality preservers: An overview. *Foods* **2021**, *10*, 249. [CrossRef]
29. Umaraw, P.; Munekata, P.E.S.; Verma, A.K.; Barba, F.J.; Singh, V.P.; Kumar, P.; Lorenzo, J.M. Edible films/coating with tailored properties for active packaging of meat, fish and derived products. *Trends Food Sci. Technol.* **2020**, *98*, 10–24. [CrossRef]
30. Kumar, K.S.; Chrisolite, B.; Sugumar, G.; Bindu, I.; Venkateshwarlu, G. Shelf life extension of tuna fillets by gelatin and chitosan based edible coating incorporated with clove oil. *Fish. Technol.* **2018**, *55*, 104–113.
31. An, S.-H.; Ban, E.; Chung, I.-Y.; Cho, Y.-H.; Kim, A. Antimicrobial activities of propolis in poloxamer based topical gels. *Pharmaceutics* **2021**, *13*, 2021. [CrossRef]
32. Fazial, F.F.; Ling, T.L.; Ahmad, A.A.A.; Zubairi, S.I. Physicochemical changes of tuna fish (Euthynnus affinis) throughout refrigerated storage condition. A preliminary study. In Proceedings of the 2018 UKM FST Postgraduate Colloquium, UKM Bangi, Selangor, Malaysia, 4–6 April 2018; p. 050002.
33. Lahreche, T.; Durmus, M.; Aksun Tümerkan, E.T.; Hamdi, T.-M.; Özogul, F. Technology. Olive leaf extracts application for shelf life extension of vacuum-packed frigate mackerel (*auxis thazard* L.) fillets. *Carpathian J. Food Sci. Technol.* **2020**, *12*, 70–83. [CrossRef]

34. Yazgan, H.; Burgut, A.; Durmus, M.; Kosker, A.R. The impacts of water and ethanolic extracts of propolis on vacuum packaged sardine fillets inoculated with Morganella psychrotolerans during chilly storage. *J. Food Saf.* **2020**, *40*, e12767. [CrossRef]
35. Duarte, A.M.; Silva, F.; Pinto, F.R.; Barroso, S.; Gil, M.M. Quality Assessment of Chilled and Frozen Fish—Mini Review. *Foods* **2020**, *9*, 1739. [CrossRef]
36. Yi, Z.; Xie, J. Assessment of spoilage potential and amino acids deamination & decarboxylation activities of Shewanella putrefaciens in bigeye tuna (*Thunnus obesus* L.). *LWT* **2022**, *156*, 113016. [CrossRef]
37. Vieira, B.B.; Mafra, J.F.; Bispo, A.S.d.R.; Ferreira, M.A.; Silva, F.d.L.; Rodrigues, A.V.N.; Evangelista-Barreto, N.S. Combination of chitosan coating and clove essential oil reduces lipid oxidation and microbial growth in frozen stored tambaqui (*Colossoma macropomum* L.) fillets. *LWT* **2019**, *116*, 108546. [CrossRef]
38. Khan, M.I.; Adrees, M.N.; Arshad, M.S.; Anjum, F.M.; Jo, C.; Sameen, A. Oxidative stability and quality characteristics of whey protein coated rohu (*Labeo rohita* L.) fillets. *Lipids Health Dis.* **2015**, *14*, 58. [CrossRef] [PubMed]
39. Roy, S.; Priyadarshi, R.; Rhim, J.-W. Development of Multifunctional Pullulan/Chitosan-Based Composite Films Reinforced with ZnO Nanoparticles and Propolis for Meat Packaging Applications. *Foods* **2021**, *10*, 2789. [CrossRef] [PubMed]
40. Shavisi, N.; Khanjari, A.; Basti, A.A.; Misaghi, A.; Shahbazi, Y. Effect of PLA films containing propolis ethanolic extract, cellulose nanoparticle and Ziziphora clinopodioides essential oil on chemical, microbial and sensory properties of minced beef. *Meat Sci.* **2017**, *124*, 95–104. [CrossRef] [PubMed]
41. Khoshnoudi-Nia, S.; Moosavi-Nasab, M. Comparison of various chemometric analysis for rapid prediction of thiobarbituric acid reactive substances in rainbow trout fillets by hyperspectral imaging technique. *Food Sci. Nutr.* **2019**, *7*, 1875–1883. [CrossRef] [PubMed]
42. Connell, J.J. *Control of Fish Quality*; Fishing News Books: Farnham, UK, 1980.
43. López de Lacey, A.M.; López-Caballero, M.E.; Montero, P. Agar films containing green tea extract and probiotic bacteria for extending fish shelf-life. *LWT—Food Sci. Technol.* **2014**, *55*, 559–564. [CrossRef]
44. Yu, D.; Li, P.; Xu, Y.; Jiang, Q.; Xia, W. Physicochemical, microbiological, and sensory attributes of chitosan-coated grass carp (*Ctenopharyngodon idellus* L.) fillets stored at 4 °C. *Int. J. Food Prop.* **2017**, *20*, 390–401. [CrossRef]
45. Grigorakis, K.; Alexis, M.; Gialamas, I.; Nikolopoulou, D. Sensory, microbiological, and chemical spoilage of cultured common sea bass (*Dicentrarchus labrax* L.) stored in ice: A seasonal differentiation. *Eur. Food Res. Technol.* **2004**, *219*, 584–587. [CrossRef]
46. Socaciu, M.-I.; Semeniuc, C.A.; Vodnar, D.C. Edible Films and Coatings for Fresh Fish Packaging: Focus on Quality Changes and Shelf-life Extension. *Coatings* **2018**, *8*, 366. [CrossRef]
47. Bazargani-Gilani, B.; Pajohi-Alamoti, M.; Hassanzadeh, P.; Raeisi, M. Impacts of Carboxymethyl Cellulose Containing Propolis Extract on the Shelf Life of Trout Fillets. *Arch. Hyg. Sci.* **2021**, *10*, 117–132. [CrossRef]
48. Song, Y.; Liu, L.; Shen, H.; You, J.; Luo, Y. Effect of sodium alginate-based edible coating containing different anti-oxidants on quality and shelf life of refrigerated bream (*Megalobrama amblycephala* L.). *Food Control* **2011**, *22*, 608–615. [CrossRef]
49. Ehira, S. A biochemical study on the freshness of fish. *Bull. Tokai Reg. Fish. Res. Lab.* **1976**, *88*, 130–132.
50. Ucak, I.; Khalily, R.; Carrillo, C.; Tomasevic, I.; Barba, F.J. Potential of propolis extract as a natural antioxidant and antimicrobial in gelatin films applied to rainbow trout (*Oncorhynchus mykiss* L.) fillets. *Foods* **2020**, *9*, 1584. [CrossRef]
51. International Commission on Microbiological Specifications for Foods (ICMSF). *Micro-Organisms in Foods: A Publication of the International Commission on Microbiological Specifications for Foods (ICMSF) of the International Association of Microbiological Societies*; University of Toronto Press: Toronto, ON, Canada, 1986.
52. Gui, M.; Zhao, B.; Song, J.; Zhang, Z.; Peng, Z.; Li, P. Paraplantaricin L-ZB1, a Novel Bacteriocin and Its Application as a Biopreservative Agent on Quality and Shelf Life of Rainbow Trout Fillets Stored at 4 °C. *Appl. Biochem. Biotechnol.* **2014**, *174*, 2295–2306. [CrossRef]
53. Volpe, M.G.; Siano, F.; Paolucci, M.; Sacco, A.; Sorrentino, A.; Malinconico, M.; Varricchio, E. Active edible coating effectiveness in shelf-life enhancement of trout (*Oncorhynchusmykiss* L.) fillets. *LWT—Food Sci. Technol.* **2015**, *60*, 615–622. [CrossRef]
54. Jalali, N.; Ariiai, P.; Fattahi, E. Effect of alginate/carboxyl methyl cellulose composite coating incorporated with clove essential oil on the quality of silver carp fillet and Escherichia coli O157:H7 inhibition during refrigerated storage. *J. Food Sci. Technol.* **2016**, *53*, 757–765. [CrossRef]
55. Tosi, E.A.; Ré, E.; Ortega, M.E.; Cazzoli, A.F. Food preservative based on propolis: Bacteriostatic activity of propolis polyphenols and flavonoids upon Escherichia coli. *Food Chem.* **2007**, *104*, 1025–1029. [CrossRef]
56. Ramata-Stunda, A.; Petrin, Z.; Valkovska, V.; Borodušk, M.; Gibnere, L.; Gurkovska, E.; Nikolajeva, V. Synergistic effect of polyphenol-rich complex of plant and green propolis extracts with antibiotics against respiratory infections causing bacteria. *Antibiotics* **2022**, *11*, 160. [CrossRef]
57. Mohammadzadeh, S.; Shariatpanahi, M.; Hamedi, M.; Ahmadkhaniha, R.; Samadi, N.; Ostad, S.N. Chemical composition, oral toxicity and antimicrobial activity of Iranian propolis. *Food Chem.* **2007**, *103*, 1097–1103. [CrossRef]
58. De Marco, S.; Piccioni, M.; Pagiotti, R.; Pietrella, D. Antibiofilm and Antioxidant Activity of Propolis and Bud Poplar Resins versus *Pseudomonas aeruginosa* L. *Evid. Based Complementary Altern. Med.* **2017**, *2017*, 5163575. [CrossRef] [PubMed]
59. Françoise, L. Occurrence and role of lactic acid bacteria in seafood products. *Food Microbiol.* **2010**, *27*, 698–709. [CrossRef] [PubMed]
60. Burt, S. Essential oils: Their antibacterial properties and potential applications in foods—A review. *Int. J. Food Microbiol.* **2004**, *94*, 223–253. [CrossRef] [PubMed]

61. Duman, M.; Özpolat, E. Effects of water extract of propolis on fresh shibuta (*Barbus grypus* L.) fillets during chilled storage. *Food Chem.* **2015**, *189*, 80–85. [CrossRef] [PubMed]
62. Kalogeropoulos, N.; Konteles, S.J.; Troullidou, E.; Mourtzinos, I.; Karathanos, V.T. Chemical composition, antioxidant activity and antimicrobial properties of propolis extracts from Greece and Cyprus. *Food Chem.* **2009**, *116*, 452–461. [CrossRef]
63. Lerma-Fierro, A.G.; Flores-López, M.K.; Guzmán-Robles, M.L.; Cortés-Sánchez, A.D.J.J.T. Microbiological evaluation of minimally processed and marketed fish in popular market of the city of Tepic Nayarit, Mexico Sanitary quality of tilapia (*Oreochromis niloticus* L.). *Tropicultura* **2020**, *38*, 1–19.
64. Bazargani-Gilani, B. Activating sodium alginate-based edible coating using a dietary supplement for increasing the shelf life of rainbow trout fillet during refrigerated storage (4 ± 1 °C). *J. Food Saf.* **2018**, *38*, e12395. [CrossRef]
65. Iñiguez-Moreno, M.; Ragazzo-Sánchez, J.A.; Calderón-Santoyo, M. An extensive review of natural polymers used as coatings for postharvest shelf-life extension: Trends and challenges. *Polymers* **2021**, *13*, 3271. [CrossRef]
66. Mahdavi-Roshan, M.; Gheibi, S.; Pourfarzad, A. Effect of propolis extract as a natural preservative on quality and shelf life of marinated chicken breast (chicken Kebab). *LWT—Food Sci. Technol.* **2022**, *155*, 112942. [CrossRef]
67. Bazargani-Gilani, B.; Pajohi-Alamoti, M. The effects of incorporated resveratrol in edible coating based on sodium alginate on the refrigerated trout (*Oncorhynchus mykiss* L.) fillets' sensorial and physicochemical features. *Food Sci. Biotechnol.* **2020**, *29*, 207–216. [CrossRef]
68. Dalgaard, P.; Madsen, H.L.; Samieian, N.; Emborg, J. Biogenic amine formation and microbial spoilage in chilled garfish (Belone belone belone)—Effect of modified atmosphere packaging and previous frozen storage. *J. Appl. Microbiol.* **2006**, *101*, 80–95. [CrossRef] [PubMed]
69. Mahmoudzadeh, M.; Motallebi, A.; Hosseini, H.; Khaksar, R.; Ahmadi, H.; Jenab, E.; Shahraz, F.; Kamran, M. Quality changes of fish burgers prepared from deep flounder (Pseudorhombus elevatus Ogilby, 1912) with and without coating during frozen storage (−18 °C). *Int. J. Food Sci. Technol.* **2010**, *45*, 374–379. [CrossRef]
70. Shon, J.; Chin, K.B. Effect of Whey Protein Coating on Quality Attributes of Low-Fat, Aerobically Packaged Sausage during Refrigerated Storage. *J. Food Sci.* **2008**, *73*, C469–C475. [CrossRef] [PubMed]
71. Sallam, K.I.; Ishioroshi, M.; Samejima, K. Antioxidant and antimicrobial effects of garlic in chicken sausage. *LWT—Food Sci. Technol.* **2004**, *37*, 849–855. [CrossRef]
72. Choi, J.-W.; Jang, M.-K.; Hong, C.-W.; Lee, J.-W.; Choi, J.-H.; Kim, K.-B.; Xu, X.; Ahn, D.-H.; Lee, M.-k.; Nam, T.J. Novel application of an optical inspection system to determine the freshness of Scomber japonicus (mackerel) stored at a low temperature. *Food Sci. Biotechnol.* **2020**, *29*, 103–107. [CrossRef]
73. *ISO 21528-2*; Microbiology of Food and Animal Feeding Stuffs—Horizontal Method for the Detection and Enumeration of Enterobacteriaceae. Part 2, Colony-Count Technique. International Organization for Standardization (ISO): Geneva, Switzerland, 2004.
74. Sagdic, O.; Ozturk, I. Kinetic Modeling of Escherichia coli O157:H7 Growth in Rainbow Trout Fillets as Affected by Oregano and Thyme Essential Oils and Different Packing Treatments. *Int. J. Food Prop.* **2014**, *17*, 371–385. [CrossRef]
75. Tang, R.; Zhu, J.; Feng, L.; Li, J.; Liu, X. Characterization of LuxI/LuxR and their regulation involved in biofilm formation and stress resistance in fish spoilers Pseudomonas fluorescens. *Int. J. Food Microbiol.* **2019**, *297*, 60–71. [CrossRef]
76. Giatrakou, V.; Kykkidou, S.; Papavergou, A.; Kontominas, M.G.; Savvaidis, I.N. Potential of Oregano Essential Oil and MAP to Extend the Shelf Life of Fresh Swordfish: A Comparative Study with Ice Storage. *J. Food Sci.* **2008**, *73*, M167–M173. [CrossRef]
77. Allam, A.Y.F.; Vadimovna, D.N.; Kandil, A.A.E.H. Functional characteristics of bioactive phytochemicals in beta vulgaris l. root and their application as encapsulated additives in meat products. *Carpathian J. Food Sci.* **2021**, *13*, 173–191.

MDPI
St. Alban-Anlage 66
4052 Basel
Switzerland
www.mdpi.com

Gels Editorial Office
E-mail: gels@mdpi.com
www.mdpi.com/journal/gels

www.ingramcontent.com/pod-product-compliance
Lightning Source LLC
LaVergne TN
LVHW070909170726
843515LV00005B/735